D0908272

Yours to Reason Why

Previous books by the author

BATTLES IN BRITAIN
ORDEAL BY AMBITION
LANDS OF SPICE AND TREASURE
SOVEREIGN LEGACY

Yours to Reason Why

Decision in Battle

WILLIAM SEYMOUR

Maps and battle-plans by W.F.N. Watson

Sidgwick & Jackson
London

First published in Great Britain in 1982
by Sidgwick and Jackson Limited

ISBN 0-283-98788-X

Phototypeset in Monophoto Century Schoolbook by
Servis Filmsetting Limited, Manchester

Printed in Great Britain by
R. J. Acford, Industrial Estate, Chichester, Sussex
for Sidgwick & Jackson Limited
1 Tavistock Chambers, Bloomsbury Way
London WC1A 2SG

To Richard and
Dorothy McArdle

---

with grateful thanks
for all their kindness and help
in the writing of the battles
fought in their country.

Contents

List of Plates

Introduction

FIELD-MARSHAL Lord Roberts wrote, 'The study of military strategy is no very exact science, leading to propositions susceptible of mathematical proof.' There are always extraneous factors to be considered bordering on the realms of psychology and, although diaries, despatches and official papers relevant to later battles help one to get into the skin of the commander, for battles fought long ago there are many facets that it is difficult to summon from the shades, even through the most diligent research.

I have confined the options in this book to a strategical and tactical level. That is not to say that they are taken solely from a look at the map; in analysing the problems confronting the commanders in the ten selected battles or campaigns, I have taken into account as many of the known factors as possible – including that most important one of all, logistics – and related them to the conditions prevailing at the time of the battle. I have tried to present the options in a straightforward and simple manner with, it is hoped, sufficient background for the reader to understand the problem but without the confusion of too much detail. Inevitably the book is speculation, and therefore not a work of scholarship, for speculation cannot be scholarship.

The idea has been to set the scene and, as the battle develops, put forward some of the options that might reasonably have been expected to confront the commanders. The decision that the commander took is then made clear, and commented upon in relation to the other possibilities. To agree with the decisions taken, or in some cases to prefer a better one, may be considered presumptuous, for there are usually unanswerable reasons to be given for and against most decisions taken in battle. All that has been done is to examine thoroughly the possible courses of action open to the commander, and, with the benefit of hindsight, form as balanced a judgment as possible. No infallibility is claimed, and readers of the book may disagree with some, or all, of the alternative solutions – that is to be expected, and indeed encouraged.

The battles and campaigns, which span almost nine hundred years, have been chosen to cover differing types of warfare, and at the same time to be those which seemed to offer the greatest number of comparatively uncomplicated options open to the opposing commanders. Although each battle has been described in outline this is not principally a battles book.

I am indebted to a great number of kind people who have given me their advice and assistance. The inspiration for the book came from William Armstrong, the managing director of my publishing firm, whose wise counsel and encouragement throughout, together with that of his managing editor, Margaret Willes, has been invaluable. Colonel Watson has once again come to my rescue with his excellent maps and battle plans, without which the options could never have been intelligently conveyed.

During my two visits to the United States to look at the battlefields of the War of the Revolution and the Civil War, I met with kindness and enthusiasm for the work wheresoever I went. In particular, I must mention my friend of many years, Richard E. McArdle (to whom this book is dedicated) who, with his son Dick, did so much preparatory work for my visits, and drove me to and around the battlefields of Chancellorsville and Gettysburg. At Chancellorsville Mr Edmund Raus put his great knowledge of the battle at my disposal and with Mr Robert Krick kindly read and commented on those chapters, while Mr Kim Holien did the same for the Gettysburg chapter. Dr Harry Pfanz, Mr Harrison and Miss Cathy George of the National Parks Service, Brigadier-General James Collins junior, Chief of Military History, and Mr Curt Johnson of the Historical Evaluation & Research Organization are others who gave me their time and assistance, and to whom I offer my grateful thanks.

Field-Marshal Lord Harding of Petherton, Sir David Hunt, Major Peter Verney and Major Nicholas Straker kindly gave me advice and help with the Anzio chapters, and General Sir David Fraser was very helpful over the Waterloo campaign. The staff of the London Library, as always, were most patient and painstaking over my demands, as also were the staff of the historical section of the Cabinet Office and the Ministry of Defence Library. I have left to the last my hard-working secretary, Miss Patricia Carter, without whose untiring efforts at deciphering my writing, typing and re-typing the manuscript, and setting out the tables, the book might still be unfinished.

WILLIAM SEYMOUR
Park House, Shaftesbury

1 The Battle of Hastings

14 October 1066

WHEN Edward the Confessor died at Westminster on 5 January 1066, the Witan, or Great Council, immediately elected Harold Godwinson, Earl of Wessex, to be their king. He ascended the throne uneasily. He had scarcely a drop of royal blood in his veins; he was not loved by the powerful Earls of Mercia and Northumbria (Edwin and Morcar), and there were two foreign claimants who had been watching and waiting for some time. But he had shown himself to be a leader of men, and a commander of above average ability (if at times a trifle too impetuous), when a few years earlier he had broken the power of the Welsh prince, Griffith ap Llywelyn.

The Norsemen had as their king Harald Hardrada, who was reckoned the greatest captain of his day. His claim to the English throne was extremely flimsy, being based on an unrecorded agreement said to have been made a quarter of a century earlier between Magnus, Hardrada's predecessor on the Norwegian throne, and King Hardicanute, the last Anglo-Danish king. Nevertheless, the weakness of his case did not deter him from preparing to invade with ruthless determination.

The claim of the other foreign contender, William Duke of Normandy, had more substance. He was a first cousin once removed to Edward the Confessor; it is quite possible that as far back as 1051 Edward – without any right to do so – had promised William the succession; and in 1064 Harold, when the guest of William in Normandy, had sworn an oath that he would not oppose the Duke's accession to the English throne. The sanctity of the oath, incidentally, and the fact that Harold was already out of favour with the papacy, were to strengthen William's cause in Rome and enable him to march under the papal banner.

Harold was undoubtedly the best man that England could have found; he had seized the throne with speed and resolution from the two foreigners who thought it belonged to them, in the face of the domestic dangers that usually accompany a coup d'état. As it turned out the two northern Earls remained loyal, and the first danger that Harold was called upon to face came from his troublesome younger brother Tostig, who had been banished from the earldom of Northumbria.

Tostig's raids on the Isle of Wight and Sandwich in late May, while in themselves total failures, had serious consequences for Harold. Convinced that they heralded an immediate threat of invasion from Normandy, Harold called out the southern fyrd and manned what few ships were serviceable in the English navy. Unable to maintain the levies for more than three months, he demobilized them on 8 September, and returned the fleet to London. He was, therefore, caught off-balance when news came that Harald Hardrada, accompanied by the egregious Tostig, had sailed up the Ouse and disembarked a large force at Ricall in Yorkshire.

But Harold was a man not easily daunted by adversity; in one of the great marches of history he, and the fresh levies he called out as he went, covered 180 miles in four days. Meanwhile, on 20 September, the Earls of Mercia and Northumbria, unaware of Harold's movements, gave battle to the invaders at Fulford, just south of York, and were heavily defeated. But Harold's whirlwind march took the Norsemen completely by surprise, and on 25 September he came upon Hardrada's army taking their ease on the banks of the river Derwent at Stamford Bridge. In a battle that lasted from midday until dusk Harold gained a decisive victory; the Vikings fled the field, leaving upon it the bodies of their King and Harold's brother Tostig. But the victory was not gained without cost, and Harold knew that somehow he must fill his thinning ranks and be ready to stake his crown in the even sterner task that lay ahead.

When Duke William of Normandy disembarked his army at Pevensey in Sussex on 28 September, Harold was either resting at York after his victory at Stamford Bridge, or, more probably – for it is unlikely that he would have lingered there for four days when he must have known that William had got the wind he wanted – on the march south. He reached London, almost certainly well ahead of his main army, during the first week in October, and there he learned of the great destruction William's army was causing to the countryside round Hastings.

The Anglo-Saxon army was composed of two elements, the house-carls and the fyrd. The house-carls were a comparatively small corps d'élite of around 3,000 professional soldiers, whose principal armament was the powerful double-handed axe; they carried long shields and wore helmets and chain mail almost down to their knees. The fyrd, or shire levies, were free men raised from each hundred and liable to serve for a period of two months annually; it was possible to assemble about 12,000 of these irregulars at any one time. They wore no armour – though some carried shields – and were armed with a curious assortment of weapons ranging from spears, axes, stone slings and javelins to scythes. They were often well led by their thanes, and were by no means the ill-disciplined rabble that thev are sometimes described as. The army had no cavalry, and precious few bowmen. It is not known how many housecarls fell at Stamford Bridge, but probably the muster in London was not much above 2,000, and the levies that had rallied to Harold's banner as he marched north had also been sadly depleted in the fight. *In London Harold was faced with two options:*

(1) He could wait there until those of his troops who had fought at Stamford Bridge had recovered from their exertions, and new levies had been assembled – perhaps a period of a fortnight, during which time he would have to be prepared to resist a Norman attack should William decide to move against him.

(2) He could march within a week, leaving some levies to follow on and gathering others as he went, in an attempt to surprise the Normans (as he had the Vikings) and drive them into the sea before they became properly established.

Harold decided on the second course, and left London probably on 12 October. It was a wrong decision, and by it he threw away his best chance of winning the battle. Harold's worst fault was impetuousness; he always longed to be at the throat of his adversary, and this was particularly so now, for he was roused to anger by the destruction of even a small part of his beloved Wessex. William had thrown down the gage; Harold accepted it without giving thought to the problems confronting his enemy. Had he done so he would have realized that these held the secret of possible success.

Meanwhile, on the other side of the Channel, William of Normandy, casting aside all disguises, was preparing to take by force what he claimed was his by right. When he had received the necessary support from his innermost council, and the blessing of the Pope for the holy mission against England that he was about to undertake, he set about, with his customary vigour, persuading the barons to raise and equip a prescribed number of mounted knights and foot soldiers, and to assist with materials and money for the building of a fleet. We do not know how large a force William transported across the Channel, but it seems most likely that it would have been around 8,000, of which about 2,000 would have been mounted knights. There was no Norman fleet of any consequence, and it is unlikely that the small craft, which were hewn by the woodcutters' axes from the forests above the town of Dives, could hold more than twenty men, so about 1,000 boats had to be constructed to carry the army and necessary non-combatants, 2,000 or so horses and a certain quantity of provisions and war materials.

At this time there was no armour for horses, but the knights themselves wore long coats of mail slit at the bottom to fall over the saddle, cone-shaped helmets fitted with nose-pieces, and neck-guards of fine material which were usually attached to the hauberk. Their weapon was a lance or sword, and they carried kite-shaped shields. The infantrymen wore leather hauberks covered with flat iron rings and quilted; their helmets were similar to those of the knights. They carried shields large enough to protect most of their body, and fought with swords, spears and short axes. The archers wore no mail and used the short bow, which had a maximum range of 150 yards.

Throughout the long summer months William's iron discipline and inspired leadership kept his large army from disintegrating while the

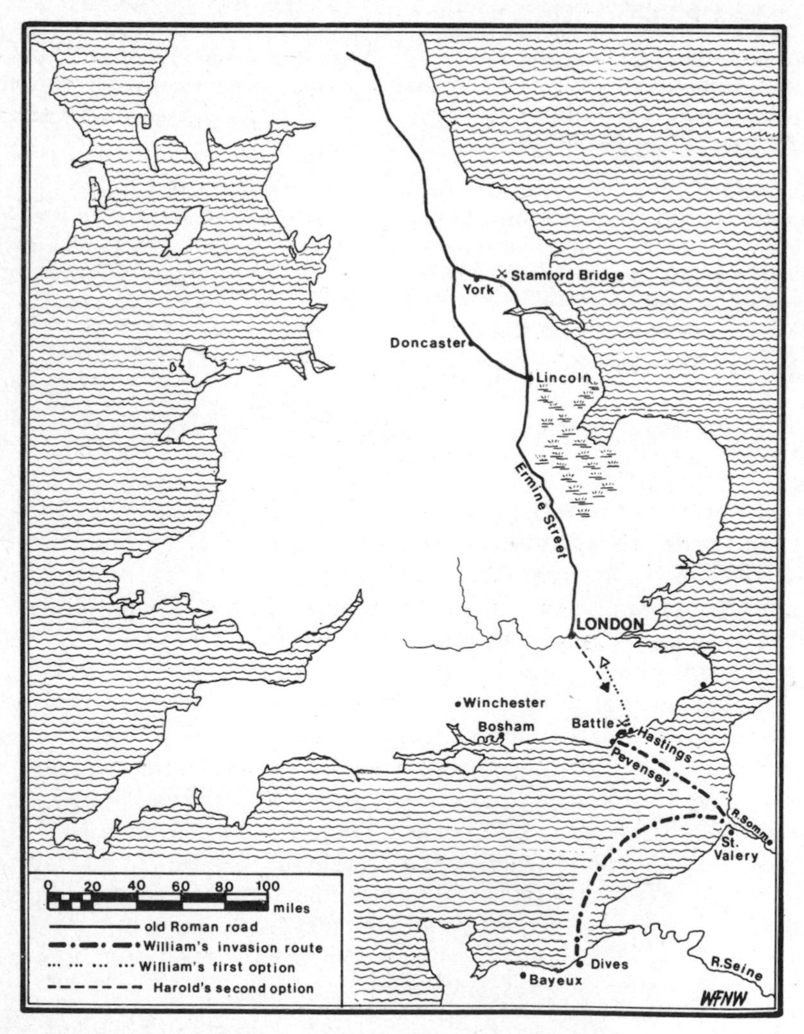

1 The Battle of Hastings: Stamford Bridge and the Norman invasion

ships were being built. At the beginning of September, when he learned that Harold had had to disband the southern fyrd, he sailed his troops round to St Valery, at the mouth of the Somme, and awaited a favourable wind to blow him to England. This at last he got on 27 September, and as we have seen he disembarked his army at Pevensey on the 28th. On the next day he removed both men and ships to Hastings; situated on a narrow peninsula and protected by the Brede and Bulverhythe estuaries, Hastings was an ideal place for a covering action should withdrawal on the fleet become necessary.

The landing had been entirely unopposed, but William would not have known the whereabouts of Harold and his army, for news of Stamford Bridge is unlikely to have reached him for a few days. Time was an important factor for him, for if he could not win a decisive battle quickly he would have to return to Normandy. He would be unable to feed his large army off the country, and by waiting too long he might risk a sea battle – for he probably did not know the extent to which Harold's fleet had been damaged on returning to London earlier in the month, after patrolling the South Coast. *There were two options open to William:*

(1) He could advance on London, leaving a small detachment to guard the base camp and transports.

(2) He could stay close to his base, and wait for Harold to come to him.

William decided to stay close to his base, and as bait for Harold – whose character he knew very well from their association on the battlefields of Normandy – he decided to lay waste and burn the surrounding countryside. He was a good commander, and well aware that the farther he moved from his base the greater became his perils and his problems. It was a bold decision, and almost certainly the right one, but for several days he was a prey to much anxiety, and the chroniclers tell us that he was agitated, morose and unsure of himself until he learned that he had calculated the risk correctly.

Harold's precipitate decision to leave London within a week of his arriving there from the north meant that he began his march with probably no more than 5,000 men. Some contingents would have met him en route, and others would have been ordered to march direct to the rendezvous. This was a hoar apple tree (such trees were often prominent land and boundary marks), which stood just beyond the southern boundary of that massive forest called the Andredsweald, probably on what is now known as Caldbec Hill, some sixty miles from London.

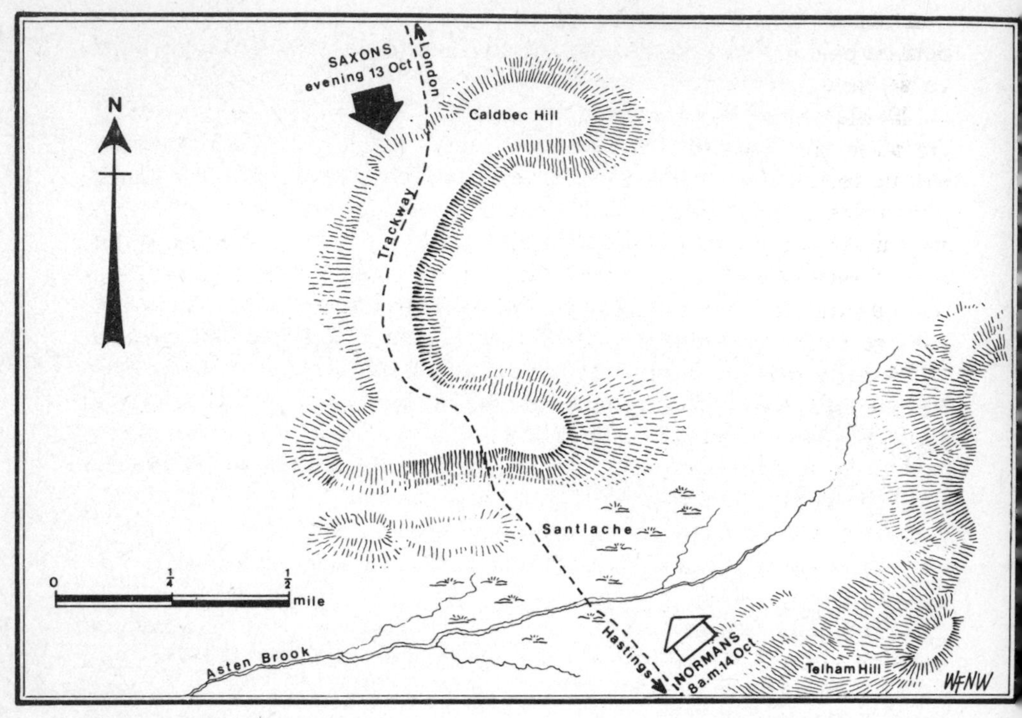

2 Hastings: the battle ground

An examination of the ground over which the battle of Hastings was fought reveals that with the exception of an inroad into the eastern part of the ridge, where Battle Abbey House was built, the shape and contours of the land must have been much the same in 1066 as they are now. But the condition of the ground was, of course, totally different. From Caldbec Hill there runs a narrow neck of land (now largely occupied by Battle High Street), and from this isthmus the ground drops away steeply on both sides – particularly to the east. At the point where this narrow ridge begins to dip into the marshy area, called in medieval times Santlache and known to history as Senlac, there is a fairly level cross ridge stretching for approximately half a mile east and west. The slope down from this Senlac ridge to the west was not particularly steep, but to the east and south the slopes were steep enough to put attacking troops at a grave disadvantage. On the Hastings side of the slightly undulating valley the ground rises gradually to Telham Hill, the summit of which is a mile from the Senlac ridge and nearly 200 feet higher; as the land – with the exception of a few trees – was open, the ridge was clearly visible from Telham Hill. However, the ground between these two high points, especially that on the west, must have been very wet and marshy even in October, for it was intersected by a number of streams and by the Asten brook. The southern slopes of the Senlac position were most probably uncultivated, rough bracken and gorse land.

Harold would have reached the rendezvous on the evening of 13 October, and much of his army, well strung out and utterly exhausted from trudging miles over rough and rutted tracks, would have trickled in throughout the night, finding their way by the light of a waning moon – it was twenty-two days old. By dawn the next day his army would still have been fewer in numbers than that of the Normans, and, as already mentioned, he had no cavalry. He could expect to receive reinforcements throughout the morning, however, and in fact when fully mustered his strength was probably slightly in excess of William's total force of 8,000. *Once again Harold was faced with two fairly clear-cut options.*

(1) He could advance to the attack.

Harold chose to defend the Senlac ridge, and it could not have been a very difficult decision even for someone as impetuous and determined as Harold. His army was tired and incomplete, so a dawn attack was quite out of the question; surprise was no longer possible, for William would have had plenty of warning of the Saxon approach, and to pit the shire levies against the Norman mounted knights in open country would have been disastrous.

(2) He could take up a defensive position on ground of his own choosing.

The Senlac position offered the finest defensive position in the neighbourhood, and it is probable that Harold was familiar with it, and fully understood the value of an offensive-defence. To invite the enemy to attack a difficult position and in so doing to blunt their cavalry weapon, and then to go over to the offensive with part or all of his force and roll back a tired and dispirited army, was his best hope of victory.

The Norman army left the Hastings camp at dawn on 14 October, and shortly before 8 a.m. the head of the column (which stretched for about three miles) would have reached Telham Hill. Here, while his army closed up and prepared for battle, William rode forward to the point now occupied by Glengorse School, where the Senlac position first became visible. He would have seen that the Saxon line extended from east to west for a little under 800 yards, almost square to the ancient trackway, thus barring his road to London. The left stretched to the area of the present primary school, and the line cut diagonally across where the east-west abbey ruin now stands, with the right flank resting on the easier slope close to the fence that now divides the park from the abbey grounds. There were no effective defence works in front of the Saxon line,

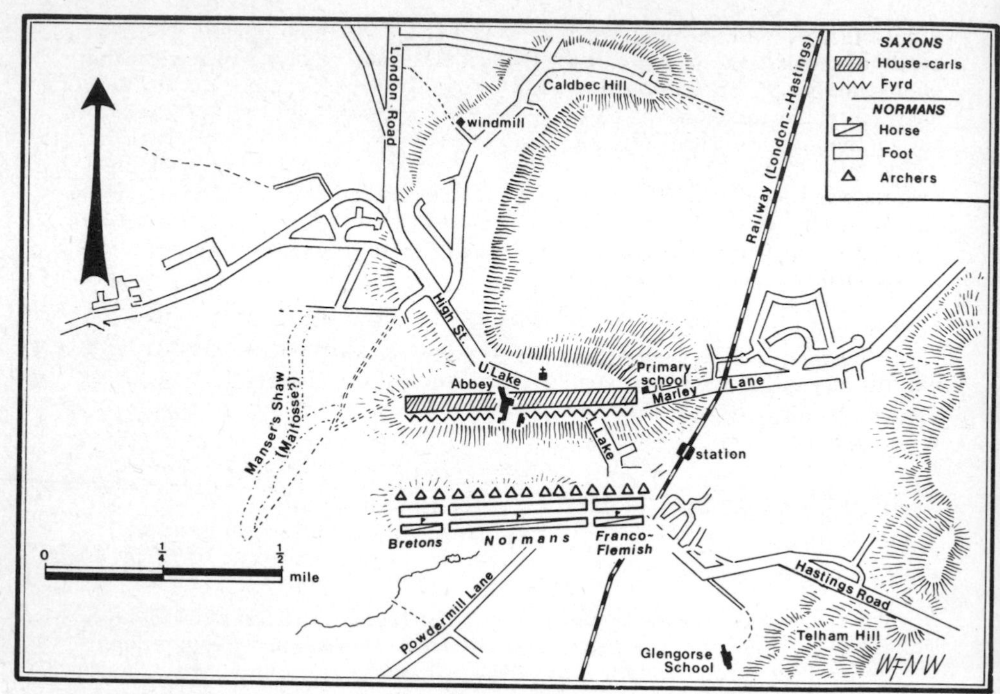

3 The Battle of Hastings: the estimated positions of the Saxon and Norman armies, and the modern town of Battle

but the steepness of the ridge at Harold's left and centre, and the marsh land to his right front, made it a natural defensive position. The close-packed ranks, even at that hour many rows deep, with the formidable house-carls in their long chain mail proudly arrayed on this famous ridge, would have left Duke William in no doubt about the magnitude of his task. *There were three options to him:*

(1) He could advance his bowmen to engage the enemy at their maximum range, and, while the Saxons were held, attempt to turn their right flank with the bulk of his army.

(2) He could carry out a frontal attack in two or three distinct phases, using his archers to soften up the enemy, then advance his men-at-arms, keeping his cavalry in reserve for the coup de grâce and pursuit.

(3) He could carry out a combined offensive using all three arms, by passing the infantry through the archers while the latter still fired, and bringing the cavalry up (or most of it) behind the infantry to press the attack home.

William chose a frontal attack (option 2), which was predictable for two reasons. Frontal attacks were the orthodox tactics of the day; it is true that Harald Hardrada in the heat of the battle of Fulford employed a most successful left hook, but the setpiece battle of those days did not as a rule offer much scope for turning the flank, and William probably never considered such a manoeuvre. Even if he had he would quickly have seen that the ground was unsuitable, for on his right the ridge was far too steep, and the boggy ground on his left flank greatly diminished the chances of success.

Having decided to attack frontally he deployed his army (probably in the valley between the present railway line and Powdermill Lane) on a three-division front. In each division there were three echelons; the first was composed of bowmen, the second of men-at-arms, and in the rear came the mailed and mounted knights. When extended the line would have covered the Saxon position with the Franco-Flemish mercenary contingent under Roger of Montgomery on the right, to attack that section of the Saxon line that lay in the area of the primary school–Chequers Hotel. In the centre came the Norman division, larger than both the two wings put together and under the personal command of the Duke; and on the left were the Bretons and men from Maine and Poitou commanded by Alan of Brittany. Their left flank rested originally immediately south of the small hillock, which is still clearly visible.

William's battle order would have served equally well for a combined attack by all arms, or the employment of each arm independently. He in fact decided on the latter course (option 2)

It is unlikely that the Norman army could have been fully assembled on Telham Hill before 8.30 a.m., and it is difficult to see how a long column could form line in less than an hour. It is therefore unlikely that William's archers advanced to the attack before 9.30 at the earliest. These bowmen made virtually no impression on the Saxon phalanx; when they came in range they proved very vulnerable to the shower of well-aimed missiles, which accounted for many of them. Moreover, as Harold was virtually without archers, the Normans were soon left with no arrows.

This phase of the battle could not have lasted much more than an hour, before William realized what little impression was being made on the enemy ranks and sent in his infantry – tough men, most of them schooled by many seasons of campaigning. Their struggle with the house-carls was a bloody one and both sides fought with great courage. Dents were made in the Saxon line, but these were quickly filled by men from the rear, and this massive assault by the men-at-arms failed to make the expected penetration through which the cavalry might sweep. What exactly happened next is uncertain, but a part of the cavalry – whether under direct orders from the Duke, or in an independent display of élan by some of the knights – passed through the infantry and went into the attack. But the boggy ground on their left and the steep slope in the centre and right gave them insufficient momentum, and losses were suffered. There then occurred a vitally important incident.

The Breton knights on the left gave way, and in their retreat rolled back their own infantry, and this exposed the flank of the Norman division in the centre, who were forced to retire precipitately. All this proved too much for some of the less disciplined shire levies positioned on the right of the Saxon line, and disregarding their King's order before the battle that on no account was any man to break ranks, they charged down the slope in pursuit of the fleeing Bretons. For a brief period the Norman situation became critical; their whole line was in confusion, and William had become unhorsed.

Harold must have been deeply annoyed to see a part of his right wing, for whatever reason, disappearing down the slope (some think that his brothers Gyrth and Leofwine may have ordered a counter-attack); he was a good enough general to realize that if William could halt the limited rout his now hopelessly vulnerable men would be massacred. He had to make a snap decision. *The two options were:*

(1) Should he take the opportunity afforded by the temporary chaos of the Norman army to order a general advance to sweep the enemy off their feet?

(2) Should he remain on the ridge and fill his broken ranks from reserves that had recently arrived?

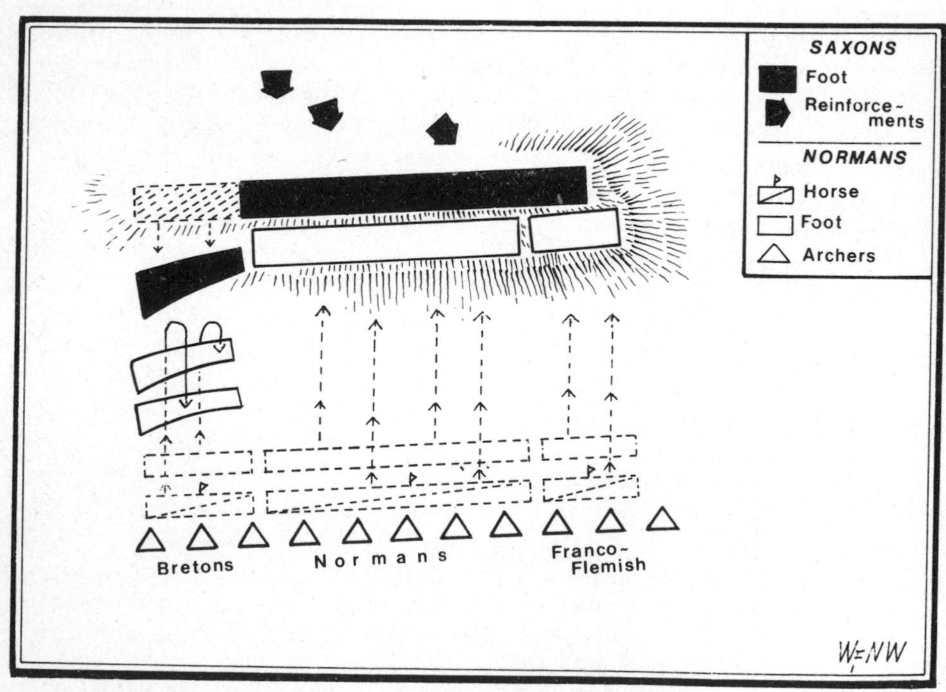

4 *The Battle of Hastings: first stage, the Saxon right pursues the recoiling Bretons*

Harold resisted the temptation to advance, and he was almost certainly right. The chance of success, if it ever was there, was fleeting and as it happened – for William was quick to remount and rally his men – would probably have been gone by the time Harold could have got his army properly under way. A considerable section of the Norman cavalry had not been committed and could have played havoc with a loosely organized charging fyrd. Harold sensibly concluded that it was too early in the day, with the enemy not greatly diminished in numbers and possessing superior close combat armour and armament, to abandon his strong position and risk all on one desperate charge.

As soon as William had remounted, and in his stentorian voice made it abundantly clear that he was unhurt and in command, he quickly swung what cavalry had not become engaged on to the now isolated and forlorn Saxon enthusiasts. Much slaughter took place around the hillock, and not a single Saxon regained the Senlac ridge. But by this time Harold's army had reached its full strength, and there were men available to fill the gap.

During the long day there must have been more than one pause in the fighting, for no troops wearing chain mail could have kept up a sustained effort over a period of almost nine hours. One such pause would very probably have come at this critical stage of the battle. William would surely have welcomed any temporary cessation in the fighting in order to formulate a new plan, for the original one had sadly miscarried. At some stage of the battle the Norman archers were resupplied with arrows, but probably at a later hour than this.

If we assume that the archers were out of action, *William still had three options open to him:*

(1) He could reform his infantry and send them in again, but this time closely supported by the cavalry.

(2) He could throw in the cavalry unsupported.

(3) He could attack with either arm or both, with a pre-arranged feigned retreat in order to lure the enemy off the ridge.

We cannot be absolutely certain whether William took the second or third option posed, but it seems more likely that it was the second, and that the Norman retreat (this time on the Franco-Flemish front) was genuine and not feigned. There would have been time, it is true, during the pause for regrouping, to give orders for such a manoeuvre (and two of the most reliable of the chroniclers – William of Poitiers and Guy of Amiens – assure us that this did occur), but a feigned retreat is a very difficult operation of war to organize and control, even in a well-disciplined army such as William's.

In any event a large proportion of the mounted knights, the corps d'élite of the Norman army, were to ride

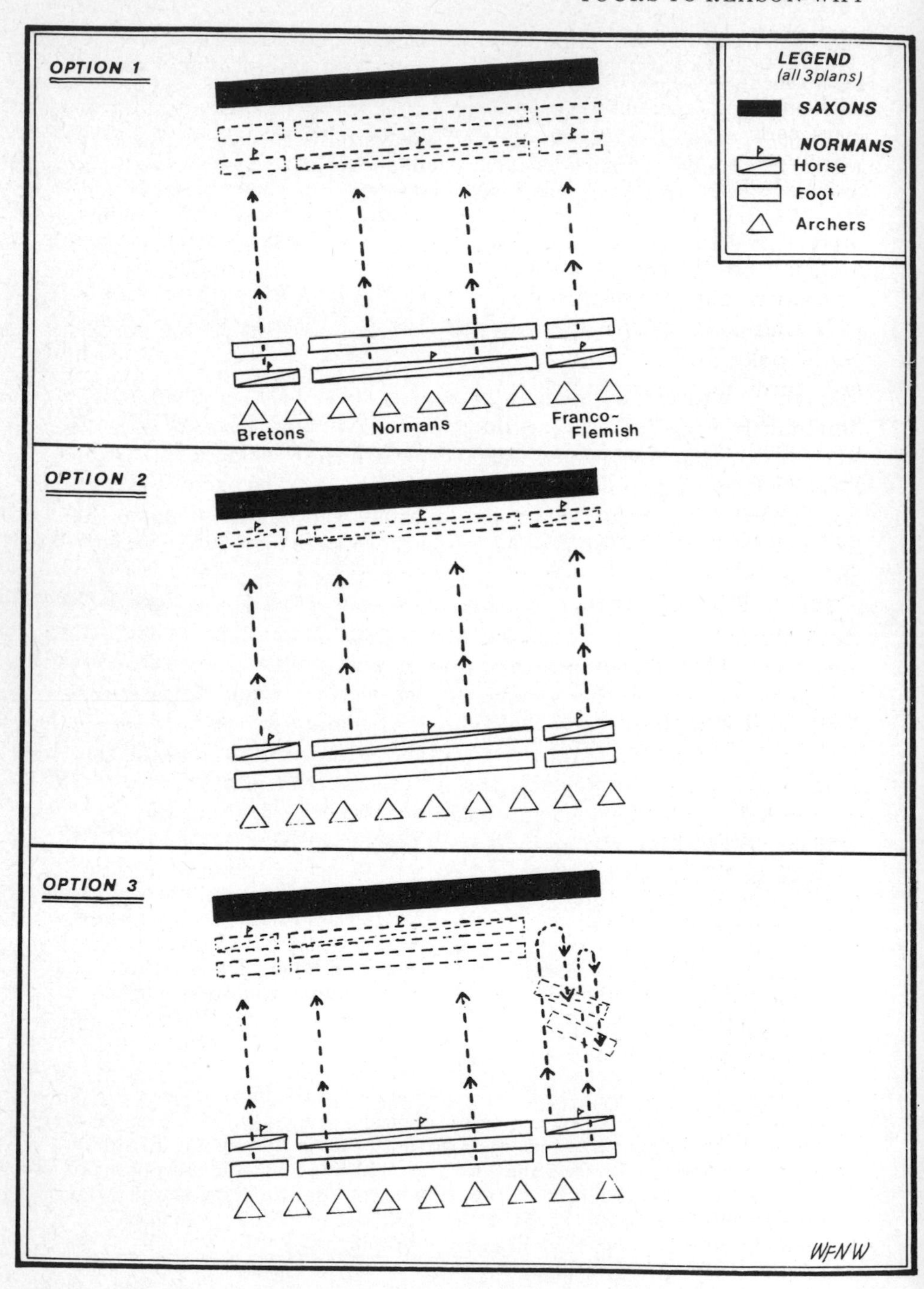

5 The Battle of Hastings: second stage, William's options

unsupported against the Saxon line – still almost intact. The annals of war record many instances of cavalry being unable to break good infantry, and the house-carls, although comparatively few in number, were probably the finest infantry in Europe at that time. To launch the knights unsupported in an uphill assault against these stalwart fighters, whose great axes could cleave man and horse at a single blow, was a decision that asked for trouble – and got it.

The day would have been more than half spent by the time the knights rode once again into the breach, and as in their previous attempt they could make little impression on the valiant house-carls, supported by the sturdy men from the shires. Men and horses floundered in vain among the dead and dying. Such persistent hammering undoubtedly weakened the Saxon phalanx, but the greatest damage was caused by a repetition of the Breton incident, this time on the left of Harold's line. The retreat of Roger of Montgomery's knights – feigned or real – once more brought the Saxon fyrd charging down the slope with disastrous consequences, for William, profiting from the first success, was quick to order up his reserve cavalry, which dealt swift destruction to these impetuous men.

It must have been somewhere around 4 p.m. when it became obvious that the unsupported cavalry attack had failed. The situation for William now appeared desperate. Not much more than two hours of daylight remained; the Anglo-Saxon army, although considerably thinned and holding less ground, still stood rocklike upon the ridge, defiant and full of fight. It had to be defeated quickly, for night might bring reinforcements from the north. For seven hours most of these men had stood steady, successfully hurling missiles and defiance against every assault – as William of Poitiers wrote, 'The only movement was the dropping of the dead; the living stood motionless.' William had to make a new plan, and this time it had to succeed.

After many hours of thrusting to the attack over steep and difficult ground, with outwardly little reward for their endeavours, the Norman army would have been very tired and dispirited. The last phase of the battle would have cost them dearly in horses, and many of the knights would now have to fight on foot. William's archers, would have been comparatively fresh and their ammunition replenished.

There were again three options open to William:

(1) A flank attack on the left with the archers holding the enemy front – similar option to the one he had had at the outset of the battle, but at this hour the chances of success were greater. Harold's right wing was now much weaker and without its commander, Harold's brother Gyrth, who

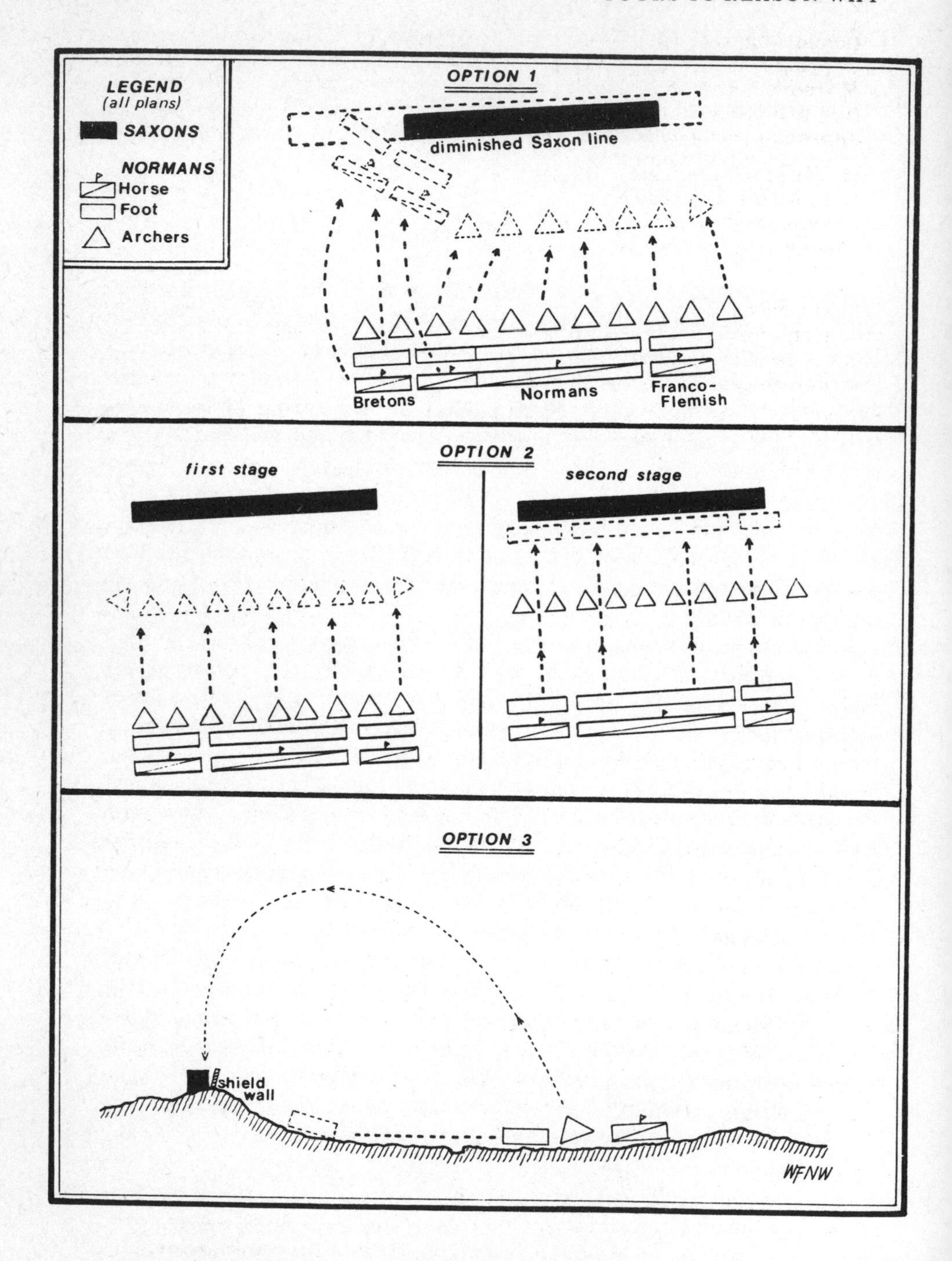

6 The Battle of Hastings: third stage, William's options

(according to the Bayeux Tapestry) had already fallen. Moreover, now that the enemy line had been shortened the approach was easier.

(2) He could send his archers in to soften up, and then advance the infantry, keeping what was left of the cavalry in reserve for an emergency or to exploit success.

(3) A fire and movement exercise, or indirect fire by the archers with the infantry moving in to the attack simultaneously, the cavalry being kept in reserve.

William decided on a fire and movement operation. It was the best tactical decision of the day, for although he might have won the battle by exercising any of the three options, the idea of getting the archers to cover the attacking troops by high-angle firing, thus allowing his infantry to come to close quarters while the enemy shields were being held aloft against falling arrows, gave his men their greatest chance of gaining a vital foothold on the ridge with minimum casualties.

As the Norman infantry passed through the line of high-firing archers in William's last great push, the solid phalanx of Saxon soldiers began to crack. The gaps made by those who had left their posts and died on the slopes could no longer be filled; the Normans were able to gain a footing on the plateau, and once upon the level ground they began to drive wedges into troops now assailed even from the sky. Both flanks were caving in; Harold's brothers had already been slain, and the enveloping pincers closed remorselessly on Harold's command post, where the house-carls still fought devotedly. Amid the chaos and carnage the English King was wounded by an arrow in the eye, and although he remained on his feet a report spread that he had been killed, and panic – always threatening at moments like this – took its grip upon the shire levies.

Somewhere around sunset on that memorable day a body of four Norman knights burst through to where Harold stood leaning upon his shield in fearful agony, and before the house-carls could come to his aid they had cut him down. Those of the valiant house-carls who remained standing closed their ranks, and under the twin standards of the golden dragon of Wessex and the banner of the Fighting Man they fought on bravely, lapped about by the enemy, who yard by yard strengthened their grip upon the ridge. The position became no longer tenable, and the levies, who had had more than enough, broke and fled, hoping that night and the forest would save them. But the Conqueror, like the good general he was, sent his cavalry in immediate pursuit. Victory was complete, and the Norman dynasty was born.

2 The Hundred Years War

The Crécy Campaign

THE date of the beginning of the Hundred Years War is usually said to be 24 May 1337, when King Philip VI of France declared that Guyenne (the duchy, which the kings of England had held as feudatories of the kings of France) had been forfeited by Edward III because of various rebellious acts. In October of that year Edward, foreshadowing his claim to the French throne, addressed a defiant letter to 'Philip of Valois who calls himself King of France'. Shortly afterwards Philip launched a massive attack on Guyenne.

The war thus begun falls into four parts: the successful invasions of France by Edward III, which terminated in 1360 with the Treaty of Brétigny; a long period of decline during which the English lost control of almost all their continental possessions; the triumphant return by Henry V; and the final eclipse of the English, rather more than a hundred years after the opening of this senseless war, when all that was left to them from their desperate endeavours was Calais.

The causes of the war are too complex to be gone into in any detail here. Immediately there were three: the disputed succession, Flanders and Scotland. But the antagonisms stretched back over two centuries. Ever since Henry II, through marriage and inheritance, found himself in possession of the whole of western France, he and his successors had held their French lands as vassals of the French king. This was the cause of much bitterness, especially as the English kings were often the more powerful, for England had been for many years a cohesive nation, whereas it needed the Hundred Years War before the colourful threads of medieval France were firmly sewn together. Thus we have a cause of constant friction, which was still at work when Edward III ascended the throne, and in 1329 went to Amiens to pay homage for the Duchy of Guyenne and the County of Ponthieu.

Edward did not go to war over the succession to the French throne, even if he did think that as a nephew (through his mother, Isabella the daughter of Philip IV) of the last three French kings he should have been chosen in preference to their first cousin, for he did not put forward his claim officially until 1340, and then principally at the request of the Flemings. Flanders was of the utmost importance to England, chiefly on account of the wool trade, but also as a useful stepping stone for any invasion of France, and Edward lent a ready ear when the Flemish merchants – who would have starved without English wool – begged him to intervene against their count, whose sympathies towards the French King had led him to suppress English imports. England's attitude towards Flanders was not dissimilar to France's towards Scotland, and Edward's endeavours to fulfil the task begun by his grandfather of conquering the Scots was constantly thwarted by the intervention of the French King.

Such in outline were the causes of the war. But what benefit did Edward hope to get from it? It is difficult to be sure, for he was not a man who spoke his thoughts freely. Self-confident and ambitious, he felt it possible that the conquest of France was not beyond his horizon. But a more limited objective – and one that would set the seal on years of acrimonious wrangling – was the complete sovereignty of Guyenne. And perhaps more than anything else an invasion of France in any form was irresistibly attractive to the chivalrous romanticism of Edward's nature, for he was as splendid in war as he was in love. Strong, courageous and with enormous vitality, at times a little impulsive but with clear judgment and quickness of perception, he was a king very capable of ruling at a time when kingship was a personal affair demanding skill in administration and diplomacy, fairness in justice, and leadership in war.

The war had been in progress for some eight years before Edward embarked on the campaign that was to win him such renown. In 1339 and 1340 the English King was in Flanders, but more concerned with diplomacy than fighting. Edward had been quick to take advantage of the conflicting interests of the Empire, and the duchies, principalities and provinces of western Europe, for this was a time of divided loyalties and much dissembling. By 1338 he had all but ringed Philip with foes, and although his alliances had a habit of collapsing almost as quickly as they were made, he was untiring in his efforts to form new ones. Shortage of money was an embarrassment in 1340, when Edward was forced to leave his wife and family in Ghent as hostages to his Flemish creditors, but his return in June to redeem them was the occasion of the great naval victory of Sluys, which gave him temporary command of the sea. Another dynastic quarrel and disputed succession in Brittany gave Edward the chance to wage war there, and in 1345 he launched a large-scale operation in Guyenne under the Earl of Derby (shortly to become Duke of Lancaster). A year later he was preparing to leave England at the head of an army numbering 15,000 combatants.

This was the largest army that a king of England had so far led across the Channel, and its terms of service had recently been changed. Edward III had found that the old feudal system was failing in its purpose of raising men-at-arms for the Crown, and in its place he substituted a system of written indentures, whereby his commanders contracted to provide a specified number of soldiers (usually of all arms) to serve as paid professionals for a given period of time. He thus provided himself with a long-service professional army, but in encouraging his vassals to raise ever larger numbers of soldiers he unwittingly gave birth to that pernicious 'livery and maintenance' system that became the curse of the Wars of the Roses.

This particular army consisted chiefly of English and Welsh bowmen. There were not many more than 2,000 men-at-arms, and perhaps 1,500 light infantrymen – skirmishers, armed with javelins or large knives. Plate armour was just beginning to take the place of chainmail, but most of the men-at-arms would still have been in mail, for the expense of equipping the necessary retinue of two armed valets and three horses was more than most knights could manage, without the additional burden of the new plate armour – although their French opposite numbers were mostly protected by plate. A surcoat with the knight's arms surmounted the steel hauberk, and the helmet was conical shaped. The principal weapon of the man-at-arms was a sword, but when he was mounted he carried a lance. However, Edward III's men-at-arms usually fought on foot, and then they might have used a halberd. The archers – if they were lucky – were issued with a short quilted coat with iron studs, those that were mounted (an innovation of Edward III) carried a lance.

It is often disputed, but seems fairly certain, that at Crécy Edward had the use of a few very elementary cannon – small-bore tubes known as *ribaulds*, which frightened his gunners even more than they did the enemy horse. But the weapon that gained Edward his victories, and until the advent of efficient cannon gave the English army the professional primacy of Europe, was the longbow. This bow (made from elm) was used in Wales as far back as 1150, but Falkirk (1298) was the first important battle in which it was used, and by then it was being made from yew – Spanish or Italian being preferred to the native species. It stood five feet six inches to over six feet in length, and fired a three-foot arrow with a triangular broad head, the fletching being of goose, or preferably peacock feather. The strings of bows were of hemp, or silk, and dressed with some waterproofing glue. An archer carried a dozen to twenty-four arrows, and a good man could discharge twelve arrows in a minute with considerable accuracy up to 220 yards.

The French army was raised from the feudal array of mounted men-at-arms and the national levy. The French recruited more mercenaries – chiefly Genoese bowmen – than the English. In theory – and indeed in practice – they could raise enormous numbers, but their feudal men did not have to fight for more than forty days outside their own province without payment, and the French treasury was usually short of funds. French knights had little regard for the common soldier, and less for a mercenary; they considered themselves the élite of the battlefield and expected to gain the victory without much assistance. The Genoese used the crossbow, which fired a quarrel fifteen inches long. It was a cumbersome, rather heavy weapon, requiring winding, ratcheting and laying; its chief advantage over the longbow was greater accuracy over a longer distance, but this was nullified by its very slow rate of fire – no more than four quarrels a minute from a good bowman.

While Edward was assembling his vast host at Porchester at the beginning of July 1346, English soldiers were heavily, and on the whole successfully, engaged in Brittany and Guyenne. In Guyenne there had been little fighting during the first seven years of the war, but in Brittany the Earl of Northampton, assisted by the gallant Robert d'Artois (a dispossessed Frenchman in the English service), had fought a number of important engagements, and for four months in the winter of 1342–43, until a truce had been arranged, Edward himself had campaigned there. In June 1345 war broke out again in Brittany, and this time Sir Thomas Dagworth sailed with Northampton. After one or two minor successes Dagworth, fighting against vastly superior numbers, won the battle of St Pol de Léon. This was not only a most prestigious victory, for it was the first battle in France in which the longbow made its mark, but it also had the effect of bringing French troops to Brittany from Normandy, and this just a month before Edward set sail.

The operations in Guyenne had had exactly the same result. Here Edward's Plantagenet cousin, Lord Derby, assisted by that battle-scarred knight, Sir Walter Manny, had since June 1345 been performing prodigies of war. Bergerac and La Réole had been besieged and taken with the courage and coolness of judgment that are the hallmark of good generalship, while Auberoche was a victory that ranks high as an example of strategical and tactical skill. So alarmed had Philip become at what was happening in the Duchy that he had sent his son John, Duke of Normandy, there at the head of an army numbering at least 20,000 men. Shortly before Edward was due to sail from England he learned that this force had invested Manny and the Earl of Pembroke in the strategically important town of Aiguillon.

Thus Edward started his own campaign with the considerable advantage that his generals in Guyenne and Brittany had spread throughout France the might and puissance of English arms. Moreover, he had hopes of confronting the French King with a fourth army. As previously mentioned, Edward's alliances tended to collapse before they could be of real value, and with no people was this more so than the Flemings; the Duke of Brabant and the Count of Hainault (Edward's brother-in-law) had already defected, and in 1345 Jacob van Artivelde, who had usurped power and allied himself to Edward, was murdered. The resulting confusion ruled out Flanders as a jumping-off base for operations against the French King, but a fortnight before Edward intended sailing news came that the men of Ghent, Bruges and Ypres, led by Count Henry of Flanders, were prepared to co-operate with him. Taking advantage of this offer Edward immediately made arrangements for a small but powerfully led force to sail in July to join with Count Henry in the English King's forthcoming offensive against Philip of France.

Edward's fleet of between 700 and 1,000 vessels carrying his 15,000 soldiers sailed down the Solent on 11 July 1346, a previous attempt made six days earlier having been thwarted by gales. Absolute secrecy had been maintained at the time of sailing, and orders for the fleet's destination were given while at sea. *Edward had five possible strategic choices:*

(1) He could sail for Bordeaux to reinforce Derby – who had been requesting assistance – defeat Duke John, relieve Aiguillon and then march against Philip.

(2) He could relieve the pressure on Aiguillon indirectly, by attempting the conquest of Normandy – the Duchy so long the possession of his ancestors, but lost to France by King John – and thereby providing himself with a firm base for future operations.

(3) He could land near the mouth of the Seine and make a dash for Paris, where the French King would seemingly be at his mercy.

(4) He could carry out a march across Normandy with open objectives depending on his progress, but with the conquest of Calais as a possible terminal prize.

(5) He could arrange with Count Henry of Flanders a joint action, whereby the two armies might meet in the area of Amiens, and together engage Philip's army with a total force more equal in numbers than if Edward had engaged on his own.

It came as a complete surprise to everyone in the fleet to learn that their destination was St Vaast la Hogue on the Contentin peninsula. It had been universally expected, and indeed Edward had hinted at it before sailing, that the army was destined to reinforce the troops in Guyenne. It is possible that Godefroi d'Harcourt, a Norman baron who had been dispossessed by Philip and joined Edward, was responsible for the choice of Normandy, for his services there would be invaluable. It is also possible – and sometimes argued – that Edward had no real plan, and was only interested in a chivalrous adventure, with perhaps Calais as a distant objective. His strategy, or rather lack of it, has also been called in question because he

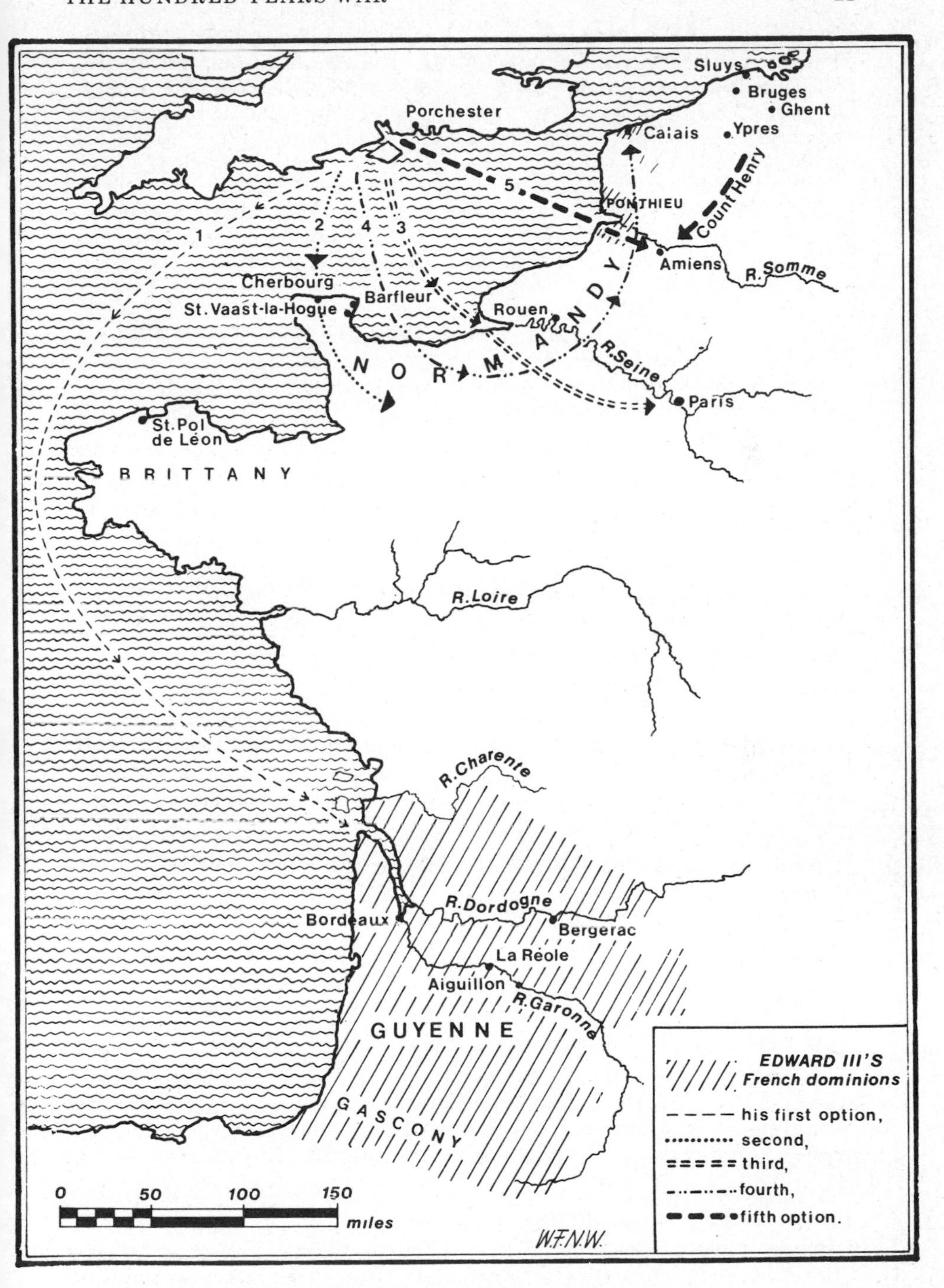

7 *The Crécy Campaign: Edward III's invasion options*

failed to direct his various armies (all fighting on exterior lines, be it noted) to a concerted attack on Philip's army near Paris – but that could never have been his intention, for two of those armies had work to do independently.

One cannot be certain of Edward's overall plan, for neither at the time nor later did he disclose his original intentions. But the grand strategy for the campaign, viewed in retrospect, had a great breadth of vision, and those historians who give Edward the benefit of believing that he had adopted option 5, and think that he had agreed to combine with Count Henry at some central point, are probably right. This might be one reason for the choice of St Vaast as the port of disembarkation, for it and Ypres (from where the Flemings started their advance on 2 August) are almost equidistant from Paris. Another reason was the short sea crossing. However, if the defeat of Philip and the taking of Paris were the ultimate objectives – which seems likely – then Edward had more chance of achieving them by taking the shortest route in the quickest time, before Philip was fully mobilized.

Assuming that Edward did have a plan – and in the opinion of the present writer he was far too good a general to launch 15,000 troops on to the continent of Europe without one – then the *chevauchée* across Normandy, more akin to mass murder than medieval chivalry, with its pleasing plunder, was incidental to the main objective.

The English army disembarked on 12 July, and Edward set about laying waste much of the Cotentin peninsula. Barfleur, Cherbourg and other towns suffered the full horrors of sack and pillage. Five days later the army set off towards Rouen, with the fleet sailing abreast of it; by 26 July they were before Caen, where Edward sent a message to the Bishop of Bayeux, who was occupying the castle, that he would spare this large town if it surrendered. The messenger was promptly imprisoned, and Edward set about storming the place. Something like 2,500 Frenchmen lost their lives, and a number of important prisoners and a vast amount of booty were taken. The bishop was left undisturbed in the castle, which was considered impregnable.

The fleet had sailed up the river to give assistance, and therefore ships were available to take prisoners and booty back to England. The crews of those ships that were not detailed to return to England decided that they had had enough, and sailed without orders. This may have caused Edward some anxiety, but unless he was to abandon the Flemings (and he learned while at Caen that they were into France and approaching Béthune) there could be no falling back to a safe base in Brittany. And so, after a strangely long (five days) period of rest and reorganization, the march was resumed on 31 July. The trail of destruction and misery continued to be appalling; villages and towns were laid waste, and the advance of the English was lit by lurid flames. On 6 August a halt was made some twenty miles south-west of Rouen, and a party was sent forward to reconnoitre the city.

Meanwhile, Philip had taken full advantage of the respite given him by Edward's leisurely looting in Normandy. He had been aware for some time that an invasion was pending, and he had set about raising a very large army – it ended up more than three times the size of Edward's. But until he got news (which he did on 19 July) of the direction that Edward was taking, he had concentrated his troops in the area of Paris. *He now had two options:*

(1) He could attempt to intercept Edward in Normandy, and offer battle with his vastly superior force. This should not be too difficult, for Edward would almost certainly make for Rouen, the capital of Normandy. It would have the additional merit of halting the destruction of towns and villages.

(2) He could remain on the right bank of the Seine, destroy all the bridges, and await Edward at Rouen.

Philip had no conception of war, which he regarded as a glorified tournament where chivalry and courage counted for more than victory. He was indecisive, impulsive and ill-advised; very rarely did he take the initiative, and only once in the campaign did he steal a march on Edward. As a general he was not in the same class as his opponent, but he was a highly intelligent man, and perfectly capable of appreciating a difficult military situation, even if he had a problem with its solution.

On this occasion he chose the passive role (option 2) and his army marched into Rouen on 2 August, which was the day the English entered Lisieux, some forty miles away. A more aggressive commander might have chosen to attack, but except for the continuing misery it brought his Norman subjects this cautious strategy was probably the best. He had gauged Edward's intentions as well as anyone could do, and he held most of the cards in a waiting game.

Edward's reconnaissance troops had got close enough to Rouen to be engaged by the enemy, and to discover that the city was held in strength. Edward had no alternative but to seek another less strongly guarded crossing, for somehow he had to get across the Seine. There followed for the English an agonizing week of long marches upstream, and constant disappointments. Every time a bridge was reached it was either down, or – as in the case of the fortified towns of Pont de l'Arche and Vernon – too strongly guarded. Edward found the time to relieve the frustrations and monotony of the march by sacking some of the principal towns on his route, but until they got to Poissy only one attempt to force a crossing

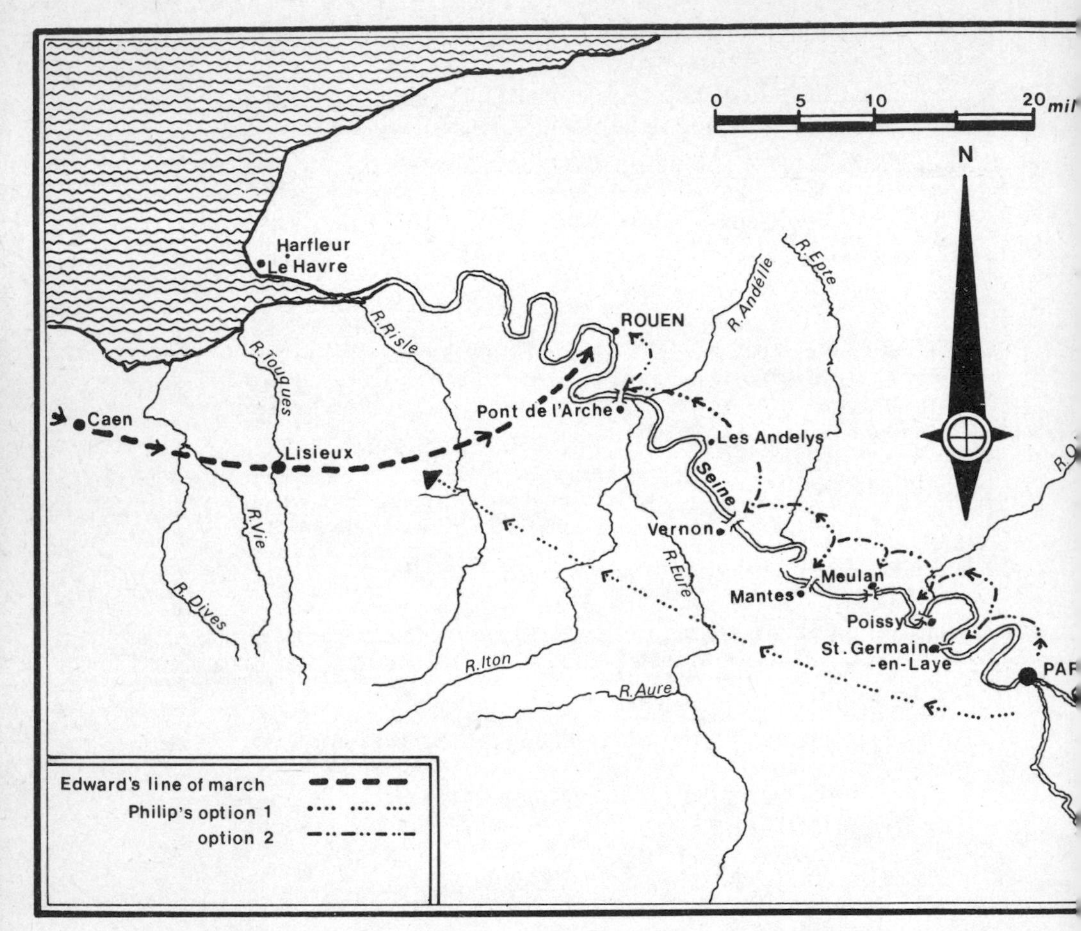

8 *The Crécy Campaign: Edward's advance through Normandy and Philip VI's options*

seems to have been made, when the Earls of Northampton and Warwick were unsuccessful at Meulan.

The French army marched more or less parallel to the English on the opposite bank of the Seine until they were east of Meulan, by which time an English detachment under Edward, Prince of Wales (later to be known as the Black Prince), was seen to be causing some havoc in the suburbs of Paris around St Germain-en-Laye. Philip had made no attempt to attack the English across the river during their march, *but now he had three options:*

(1) He could fall back on Paris with his entire army.

(2) He could hold the river at Poissy, and defend Paris in the suburbs of St Germain and St Cloud.

(3) He could keep a holding force in the western suburbs in case of an emergency, but use his main army to engage the English, when they were crossing the river and at a disadvantage.

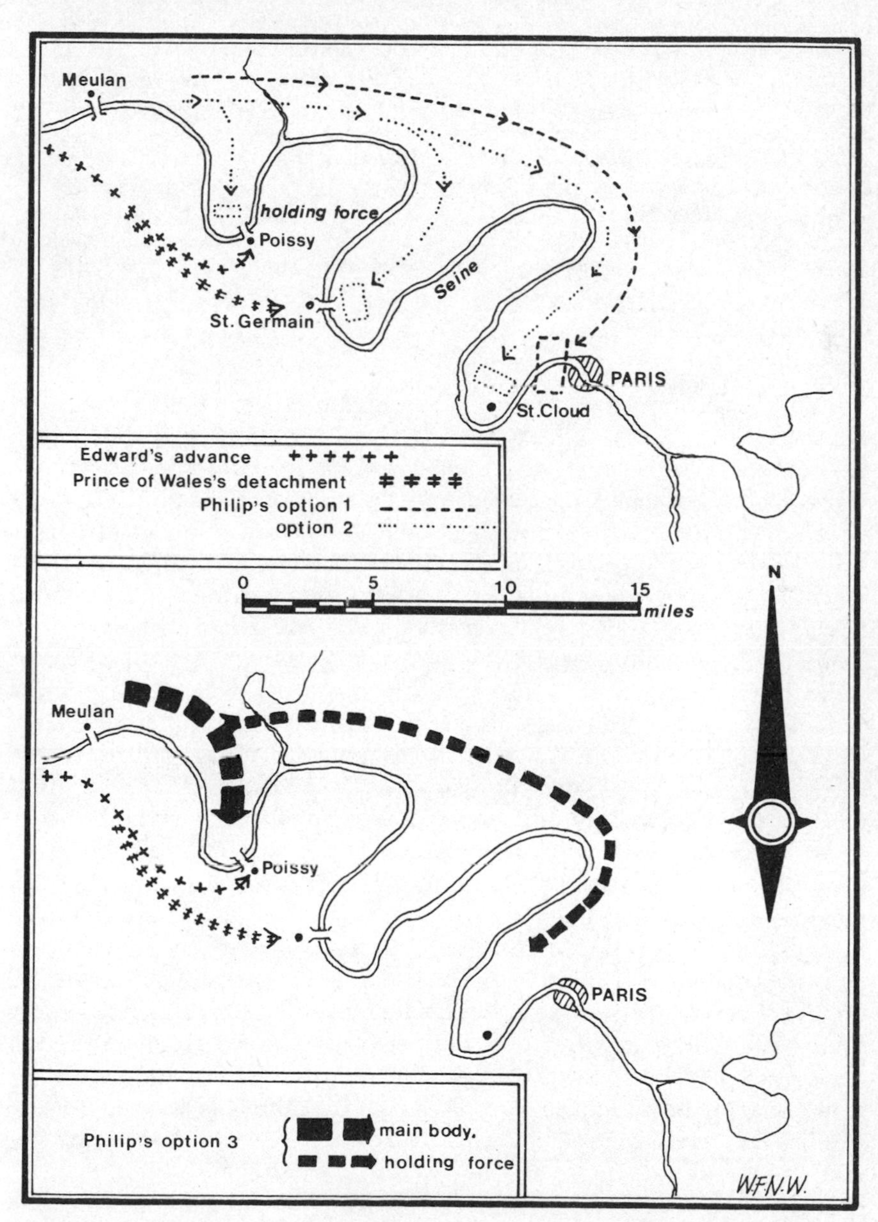

9 *The Crécy Campaign: Edward's march on Paris, and Philip's three options*

Philip fell back on Paris with his entire army (option 1), and appeared to be gripped with indecision, for he marched from one side of the capital to the other without ever engaging the enemy. He had ordered the bridge at Poissy (which Edward reached on 13 August) to be destroyed, but only a small force was left to guard the crossing, and this was soon defeated by an English advance guard, who crossed the river in boats. It took the English three days to repair the bridge sufficiently for the army to cross, and, except for one short engagement with some French troops, thought to have been on their way into Paris from Amiens, they were allowed to complete the task unimpeded. All that Philip did was to send a challenge to Edward for a meeting between the two armies on some mutually agreeable ground. Such acts of chivalry – and those challenges to personal combat – were not unusual between monarchs, but seldom accepted. Philip would have done much better not to have panicked over a few fires in the suburbs of Paris, but to have launched a full-scale attack at the river crossing.

The Seine was only one of two great rivers that Edward had to cross, and his situation was now more desperate than ever. Sandwiched between the Seine and the Somme, with almost certainly no news of his Flemish allies and a vast enemy host breathing down his neck, he had virtually no choice but to get across the Somme, and quickly too. To have accepted the French King's challenge might have been chivalrous, but with his numbers disastrous, and to have sent an immediate refusal could have been equally disastrous. And so Edward resorted to subterfuge; he either accepted a rendezvous to the south of the Seine, or more likely delayed his answer until his army was at Grisy (the chroniclers differ), but the result was the same. He had stolen a march of sixteen miles before Philip learned the position. There followed two remarkable marches. Edward struck north, and marching at a great pace for more than sixty miles scarcely deviated from a straight line – Grisy, Auneuil, Beauvais, Summereux, Poix, Camps and Airaines, where the army arrived on 21 August.

Edward's plan had been to cross the Somme somewhere between Amiens and Abbeville, for although those two towns would be garrisoned there were smaller places where the crossings would be unguarded. At Airaines he would be well situated for this purpose, for there were bridges at Picquigny, Longpré and Pont Remy, and although the march had been exhausting, with no time for looting, the river would seem to be no longer a problem. But his plan was doomed, for the English had no sooner reached Airaines than they learned with surprise and considerable chagrin that the French, marching parallel and a little to the east, had entered Amiens and were across the Somme. They had marched seventy-three miles in five days – a magnificent feat, and easily King Philip's finest manoeuvre.

The English situation had become almost desperate. Food was becoming scarce, the French were now in command of the crossings, and unless Edward could cross the Somme his army seemed certain to be destroyed. He had perhaps three options, and all of them must have seemed fairly forlorn chances, even to a man of Edward's self-assurance. *These were his options:*

(1) He could try to force one of the crossings between Amiens and Abbeville.

(2) He could retrace his steps and attempt to cross the river much higher up – an operation that had proved successful in crossing the Seine.

(3) He could march west in the direction of Oisemont in order to conceal an attempt to cross the river near its swampy estuary between Abbeville and the sea.

Edward chose the third option. It is true that he did attempt to force the bridges at Picquigny, Longpré and Pont-Remy, but it was more in the nature of a strong probe with a force commanded by the Earl of Warwick, and when the Earl was engaged and repulsed at all three crossings Edward did not try to reinforce failure. To retrace his steps and try to cross the river higher up could have had no appeal to an army that had already marched itself almost to a standstill, and this time Philip was unlikely to refrain from attacking. The third option offered the best chance of success, if the intention could be concealed, even briefly, from the enemy.

Whether or not Oisemont was sacked, as Froissart states – and it seems unlikely that time could have been spared for this – the retreat to the estuary was certainly well inland from the river, and directed first on Acheux and then on Boismont. To find a ford was of vital importance, and the locals proved stubbornly loyal to Philip; but at Mons a prisoner, named Gobin Agache, was found willing to sell his country for a large sum of gold, and he disclosed that the only practicable ford was situated at Blanche-Taque.

Philip's indecisiveness in Paris had been atoned for by his splendid march to get behind the Somme before Edward could make the crossing. He was now back in the commanding position that had been his all the way along the Seine from Rouen to Poissy. *Like Edward, he also had three options:*

(1) He could continue his cat and mouse tactics of watching the struggling Edward from the right bank of the Somme, and wait on events.

(2) He could cross the river and give battle with the greater part of his army, leaving a skeleton force to guard the crossing places.

(3) He could make certain that he had sufficient troops on the right bank to hold all the crossings, and go in pursuit of Edward with the rest of his army.

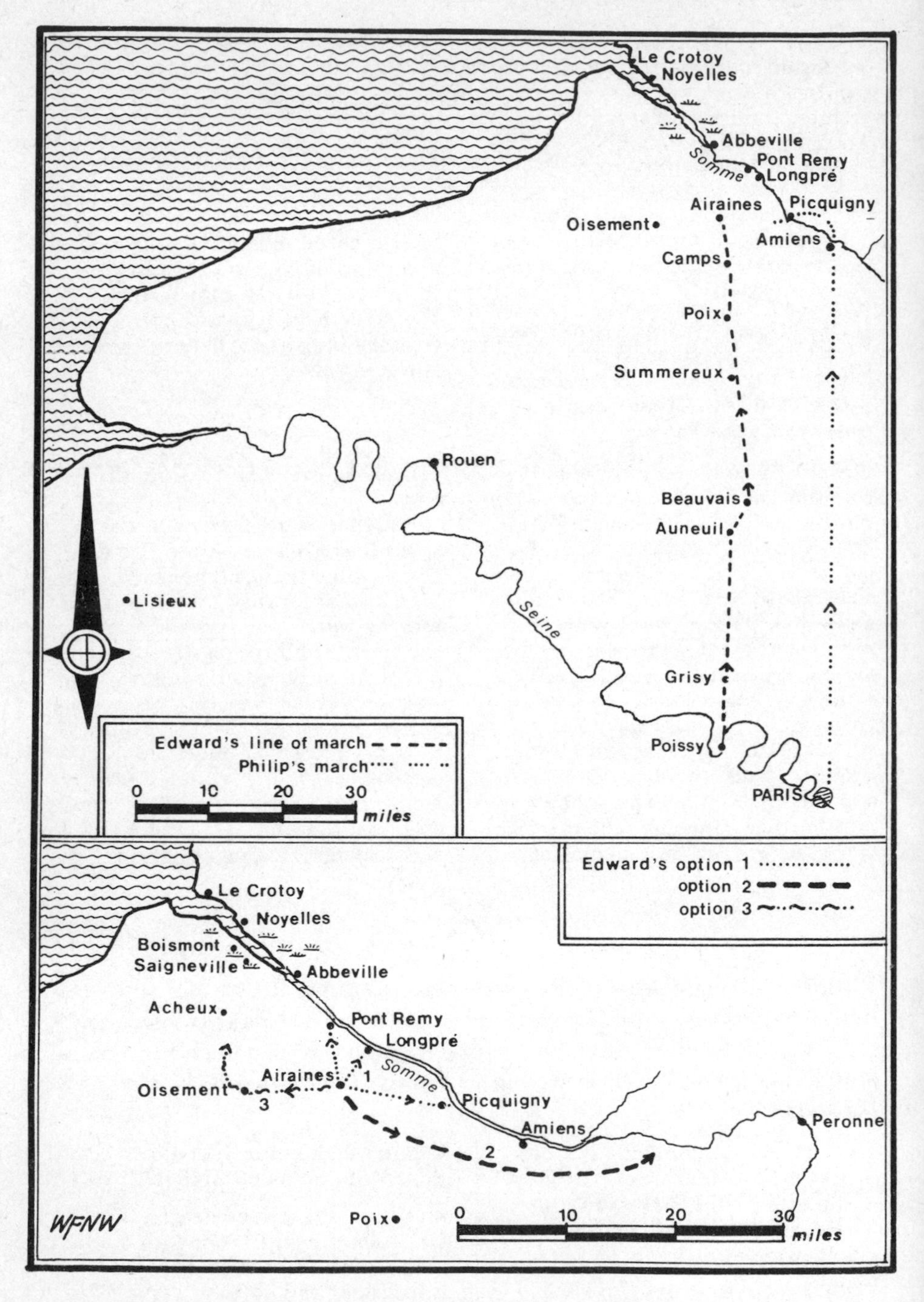

10 The Crécy Campaign: Edward's and Philip's marches to the Somme, and Edward's options for crossing

This time Philip took the initiative (option 3); after the English had been repulsed in their attempts to force the bridges, and had just moved out of Airaines, he arrived in that town having crossed the river (probably at Picquigny) at the head of a large force. To hold the crossings strongly meant dividing his army, but it must have been the right decision to try to bottle up the English.

Edward's crossing of the Somme is among the most exciting events of English military history. The safety of his entire army depended upon the word of a traitor, and the speed of an ebbing tide, for the French host was closing on him, and Agache had said that the causeway – a belt of firm marl amid the green mats of sinuous slime that then comprised the Somme valley – was passable only at low tide. The moon was almost full on the night of 23–24 August, and the army began its march from Boismont in the early hours of the morning. Saigneville marked the edge of the marshy estuary – over 2,000 yards wide at this point – and when the leading troops of the long column reached the steep bluffs that bounded the marsh there stretched before them a broad shining flood of silver, for the tide was still high and the moon was still up.

The army had marched in single column, with Warwick commanding the van, and the sun was beginning to rise before the last man had reached the edge of the flood-water. The tide was starting to ebb, but four hours were needed before the treacherous swamp could be waded, and Philip was narrowing the gap all the time. Nor had he left the ford unguarded. Godemar de Fay with 3,000 soldiers was ready on the far bank to deny the English a safe landing, so Philip had every reason to be confident of victory.

It was about 10 a.m. when the first English archers and men-at-arms splashed into the water; wading knee deep they came within a few hundred yards of the farther bank before de Fay's Genoese bowmen released their quarrels, and his mounted men rode into the river. There was a very brisk engagement, and although the French were outnumbered it cannot be said that Philip had miscalculated, for de Fay had the great advantage of defending a crossing against troops having to debouch from a narrow column in the middle of a river. De Fay might well have prevailed but for the intense and destructive fire of the longbow which here – as throughout the campaign – proved decisive.

The English van had soon pushed the enemy back, and formed a bridgehead while the rest of the army crossed. But it was a very close-run thing, for as the last troops were crossing the causeway the French appeared from the direction of Saigneville. Edward's rear division was in no position to save the bulk of the wagons, which had just begun to cross, and much booty fell to Philip. But the army was safely across, and Edward was now in his own county of Ponthieu, and with a good chance of escape to friendly Flanders, or England if needs be. Moreover, Philip had lost the tide, and anyway his army was too large and cumbersome to make full use of the ford.

The English camped on the night of 24 August five miles north of the river near Forest l'Abbeye; raiding and scavenging parties were sent out, and Noyelles and Le Crotoy were both taken. Edward, knowing that the unwieldy French army would have to retrace its steps to Abbeville to recross the river, realized that he was now a good day ahead of Philip. This meant that his retreat was almost certainly secure, for he could cross the Authie at Ponches well ahead of the French, and the Roman road stretching away to the north would be open to him. *That Friday the 25th, Edward had three options:*

(1) He could march north-east to carry out the original plan of joining forces with the Flemings.

(2) He could stand and fight, although he could not be certain on past performance that Philip, in spite of superior numbers, would accept battle.

(3) He could march north with a view to taking Calais, or making the short Channel crossing to Dover.

Edward must have been tempted to adopt the third option, and march for Calais and home, because on the Friday it was the only course he could be sure about. He had almost certainly received no recent news from Count Henry, and as it happened had he tried to make contact with his allies he would have found it difficult. The Flemings, having been forced to raise the sieges of Béthune and Lillers, and being without certain knowledge of Edward's position, had become discouraged and fallen back to the line of the Lys round Estairs. And if he were to offer battle to a vastly superior force (option 2) a good defensive position was essential, and at the time he could not be sure of finding one. However, Edward was a man who preferred the pursuit of glory through the hazards of battle to running for home with nothing to show but the crushing burden of his debts. He chose to stand, and the outcome of the contest vindicated his decision.

On 25 August Edward ordered the army, less his own division, to march by the track that skirted the eastern side of the great Forêt de Crécy; his own men marched through the forest keeping in touch with the foraging parties to the west. At the northern edge of the forest there stands the village of Crécy-en-Ponthieu, and just beyond the houses Edward's keen tactical eye discerned a likely defensive position. Immediately to the north-east of the village there stretches a ridge of some 2,000 yards to the

village of Wadicourt; at the Crécy end of this village the fall of land is steep, but at Wadicourt the decline is very gentle. On the forward slope of the ridge are three terraces – possibly sites of ancient vineyards – and below is the undulating valley known as La Vallée aux Clercs with an almost inconspicuous rise at its south-east boundary. The fertile fields of Ponthieu are cut by winding streams, and in particular the tiny river Maye rises below the village of Crécy. It was this Crécy–Wadicourt ridge that Edward decided to defend, making his command headquarters in the windmill that stood some 600 yards from Crécy.

The English army that had landed at St Vaast in July would have been reduced by now to around 12,000 combatants – perhaps a little less than 2,000 men-at-arms, 8,000 archers, and 2,000 spear and knifemen. The ridge that Edward intended to hold was full long for the size of his army. *In making his tactical dispositions he had five options.*

(1) He could extend his army to cover the whole front with all three divisions (or 'battles') up: the archers and spearmen in the van, positioned in chequer formation to give a better field of fire; the men-at-arms fighting mounted, or dismounted, and what remained of the baggage wagons placed in rear.

(2) He could adopt the same extended coverage, but with the bowmen stationed between the divisions in wedges to provide enfilade fire, and forming a bastion between the divisions.

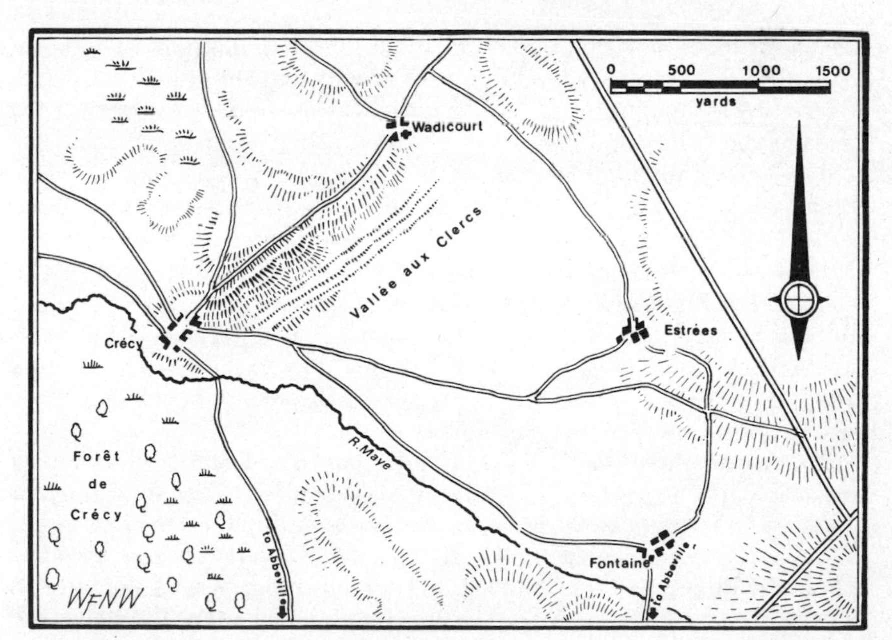

11 Crécy: the topography of the battlefield

(3) He could hold a lesser front in greater depth by leaving small gaps between the villages, and possibly using the wagon park as additional protection to the more vulnerable left flank.

(4) He could hold a still further reduced front of, say, 1,500 yards with a division in reserve, and bowmen protecting the front of all three divisions and the gap between the leading two.

(5) He could hold the same reduced front, but place the majority of the bowmen on each flank of the forward divisions in wedge-shaped formation to give enfilade fire, possibly leaving a gap where the terraces formed a natural defence, which could be covered by the bowmen.

N.B. The decision concerning the placing of the wagon train depended to some extent on whether the men-at-arms were to fight mounted or dismounted. If they were dismounted, the wagons had to form a laager to protect the horses.

Edward adopted option 5. We cannot be absolutely certain that he left the terraces unguarded, but it would have seemed reasonable with his limited manpower, and with the archers in a position to cover the gap, as well as to funnel the enemy knights into a frontal attack on the rows of English men-at-arms, who fought dismounted. The wagon park was placed in the rear, immediately alongside the small Crécy Grange wood.

The right-hand division, nominally under command of the young Prince of Wales but with the Earls of Warwick and Oxford and Godefroi d'Harcourt in support, was placed a little way down the slope within 300 yards of the valley and just above the Maye. The Prince had some 800 men-at-arms and 2,500 archers and infantrymen. The left division, commanded by the Earls of Northampton and Arundel, was a little higher up the slope with about 500 men-at-arms, and the same number of archers as the Prince. The flanks of both divisions fell short of the respective villages. Edward himself commanded the reserve division, which was stationed in rear slightly right of centre. The three cannon seem to have been fired from the left flank. The archers used the time given them by the tardiness of the French to dig potholes in front of their positions to hinder the enemy horses.

Meanwhile the French King, having recrossed the Somme on 24 August, spent the next day regrouping his scattered army. He seemed in no hurry to pursue, but it must be remembered that the French army, unlike Edward's close-knit, well-disciplined troops, was a conglomeration of soldiers, badly organized and owing allegiance principally to their feudal chiefs. It did, however, greatly outnumber the English army – probably by at least three to one.

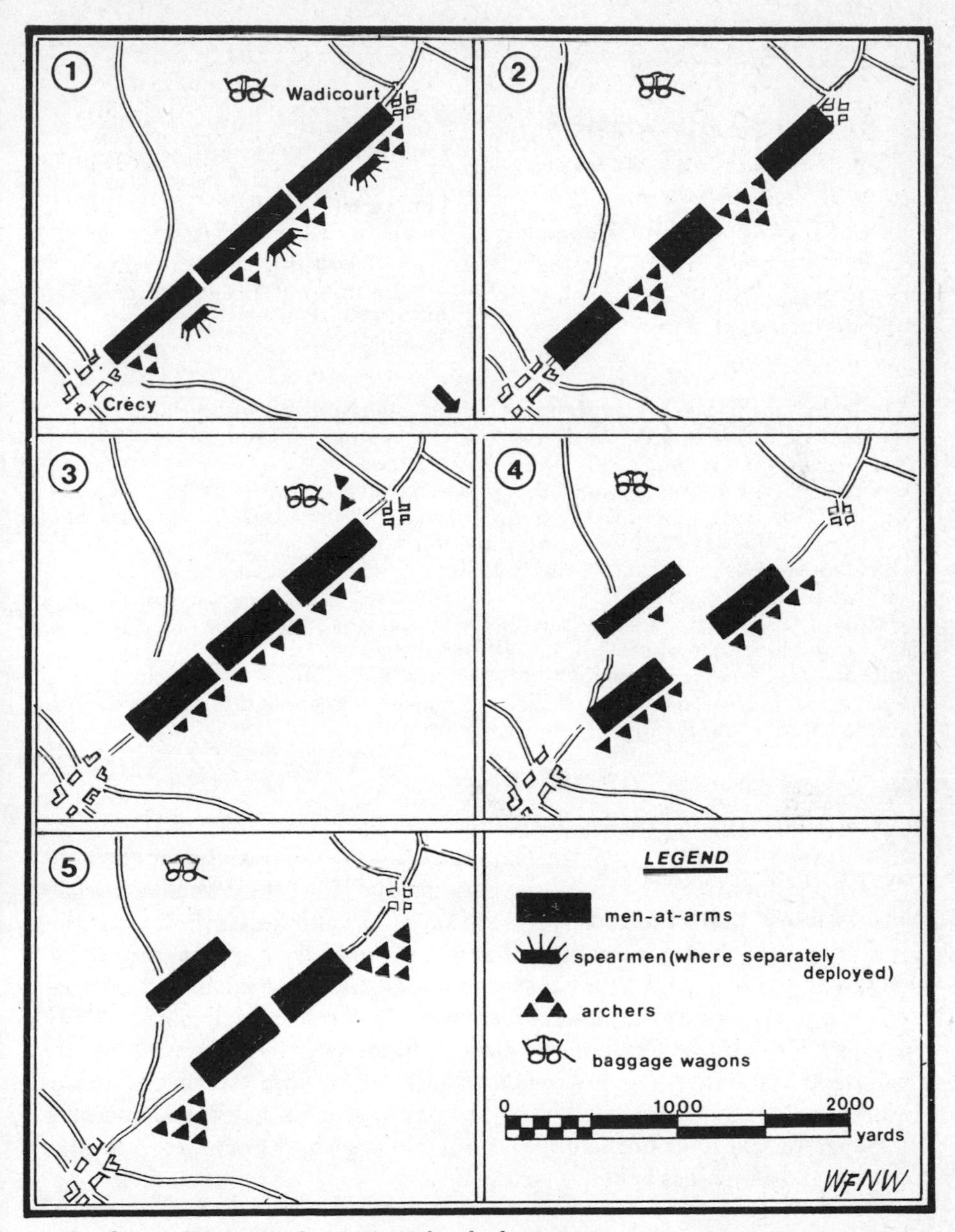

12 Crécy: Edward's five options for deployment

When Philip did at last get his host on the move, the leading elements appear to have headed for Noyelle (throughout the campaign Philip's intelligence was lamentable); on receiving correct information from scouts he had to make a detour. It was not, therefore, until the late afternoon that this cumbersome force came into view of the English position. *Philip was faced with several options.*

(1) He could attack at once.

(2) He could delay his attack until the morning.

And having made that decision he could:

(1) Advance his whole army into a frontal attack.

(2) Attack frontally with the bulk of his force, but send a strong contingent round behind the left flank of the English – he would have seen that a straightforward flank attack would have been hindered by the village of Wadicourt.

Philip decided to wait until the next morning, and this was certainly the correct decision. A battle started in the evening could not be concluded in daylight; his troops had been marching for much of the day, and it was obvious that the English meant to give battle and would not slip away. However, he reckoned without the enthusiasm and indiscipline of some of his knights, who without orders began the fight that evening, Saturday 26 August. When Philip reached the leading troops he considered it too late to halt the proceedings and ordered a general advance.

By then he probably had no choice but to attack frontally with his whole force, and in any event turning the flank was a manoeuvre seldom practised in large-scale medieval battles – it was not considered to be chivalrous. An attack from the rear must have done damage, although Edward's reserve division could probably have dealt with it.

The French advanced with the Genoese crossbowmen in the van. When they came within 200 yards of the English divisions they were met with the heaviest hail of arrows yet seen on a battlefield. Exhausted from humping their heavy weapons all day, and partially unprotected while rewinding their bows, their ranks shivered like wheat shaken by the wind, and those who did not crumple to the ground soon broke and fled. But the chivalry of France was close behind, and calling the Genoese traitors they rode them down as they pushed forward to come to close grapple with the English. They too were pierced by arrows, and the carnage of the field became appalling. Struggling horses were everywhere, and shapeless heaps of writhing men lay on the ground marked by a forest of feathers. Added to everything was the fog of powder smoke and the hideously frightening noise of Edward's cannon.

The fighting was particularly heavy on the English right, where hundreds of the enemy had rushed to swift destruction. Here Philip's ally and brother-in-law, the blind King John of Bohemia, lay dead, tied to the knights he had ordered to lead him into battle. Such was the pressure on this division that Warwick feared for the life of the Prince of

Wales. But Edward refused to send reinforcements, saying, 'Let the boy win his spurs.' The lilies of France and the banners of the French nobility were carried forward with great devotion in charge after charge (some say as many as fifteen), and every time they were shattered and scattered. The sun went down and dusk crept over the field, but it brought no cessation to the fighting, for the battle lasted until dark. Then the French left the field, and the English lay down where they were to sleep.

The next morning was foggy, and no pursuit was ordered. But a reconnaissance force under Northampton met and drove back a strong contingent coming from Rouen too late for the battle. The day was spent in clearing the wounded and dead from the battlefield. The young Prince Edward, who had distinguished himself in the fight, was greatly moved by the gallantry of blind King John, and decided to adopt his crest and motto – the three plumes and *Ich dien* have ever since been used by princes of Wales. In all 1,542 French knights and noblemen lay dead, and thousands more of the common soldiers. The English losses were very light. Truly it had been a great victory, and the French army no longer existed. Sir Winston Churchill, writing before the Second World War, said, 'This astounding victory of Crécy ranks with Blenheim, Waterloo, and the final advance in the last summer of the Great War as one of the four supreme achievements of the British Army.'*

Edward, like everyone else, must have been astounded at the extent of his great victory, and in the light of it he could not have found it easy to decide what steps to take next. His men had come a long way from St Vaast, but they were now in great heart, although getting short of military stores and equipment. *Edward on the morrow of the battle had four options:*

(1) He could take advantage of the fact that the way to Paris was almost certainly open, and the citizens, who earlier had shown signs of unrest at the futile inactivity of their king, might be in no mood to resist.

(2) By now he almost certainly had news of the whereabouts of the Flemings. He could carry out his original intention, and march to join them.

(3) He could march north and attempt to take Calais.

(4) He could sail for home. Boulogne and Calais were not in English hands, but the nearby port of Le Crotoy was.

Edward decided to march for Calais (option 3). It must have been a hard choice between that and Paris. The other two options probably had little appeal. Although he had won a magnificent victory, he would have been returning to England without any tangible results from the campaign,

* *A History of the English-Speaking Peoples*, Vol. I.

and he was now able to send to England for heavy cannon and other military stores. The original object of joining forces with the Flemings was no longer valid, for he had defeated the French King with his own small numbers. He did, however, make use of them later in the siege of Calais.

But Paris must have been very tempting. What probably decided him against the attempt was not knowing the size, or the whereabouts, of the Duke of Normandy's army, which had been summoned (and that he probably would have known) from Guyenne. Had the city not accepted the call to surrender, Edward might have been faced with a long siege, and the need to defeat the Duke of Normandy's army. It is difficult to fault Edward's decision to leave Paris alone, for he could not fully appreciate the chaotic conditions prevailing throughout France at this time, nor could he have known that the Duke of Normandy had left a substantial part of his army in Guyenne. As it was Paris might have fallen to the English more quickly than Calais. And had Paris been captured, the battles of Poitiers and Agincourt might never have been fought.

The siege and fall of Calais is one of the great epics of the Hundred Years War; it highlighted Edward's clear judgment, confidence and tenacity of purpose, while underlining the procrastination and pusillanimity that had been the principal characteristics of Philip's whole campaign.

The English arrived before the town on 4 October, and because Edward intended that henceforth Calais should be a part of England he did not want – even had he been able – to flatten it by assault. He therefore, with the help of the Flemings, began a blockade. The Governor, Jean de Vienne, had emptied the town of all who could be of no assistance, which, together with the fact that the English navy did not have complete command of the sea and French vessels were twice able to run the blockade, inevitably prolonged the operation. But for six months Philip did nothing; it was not until March 1347 that he bestirred himself to organize a very large (perhaps 50,000 strong) relief force, and not until 27 July did that relieving army arrive within striking distance of Calais. By this time Edward had received reinforcements from England and Guyenne, but he would still have been outnumbered had Philip forced a battle. However, the French King, having raised the hopes of the starving garrison, contented himself with issuing one of his challenges to Edward, and when that came to nothing he burnt his tents, and leaving all his stores and equipment slunk away with his entire army on the night of 1–2 August. The next day the garrison threw down the royal standard in disgust, and surrendered the town.

The following month, at the instigation of Pope Clement VI, who deplored the pain and suffering that the war had so far caused, France and England arranged a truce, and on 12 October 1347 Edward sailed for home after an absence in France of fifteen months. In 1348 the Black Death swept across Europe and England, inflicting far greater losses than had been suffered in any war, and effectively putting an end to hostilities for a while.

The principal prizes of the Crécy campaign were Calais and confidence. For two centuries Calais was to be the gateway for Englishmen into France; a piece of English territory which helped the English navy to control the Narrow Seas, and gave English merchants safer and shorter communication with continental markets. Before Crécy the English military machine was considered, if not contemptible, certainly of little consequence. From now on all Europe held Edward's superbly armed and well-trained troops in considerable awe, and Englishmen learned to be proud of the prowess of their fighting men.

Edward III, like his grandfather, was a great soldier-king, a general of wide horizons, who rejoiced in the intensity of risk. He continued to campaign at sea, in Scotland and in France, but after the fierce and successful sea-battle against the Castillian fleet off Winchelsea in 1350, his place as the foremost warrior of the Hundred Years War was to be taken by his son Edward, Prince of Wales, under whose inspiring leadership the English army was to continue to win great glory. Philip VI died in August 1350. It could be said of him that as a soldier he lacked the skill to plan, and the manhood to dare.

3 The Hundred Years War

The Agincourt Campaign

THE ravages caused by the Black Death, although reducing the number of soldiers available for garrison duties (and the English were hit harder than the French by the plague), did not put a halt to the fighting for very long. Fighting in varying degrees of intensity took place at sea and on land for the whole of the 1350s, and on 19 September 1356 the Prince of Wales (the Black Prince of history) defeated a large French army at Poitiers, and took prisoner their King, John II – known as the Good. By 1360 both sides – particularly the French – had become exhausted, and the Dauphin and the Prince of Wales were glad to negotiate the Treaty of Brétigny.

The terms of the treaty bore heavily upon the French. It is true that the English agreed to reduce the vast sum previously demanded for the ransom of King John, and that when part of it had been paid, Edward III allowed him to return to France.* But the figure of 3 million gold crowns was still more than the French could pay, and their monarchy was shamed by the loss of Guyenne, Ponthieu and many other districts, which were to be ceded by right of conquest and in full sovereignty. Furthermore, Calais was a loss of great strategic importance. In return Edward was to renounce his claim to the French throne. By now this claim, which was almost certainly originally made to satisfy Flemish demands, had become an important reality to the English King. In fact there was no formal renunciation – that had to wait until 1802. Nor for that matter was French sovereignty over the ceded territories officially renounced. Nevertheless, the treaty was both humiliating and damaging.

There followed nine years of peace, and then between 1369 and 1396 almost thirty years of sporadic fighting, some of it heavy. Edward III and

*King John had to leave three sons as hostages; one, the Duke of Anjou, broke his parole, whereupon John, with great chivalry, returned, only to die in London a few weeks later.

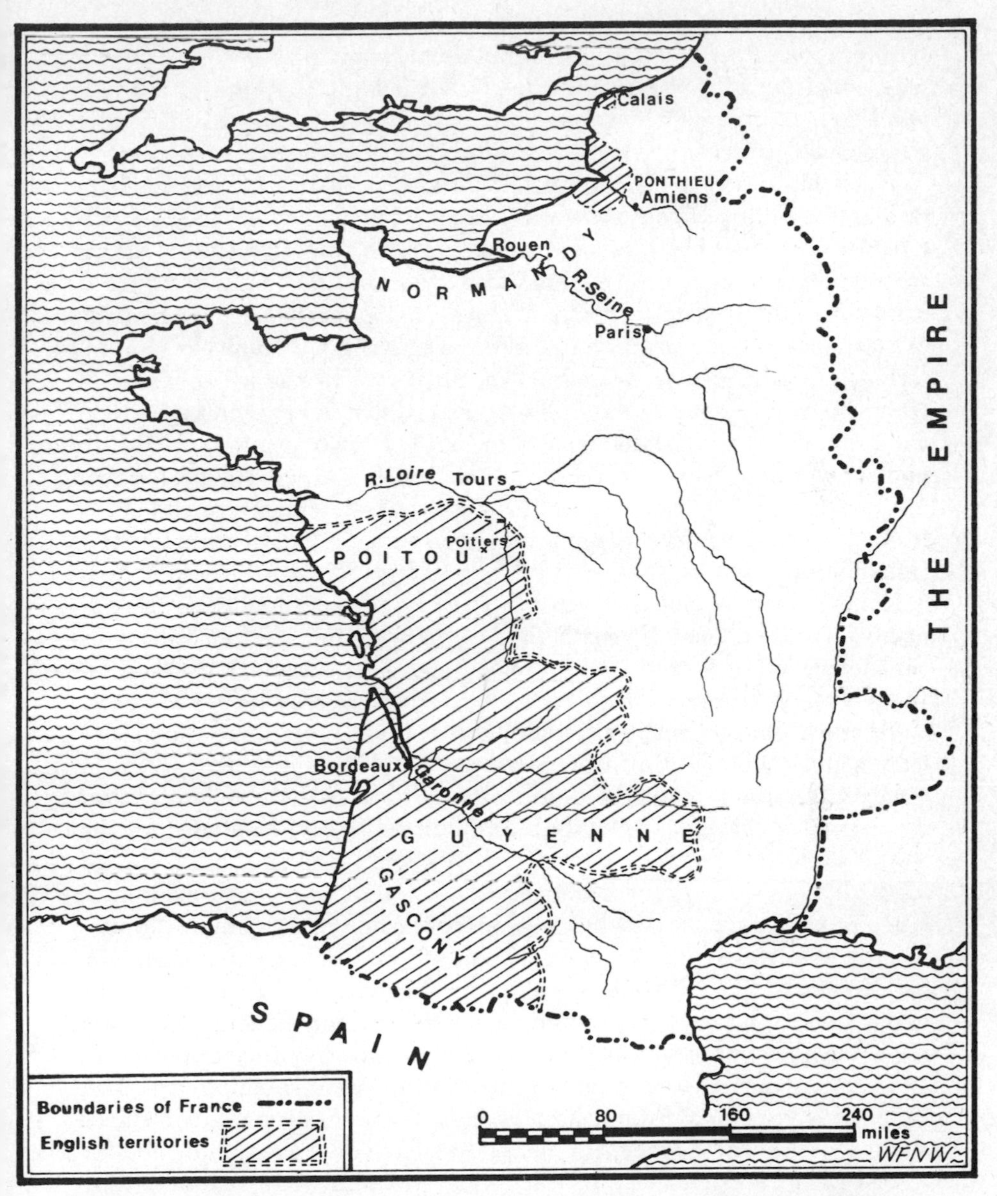

13 The Agincourt Campaign: France under the Treaty of Brétigny, 1360

his eldest son died in 1377, and three years later the French King, Charles V, also died. In Charles the French had found a good king, who understood the needs of the time. He picked sound generals, and in particular Bertrand du Guesclin, and instructed them in the type of warfare most likely to succeed. By Fabian tactics, and a determined refusal to meet the English in open battle, no matter what their strength, the French regained towns, fortresses and some territory. It was a bad period for English arms, not through any lack of military efficiency, but simply because they had not the men to deal with widespread guerrilla tactics; also the balance of sea power was changing.

When King Richard II eventually gained control of his Council he favoured a policy of peace towards France (indeed he married secondly Charles VI's daughter), but his endeavours in this direction were not wholly successful, because Charles, a weak and pathetic creature of unstable mind, was no better than a pawn between the warring factions of Armagnacs and Burgundians. When the King was sane the Duke of Orléans (leader of the Armagnac faction) was in the ascendant, and when he was insane Duke John of Burgundy, whose party had the support of the Parisians, held power. The two dukes had totally opposing policies; both were determined to rule France, and both wooed the English King – by now Henry IV – with substantial territorial offers. It was an opportunity to re-enter France that no English king of that time could resist.

Henry IV would dearly have liked to cross the Channel, but problems at home, and latterly ill health, made it impossible. But by the time his son succeeded in March 1413 the Welsh had been subdued, the Scottish King was a prisoner in the Tower, and it did not take Henry V long to bring the Lollard* conspiracy under control. The new King was a man of action and ambition. He came to the throne when twenty-five years old, but he was already a complete man with much military and administrative experience. He was of slight build, but very strong and athletic; his thin cadaverous face with its high cheekbones and long nose was set alight by the eyes – gentle and smiling when at ease, but flashing fire when he was angered. In religious matters he bordered upon the fanatic; in battle he was courageous; in affairs of state he was masterful. The brilliance of his military leadership and his fearless justice compelled widespread admiration. He could be courteous, and although he was at times ruthless, he was less cruel than many rulers of those days.

While appearing to be deeply interested in the respective territorial offers of the two rival French parties, Henry quietly prepared for a large-scale invasion of France, for he knew that the terms he demanded, which were in excess of the Brétigny settlement, were unlikely to be accepted by the Armagnacs, who were now in control of most of France, and could not be met by the Burgundians. His ambitions towards France were much the same as those of his great-grandfather, but in addition he

* Lollard was a Flemish word usually applied in a pejorative sense to a Flemish sect that had lapsed into heresy, and had come to be associated with the followers of John Wyclif, whose writings and teachings towards the end of the fourteenth century had been condemned by the Church.

was negotiating with the Armagnacs for the hand of Charles VI's daughter, together with a huge dowry. He undoubtedly believed in his claim to the French throne – although he was prepared to await the reversion on the death of Charles – but above all else he had an abiding desire to unite the princes of Europe in a crusade to recover the Holy Places.

The mood of the country favoured war, and in November 1414 Parliament voted a large subsidy. But it was not large enough, and had to be supplemented (not for the first or last time) by interest-free loans from the rich abbeys, prelates and nobles. In April 1415, at a Council of War, Henry appointed his brother, John Duke of Bedford, regent in his absence, and issued indentures for the invading army; the shire levies were also called up for defence duties in case of trouble from Wales or Scotland. There had been little change in arms and armament since the great expedition of Edward III's time; plate armour had replaced mail (and was very heavy, weighing 66 lb. and more), and Henry had a number of cannon, although they were principally siege weapons.

Henry crossed the Channel with an army smaller by a third than his great-grandfather had commanded in 1346, but this expedition was otherwise on a far more imposing scale. The King was a master of administration, and from Porchester Castle he had spent most of July organizing and supervising the arrangements for his vast host. In addition to the fighting troops – 2,000 men-at-arms, 8,000 archers, 65 artillerymen and more than 20,000 horses to meet their needs – there were a great many ancillary troops; their exact numbers are not known, but they were considerable. Chaplains were an important part of the army, and for the first time surgeons were in attendance. Large numbers of miners, armourers, masons, carpenters, butchers, drovers, fishmongers and turners comprised this medieval 'B' echelon. Then there were numerous wagons, and so as to start the troops off on good English fare there were cattle and sheep on the hoof. It was on the afternoon of Sunday, 11 August that the vast fleet of 1,500 vessels sailed past the Isle of Wight and made for the open sea.

Throughout this campaign we are struck by the many similarities to that of Crécy seventy years earlier. The first was the secrecy that surrounded the sailing. Henry, like Edward III, did not disclose the fleet's destination until they were at sea. On this occasion, however, there was little likelihood of a southern landing, and the French were well aware that Normandy, which had not been included in the Treaty of Brétigny, was Henry's first objective – but it could be reached from almost any north coast French port. Protective measures of a sort had been taken at every likely place from Brittany to Calais; and Harfleur, Henry's chosen starting point, was not found unprepared.

The fleet came to anchor on the evening of 14 August at the Chef-de-

Caux on the Seine estuary some three miles west of Harfleur, where Le Havre now stands. Embarkation of all the horses, wagons and cattle in England had not been too difficult, for there were jetties and ramps, but here there were just the shallows, and disembarkation took three days. Clearly Harfleur would not be easy to take; situated at the junction of the Lézarde and the Seine its harbour was negotiable only at high tide, and iron chains, stakes, and even trees blocked the channel. The town was protected by strong walls and many towers and barbicans, and the three gateways by ditch, drawbridge and portcullis. Outside the walls ditches had been dug and strongpoints constructed with fascines and tree trunks. To the north-east the land was flooded. The garrison had recently been reinforced by Raoul, Lord of Gaucourt, and was in good shape and well provisioned.

Henry's first task was to invest the town completely. He sent his brother, Thomas, Duke of Clarence, whose troops had to make a detour of the northern flooded side, to take up a position to the east of the town. When this was done Harfleur was surrounded, and the King was faced with the difficult task of capturing it. Obviously no assault could be made without first breaching the walls, and the strength of the place with its wide protective ditches made this an extremely difficult and hazardous operation. *Henry had three options:*

(1) He could starve the garrison into surrender.

(2) He could inch his way forward with spades, and lay mines.

(3) He could attempt to breach the walls by cannon fire.

The simplest method would have been to starve the garrison, but if Henry wanted to achieve anything beyond the capture of Harfleur, he could not afford the time needed to starve a well-provisioned garrison, who could also expect a relieving force. To breach the walls with his artillery (battering rams could not be used on account of the wide ditches) meant positioning the cannon dangerously close to the manned towers, so Henry decided to mine the walls – option 2.

It proved to be a wrong decision, for the French had miners who were more skilful than the Englishmen; they dug parallels and let off countermines to such effect that Henry was forced to abandon this scheme, and resort to his artillery. The guns were set up in the face of determined opposition, and losses were suffered. Throughout the long siege Henry never rested; completely fearless, he was everywhere planning, directing and encouraging. On 3 September he wrote to Bordeaux asking for guns and wine to be sent, but it was not a despondent letter, for in it he outlined his plans for a march on Paris, and then to Guyenne.

The siege lasted almost six weeks. The English gunners hurled stones of 500 lb. at the massive walls, but no sooner was a breach made than it was repaired. Day and night the bombardment continued, with catapults supplementing the guns. Gradually the walls began to crumble, and the incendiary balls (lighted tar-covered stones) made parts of the town an inferno. No help came for the garrison; the Constable d'Albret, on the other side of the estuary at Honfleur, was unable to cross, and the Dauphin, with an army at Rouen, failed to move. On 22 September the garrison surrendered. The terms were not harsh. Throughout this campaign Henry was to enforce the strictest discipline on his troops, and, in contrast to the *chevauchées* of Edward III, looting and burning were permitted only in exceptional circumstances under Henry V.

The English army had arrived at Harfleur in a somewhat weakened condition, for many of the men had been on board ship for some days before embarkation, and they were therefore in poor shape to resist the heat and unhealthy fetid air of the salt marshes that drained the town of Harfleur. Added to this they could not be prevented from eating great quantities of unripe fruit and drinking a lot of new, rough wine. Dysentry became a threat to the whole army, and the casualties from it in all ranks were so serious that Henry found himself with not much more than 8,000 soldiers on the active list by the end of the siege, and of these about 2,000 had to be left to garrison Harfleur. The probable strength left to him for any fresh endeavour was therefore 1,000 men-at-arms and 5,000 archers. The French could shadow this small force with a vastly superior army. On the conclusion of the siege *Henry was left with four options:*

(1) He could carry out his original intention, as expressed in his letter of 3 September to Bordeaux, and march on Paris.

(2) He could take what was left of his army home, leaving a garrison in Harfleur.

(3) He could establish a Harfleur pale on the lines of Calais, which would mean keeping the army in France to extend and garrison the boundaries of the pale.

(4) He could 'show the flag' by way of a *chevauchée* across Normandy to Calais, trusting that the French would not attack him en route.

The first option, although no doubt appealing to Henry, was probably never discussed in the war council that he held, for with a vast French army waiting to pounce and the lateness of the season, the chances of success were virtually nil. We are told that the majority of the council pressed strongly for the second option, and that there was some support for the third. But Henry overruled the more cautious veterans, and with the support of some of the younger nobles he decided to adopt the fourth option and march for Calais. To return home with only Harfleur to show for all the casualties and expense would mean a great loss of prestige, while a triumphant sortie across Normandy would raise morale at home and bring further despondency to France.

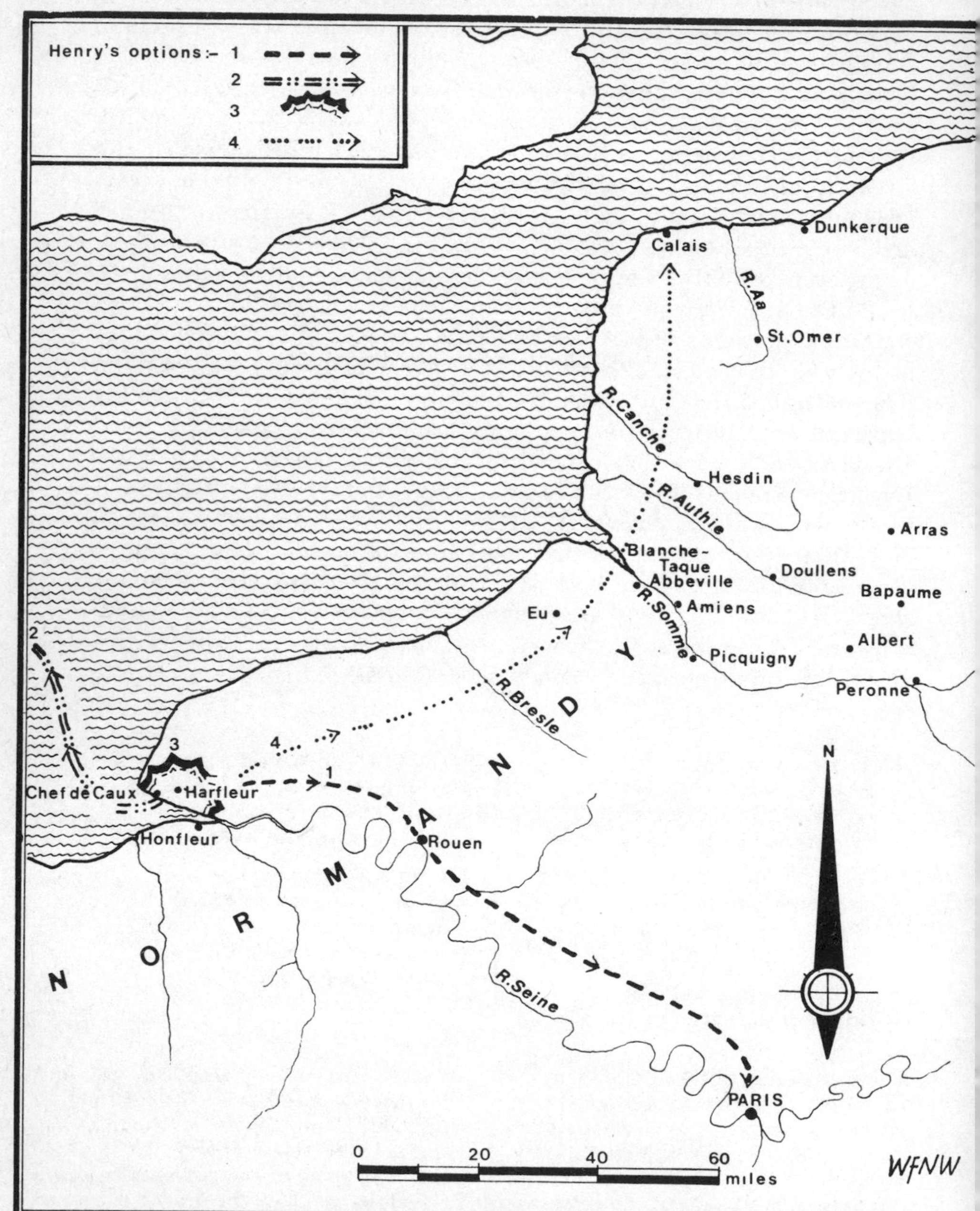

14 *The Agincourt Campaign: Henry V's options after taking Harfleur*

The risk that Henry took in deciding to trail his coat across Normandy in front of the French army was certainly great, but perhaps not so great as many historians have made it out to be. The French were suffering from the after-effects of civil war; their nation was still divided and there was little sign that invasion had united them, and above all their leadership was weak and suffered from divided control. But Henry made the risk greater by challenging the Dauphin to single combat, and while waiting eight days for the inevitable refusal failing to send forward a strong force to seize the crossing of the Somme at Blanche-Taque. Too late he instructed the garrison at Calais to do this for him.

It was probably on 8 October that the English army, marching light and without cannon or heavy wagons, set out for Calais, hoping to cover the 150 miles in eight days. There followed an operation similar to Edward III's march from St Vaast, only this time there was just one major river to cause trouble, for the Béthune and Bresle were hardly likely to present difficulties. Henry could only guess at the dispositions of the French army, but he could reasonably hope to cross the Somme without interference, for as far as he knew Constable d'Albret and Marshal Boucicaut were still at Honfleur, and the Dauphin at Rouen was probably still awaiting the main army moving up from Vernon.

But the French did now much the same as their forebears had done seventy years before – they got behind the Somme ahead of the English. We do not know exactly when Boucicaut left Rouen with a strong advanced guard (he and d'Albret had arrived there some days before Henry left Harfleur), but it came as a considerable shock to Henry to be informed by a prisoner taken at Eu that the ford at Blanche-Taque was heavily guarded. *He summoned a council, and presented them with three options:*

(1) The army could attempt to force the crossing at Blanche-Taque, which lay six miles distant.

(2) They could march east in the hope of finding a crossing farther upstream.

(3) They could retrace their steps to Harfleur.

The third and safest option was too shaming for the knights to consider seriously. To force the river at Blanche-Taque was an attractive plan, for the success of 1346 had not been forgotten, but the council were informed that by the time the army could cover the six miles the tide would be full and a long wait necessary. It was therefore decided, with considerable misgivings, to march east. It is difficult to criticize this decision without more detailed information. One account says that the Somme estuary was flooded and impassable, but if this was so why were Boucicaut's troops guarding the ford? And if it were not so, we have no information – though Henry may have got it from the prisoner – as to whether the French held the crossing in such strength as to make it impossible to

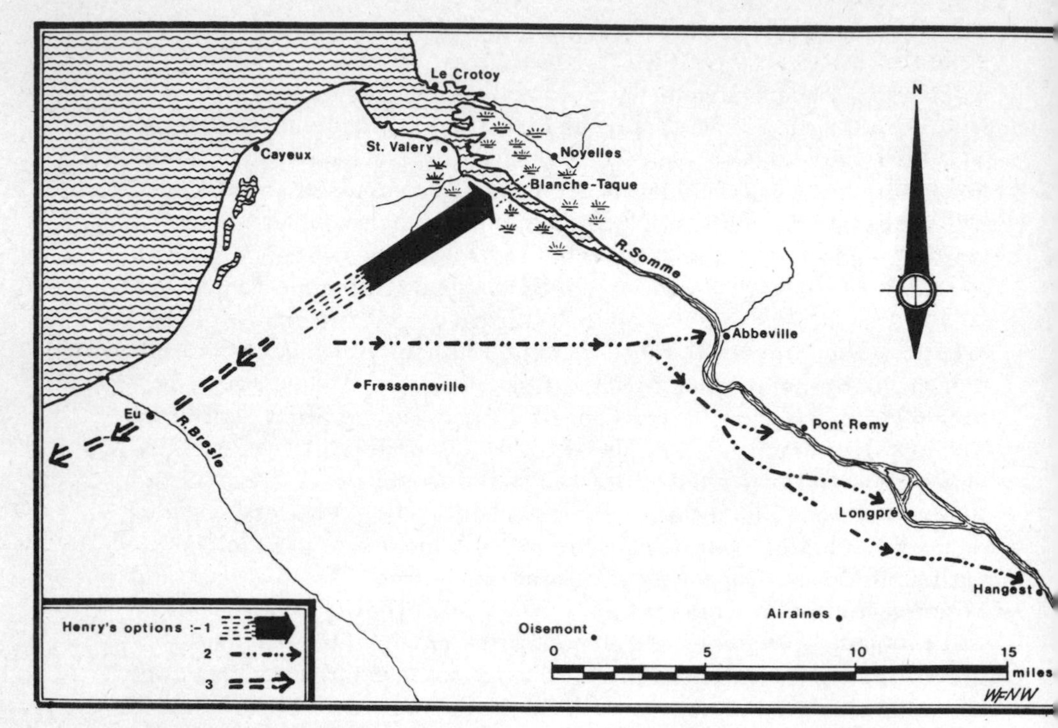

15 The Agincourt Campaign: Henry's options on reaching the Somme

force. But to march up the Somme, with a large French army playing cat and mouse along the opposite bank, would seem to have been a more hazardous operation than forcing the river, had this been possible.

On 13 October the English army altered course for Abbeville, where they found the bridge well guarded; from there they passed through places whose names had been so familiar to Edward's troops – Pont Remy, Longpré, Hangest, Picquigny and Pont-de-Metz – but the French army blocked every crossing in strength. Leaving Amiens to the north, they headed back to the river at Corbie, where the French made a sortie across the bridge, and a short engagement ensued. Learning from a prisoner that this was only a prelude to more serious attacks, Henry ordered every archer to cut a six-foot stake, to sharpen it at both ends and to be prepared to stick it in the ground at the approach of enemy cavalry. But no further attacks developed, and anyway soon after the Corbie affair Henry cut across the loop in the river and rejoined it near Nesle. Here the inhabitants at first showed fight, but the threat of destruction by fire quickly changed their intentions, and it is said that

in order to be rid of the army they informed the King that the river could be forded at Béthencourt and Voyennes.

By leaving the French army to march round the loop of the river Henry had gained valuable time and could reach the two fords well in advance of any serious opposition. But the crossings were not easy, for although there was only a small garrison at each place, the causeways leading to the river had been destroyed. Two hundred archers managed to wade across and drive off the defenders, but it took half the day to build up the causeways with trees and timbers from houses to a sufficient strength to enable the army to make the crossing. By the late evening of 19 October the whole army was safely across, and the weary troops billeted at Athies. The dangers that faced them were still considerable, but every English soldier must have slept more soundly that night from knowing the Somme was behind him.

The French army, meanwhile, had shown singularly little initiative since the advanced guard's burst of energy in racing the English to the Somme. The reason for this is probably that the main army did not cross the Somme at Amiens until the English were in the region of Boves. Before setting out from Rouen the Dukes of Berry, Orléans and Bourbon had persuaded Charles VI, and his equally pathetic son, that it would be unwise for them to risk capture by leading the army, and the King had agreed that the command should go to the Constable d'Albret – which meant in fact divided leadership, for the Dukes considered him their inferior. It was not long before an argument developed between the two soldiers, d'Albret and Boucicaut, who favoured allowing Henry to march to Calais and embark, and the royal Dukes, who were determined on a direct confrontation with the English army.

It can be assumed that while the English were crossing the Somme on 19 October, still in ignorance of the whereabouts of the French army, the latter were marching into Péronne. We know that on 20 October the French sent three heralds from there, who being brought into Henry's presence announced that their lords had assembled to defend their rights, 'and they inform thee by us that before thou comest to Calais they will meet thee to fight with thee, and to be revenged of thy conduct.' And when the heralds asked which road he would take, Henry replied, 'Straight to Calais.' Having further assured them that if the French disturbed his army they would do so at their peril, he dismissed the heralds with a hundred gold crowns each, and made plans to resist the attack to be expected on the morrow.

In order to fulfil their promise to give Henry battle before he got to Calais, *The French had four options:*

(1) To attack him in the defensive position he had hurriedly assumed after the dismissal of the heralds.

(2) To take up a strong defensive position themselves on the Péronne ridge, for they had the English well bottled up between the fortified town of Péronne and the rivers Somme and Cologne.

(3) To allow the English to proceed to Calais and attack them from some vantage point while on the march.

(4) To allow the English to proceed to Calais and block the road farther north.

The French, if their two forces had been joined by now – and there seems no reason to believe they had not been – must have been successful with any of the first three options, for they outnumbered the English by perhaps as much as four to one. Probably the second would have been the best choice, for the English must certainly have tried to fight their way through.

But the French decided to let the English proceed on their way to Calais (option 4), presumably intending to block their path farther north. It was an inexplicable decision, for they were unlikely to find another such excellent defensive position so effectively barring the route, and it can be attributed only to there still being a divided council on strategy.

On 21 October, with a strong wind blowing cold rain into their faces, the English army resumed their march. Spirits that had temporarily risen after crossing the Somme had begun to sink again in the face of the weather and an hourly expectation of attack. Some enemy had appeared as the army passed to the east of Péronne, but they did not attempt to interfere with the ragged columns. However, where the road to Albert (which the English were following) crossed the one to Bapaume, tracks clearly showed that a multitude of men had recently passed, heading north. Henry was so certain that he would be attacked on the march that he threw out a strong protective screen to his right. But no attempt was made to interfere with this weary English army as it trudged in deteriorating weather through Albert, Acheux, just east of Doullens, to Lucheux and Frévent. The French contented themselves with marching parallel some eight to ten miles to the east.

But by the time the advanced guard reached Blangy, scouts had been reporting the presence of French patrols, and it was obvious that the armies were at last converging. In fact, the French had crossed the river Ternoise ahead of the English, and from the heights to the north of the river valley at Blangy the advanced guard, commanded by the Duke of York, gazed down in awe at a massive army of magnificently mounted men – the pride of the French nobility. On seeing the enemy the French began to deploy for battle, but before Henry could consider a defensive position they resumed their march towards Tramecourt. Passing to the east of the Tramecourt wood the French swung left-handed, and, taking up a position between Tramecourt and Agincourt, they completely blocked the road to Calais. Henry advanced his army to an open expanse

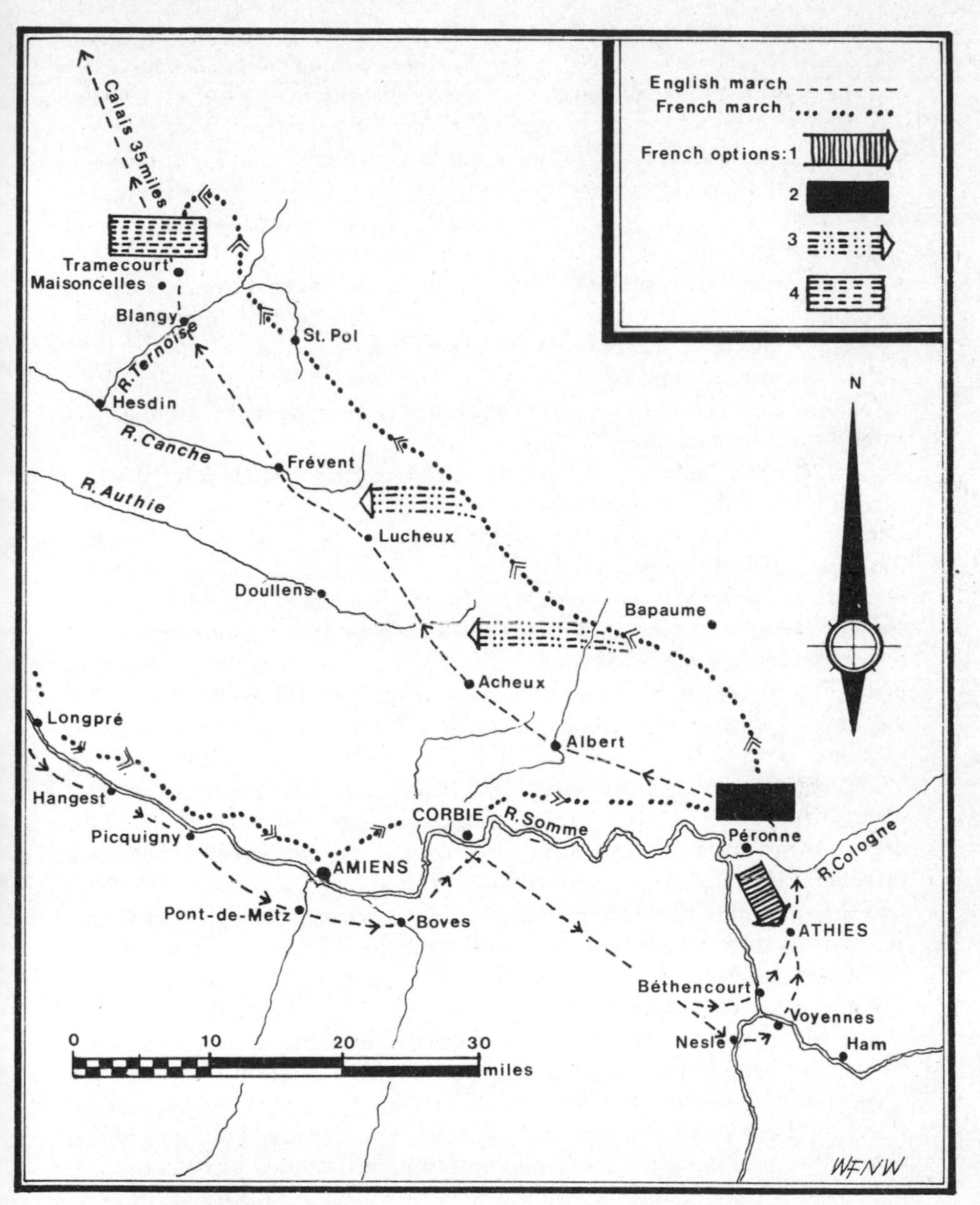

16 *The Agincourt Campaign: the French options after the English had crossed the Somme*

beyond the village of Maisoncelles, where he halted about a mile distant from the French host. Now there could be little doubt that a battle would be fought the next day – 25 October, the feast of Saints Crispin and Crispinian.

The night came early. A wind-tossed sky brought dark clouds heavy with rain, which poured upon the wretched soldiers of both armies. The English had marched 260 miles in seventeen days; it is true that most of the men were mounted, but some of the archers were not, and all were tired, hungry and very afraid of what the morrow would bring. Even Henry's usual confidence momentarily deserted him, and we learn that he sent messengers to inform the French that he was prepared to hand back Harfleur and pay for the damage, in return for a safe passage to Calais. But now we find the French commanders – even perhaps d'Albret – resolute for battle, with all doubts swept aside, and the knights busy dicing for the spoils and ransoms that would soon be theirs. There was to be no last-minute reprieve for the English, and once Henry knew it he quickly regained his steely composure; nor did he lose it throughout the next day in face of fearful odds.

Early in the morning the armies prepared for battle. The French position was not well chosen. They needed space in which to manoeuvre their great host, and could not find it between the woods that bordered Tramecourt and Agincourt. The arena chosen for the coming contest was a rectangular shaped wheat field flanked on either side by woods. At the northern, or French, end it was at its widest – about 1,200 yards – and it tapered to little more than 900 yards at its narrowest point. There was a slight dip between the two armies, which faced each other at a distance of a little over 1,000 yards. The French had the advantage of numbers (somewhere between 20,000 and 30,000 men to Henry's mere 6,000, or perhaps slightly less), but little else, and as it turned out numbers proved a liability. Squeezed into a narrow frontage, the French men-at-arms jostled each other for room, and were to become an unwieldy, uncoordinated and uncontrollable mass of steel robots. Nor was the ground to their advantage, for immediately to their front it had been badly churned up by horses, and with the night's rain had become something of a quagmire.

Henry had little choice in his deployment, for he had too few men to permit a reserve. All three divisions were therefore in the front line with the men-at-arms (dismounted) four deep. Each division had – as at Crécy – its wedge-shaped bands of archers on the divisional flanks, and in addition there was a strong force of archers posted on each flank of the army. But at this battle, unlike Crécy (where potholes had been dug), each archer carried his six-foot stake with which to repel cavalry. The French held the strategical initiative and could dictate the battle. They might well have gone into the attack the previous day, when the English were obviously unprepared, but it seems that they remembered the lesson of Crécy, and were not so foolish as to start another major battle late in the day.

Now, on the morning of the 25th, they had two options:

(1) They could remain on the defensive, and wait to be attacked.

(2) They could send the mounted knights forward in a great charge to sweep the English off the field.

The Constable d'Albret, ever a cautious man, was the principal protagonist for a defensive battle. In the past, although nominally the army commander, he had not always been allowed to follow his chosen path, but on this occasion his decision, though debated, was not disallowed. It was certainly the right choice, for the English had become desperate. Unlike their opponents they could not wait as much as a day, and for their small army to advance to the attack over the churned up ground would bring them into imminent peril. Moreover, whereas the longbow was a superb defensive weapon, it was of less value in a frontal attack. This decision having been agreed, it would have seemed logical for d'Albret to draw up the army with the crossbowmen at least in line with, if not in front of, the men-at-arms. But this was too much for the French chivalry to endure, and there was an undignified and indisciplined scramble on the part of the dismounted knights to have the honour of forming up in the front line. This completely nullified the French chances of fighting a successful defensive battle, for the crossbowmen found themselves behind the knights, and the cannon on the flank had no room, nor a field of fire.

For four hours on that October morning the two armies faced each other. The rain had stopped, but little tufts of soft cloud drifted in the wet wind; the wheat field was sodden and the trees of the wood dripped with moisture. Henry, well aware of the disadvantages of attacking, became increasingly anxious as the French stood around relaxed and chatting, showing no sign of being prepared to begin the battle. By 11 a.m. he could wait no longer, and to those immortal words, 'Avaunt banner, and this day, St George, thy help', the English army – after the soldiers, following their usual custom, had prostrated themselves on the ground and kissed the earth in token of their unworthiness – moved forward. They marched slowly, for the going was heavy for the armoured knights, and crossing the Agincourt–Tramecourt track they halted just within bowshot of the enemy.

Directly the archers had placed their stakes in position they opened up the usual withering fire. At this extreme range against plate armour the damage they did was not considerable, but it had the desired effect of provoking the enemy, and the mounted knights on the flanks were ordered to ride the archers down. As they rode steadily forward towards the inexorable firing line their array at first was perfect, for their numbers were not great and total confusion had yet to set in. But the combination of stakes and arrows drove them back upon the hopelessly crowded French dismounted knights, who were advancing close behind.

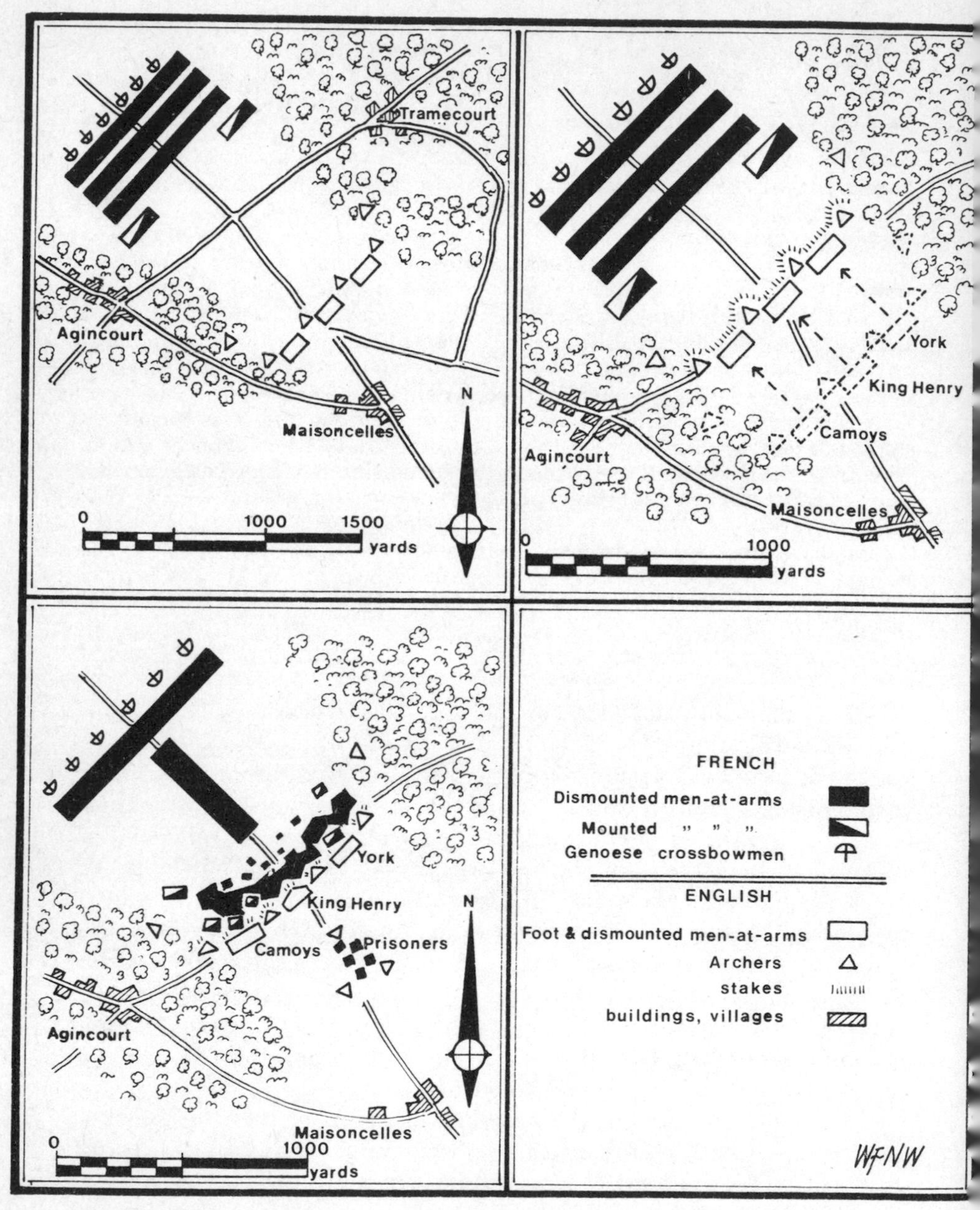

17 The Battle of Agincourt

Somehow large numbers of them struggled through, and making straight for the English men-at-arms presented – as at Crécy – a perfect enfilade target; moreover, they were so tightly jammed that they were scarcely able to use their weapons. All they could do was to sink down in tangled heaps.

The second line of French knights advanced in column, hoping thereby to present a smaller target, but the effect of the enfilading fire was just as damaging. Nevertheless, the English front lines began to bend under the sheer weight of numbers, and for a time the hand-to-hand grapple was savage. But now the archers, seeing the opportunity of doing deadly work among the solid mass of almost immobile French knights, threw down their stakes and picked up heavy weapons with which they pummelled the enemy to death. The ground was so thick with the bodies of the fallen that men were stepping on dead and wounded alike; those who were not bashed to death suffocated in the mud beneath their comrades.

The fight had lasted hardly more than half an hour, and of the French battle order only the third line of mounted warriors was left still intact. These men were clearly shattered by the turn of events, and some even started drifting away. When Henry saw that they made no move he allowed his men to start rounding up prisoners. This pleasant and lucrative task had been under way some two hours when the King thought he discerned the French horsemen about to enter the fray. It was a desperate moment. Hundreds of fully armed prisoners were milling around in the rear, and the French cavalry threatened the front. Henry was not prepared to risk losing all that he had gained through a stab in the back at this decisive stage of the battle, so he gave the controversial order that all the prisoners – save those of highest rank – were to be slaughtered. The men-at-arms were reluctant to lose their hard-earned prizes, and 200 archers were detailed for the gruesome task, which ceased only when the King saw the French knights trotting off the field in despair and disillusionment.

The French battle casualties were enormous – somewhere in the region of 8,000 – and among them were many of the highest-ranking nobles in the land. The commanding general d'Albret and the Dukes of Brabant, Alençon and Bar lay dead upon the field, as did eight counts and the Admiral of France (Jacques de Châtillon). The English lost only between 400 and 500 men. Their most distinguished casualty was the King's cousin, the Duke of York (suffocated); his brother, Humphrey Duke of Gloucester, had been wounded, but Henry, always in the thick of the battle and conspicuous by the gold crown studded with rubies, sapphires and pearls that encircled his helmet, received nothing worse than a dent in that helmet. As at the battle of Crécy, the English line never moved throughout the fight, and once again fortune favoured not the side with the largest numbers but the one whose army was better trained and better disciplined. But, unlike at Crécy, the issue this time was decided in the first half hour of fighting.

The situation facing King Henry after this battle was in many respects similar to that which faced Edward III after Crécy. He seemingly had Paris at his mercy – even more so than had Edward, for this time the French had no other army to put against the victors. Moreover, with a King who was half mad, and a Dauphin who was half dead (he died two months later), the Parisians might have welcomed a whole man. On the other hand while Agincourt, like Crécy, was a great battle won, Henry did also have Harfleur as a tangible reward for long marching and hard fighting, whereas Edward needed to bring back Calais as a gift to his subjects. Nevertheless, on the morrow of victory *Henry had three options:*

(1) He could march on Paris.

(2) He could march to Calais and embark for home.

(3) He could enhance his glory by taking some strongholds in the Calais pale, such as Ardres.

The King, bowing to pressure from his council, decided to march for Calais and home. He is said to have greatly desired option 3, for he felt it would not require much effort to gain this added prestige for the campaign. But the deciding factor must have been the condition of his army, and of their equipment. Two months of campaigning and almost 300 miles of marching must have played havoc with feet, boots, horses and clothes – not to mention the shortage of rations and prevalence of disease. Whatever Henry may have wished, he obviously made the right decision.

Henry V went back to France on 30 July 1417. This invasion was on a grander scale than the Agincourt one two years earlier. France was still in a state of chaos, but in spite of this internal turmoil Henry had set himself a most formidable task, for almost every town on the way to Paris was fortified, and there were many that could not be bypassed. During the next three years this extremely competent English King was to show leadership, strategic planning and a logistical flair unsurpassed by any of his predecessors.

The siege of Caen lasted a fortnight, and that of Rouen no less than five months. In September 1419 the Armagnacs (now known as the Dauphinists) murdered John, Duke of Burgundy, and his young successor quickly made an alliance with Henry; for the price of revenge he was prepared to sell his country. Henry's terms for such an alliance were harsh and humiliating. By the Treaty of Troyes the French Crown was to pass to Henry and his heirs for ever after the death of Charles VI, and meanwhile Henry would have the regency. The Dauphin (later to become Charles VII) was cut completely out of the succession. Henry was to marry Charles VI's daughter, Catherine, and each country would be subject to its own laws under a united Crown, with a treaty of mutual defence.

But Henry's campaigning days were not quite over. By the terms of the Treaty of Troyes he had been given a free hand to operate against the Dauphin, who held most of France south of the Loire, save Guyenne. Sailing once more for France in June 1421, Henry was to have his most arduous campaign; indeed, had the Dauphin shown a more adventurous spirit matters could have gone badly for the English. But he consistently refused open battle, and Henry made headway, capturing many towns and strongpoints. But when he died in the castle of Vincennes on 31 August 1422, much of France was still unconquered.

Thirteen years after his death most of Henry's work had been undone. Helped by the Maid, Joan of Arc, and new methods of offensive warfare, against which English tactics proved totally inadequate, the French – and in particular their King, Charles VII – gradually regained confidence. By August 1450 the whole of Normandy had gone; during 1452 and 1453 England's last military hope, John Talbot, Earl of Shrewsbury, did his best in Guyenne against a resurgent French army fighting for national independence. His defeat and death at the battle of Castillon in July 1453 meant the loss of Guyenne, and when Bordeaux surrendered on 19 October England was left with just a precarious foothold in Calais, and the Hundred Years War was over.

It had been the longest, hardest and least rewarded of all medieval wars. Both contestants were the losers. The English nation was almost bankrupted; territorial possessions had vanished, and early military predominance and glory had been somewhat eroded by improved gunnery techniques. At the cost of fearful suffering, and an appalling destruction of rural and urban wealth, the French were to obtain national cohesion through exchanging the upheavals of feudal anarchy for the oppressions of absolute monarchy. Moreover, it was to be some years before they regained commercial prosperity. But the courage and endurance of the soldiers of both nations command the admiration and respect of posterity.

4 The Third Civil War

July–September 1650

FROM 1629 King Charles I ruled England for eleven years without a Parliament; during that time many of his subjects became increasingly discontented with government through a council of courtiers, who were often inept, and sometimes corrupt. This determination to rule without Parliament presented fiscal problems that were eventually insoluble, but in the meantime Charles's enforcement of tonnage and poundage, a manipulation of customs duties, and an agreement with Spain to transport bullion in English ships for the payment of Spanish troops, enabled him to exist amid increasing unpopularity, which soon amounted to hostility. There was also grave religious discontent, and Charles's appointment of William Laud – an Arminian – to be Archbishop of Canterbury opened the way to a doctrinal split in the country. But it was the reintroduction of ship money and the extension of this tax to inland counties that caused the greatest furore.

His subjects' refusal to pay the tax, and the first of the two Bishops' Wars (brought about by Charles's attempt to impose the new Prayer Book upon Scotland), caused the King to cease his personal rule and call a Parliament in April 1640; but it was soon dissolved, and in November what became known as the Long Parliament assembled. In 1641 this Parliament, under the leadership of John Pym, forced the King to assent to many important – and to Charles unpalatable – measures, and in November of that year that most damaging indictment, the Grand Remonstrance, passed the Commons by a narrow margin. Gradually the two parties – Royalist and Parliamentarian – took shape, and the country clattered inexorably towards civil war. Early in January 1642 Charles came to Westminster accompanied by 300 troopers to arrest five members of the Commons, who were the ringleaders in a move to impeach the unpopular Roman Catholic Queen. But they had escaped by the river. The situation for the King had now become desperate, and on 10 January he and his Queen slipped furtively out of London. He was not to return to his capital for seven years, and then as a prisoner on trial for his life.

The first important battle of the First Civil War was fought at Edgehill on 23 October 1642. There were to be seven major battles fought in England, and four fought by the Marquess of Montrose in Scotland, besides twenty-eight smaller battles and sieges, before the First Civil War ended on 24 June 1646. King Charles had given himself up to the Scots in May of that year, and there then followed endless disputes and intrigues, with Charles playing off the Scots, the English army and Parliament in turn against each other. The Scots returned the King to Parliament, but so successful had his intrigues been that in March 1648 the Second Civil War broke out with a revolt in South Wales. The most serious engagement in this short-lived affair was fought at Preston, where Oliver Cromwell defeated the Scottish army in August. The Army and Parliament had fallen out, and by now the Army had triumphed. There could be no compromise with the King, who was brought to trial and executed in January 1649.

The Prince of Wales had escaped to the Scilly Isles, then to the Channel Islands, and finally to France in the spring and summer of 1646. He was eighteen years of age when, immediately after his father's execution, the Scots proclaimed him Charles II. The new King arrived in Scotland on 23 June 1650, but it had taken the Scots eighteen months to get him to their shores on the terms they demanded. Eventually at Breda on 1 May he had signed an agreement to take the Covenant, to embrace Presbytery himself, to enforce it upon his English subjects and to root out episcopacy. It was a nasty piece of hypocrisy practised more by the Presbyterians than by the King, for they very well knew that they had imposed upon Charles conditions that he was incapable of accepting in full sincerity.

For some time before Charles arrived in Scotland, the Scots had been causing anxiety to the Council of State in London. They realized that invasion was a possibility, and the idea of a pre-emptive strike seemed attractive. The first problem was who would command this strike force. Thomas Fairfax was still nominally commander-in-chief of the English army, but he was strongly averse to a national (as opposed to a constitutional) struggle, and despite Cromwell's pleadings he declined the command. Cromwell had been home only a few weeks from chastising the Irish Catholics, and was the natural choice to succeed Fairfax. He set off for Scotland on 28 June 1650 and crossed the Border on 22 July at the head of 10,800 foot and 5,000 horses. He had with him Generals Lambert, Fleetwood and Whalley, and Colonel George Monck, a renegade Royalist, whose recent fighting in Ireland had greatly impressed Cromwell. He was given a new regiment, which eventually took its name from the town of Coldstream. Although there were some raw recruits, these had been integrated into existing regiments, and the army was well disciplined, well trained and homogeneous – almost everything that the Scottish army was not.

The arrival of Charles in Scotland had been greeted with great joy, but the Marquess of Argyll, whose power in the Committee of Estates was now immense, kept the young man tightly trammelled according to the best Calvinist and Covenant principles. He was not to play any active part in the army now being hastily raised, whose commander was the experienced and capable General David Leslie. In June 1650 the standing army in Scotland was 3,000 foot and 2,500 horse, but by an Act of Levies large numbers of fresh recruits were called up. By the time the two armies met in battle the Scots could muster 16,000 foot and 6,000 horse. However the advantage of superior numbers was sadly offset by the quality of both the officers and the men that Leslie had to command. Power rested firmly with the Kirk, and many Scottish cavaliers had been excluded from further command after the disastrous battle of Preston in 1648; preferment was now on religious rather than military merit. For this and other reasons very few Highlanders were among the newly raised levies.

There was not a great deal of difference in the arms and armament of the two sides – so much so that distinguishing emblems were often worn in battle. The musketeers were mostly armed with the matchlock (although there were a few early flintlocks in use), which was a cumbersome, inaccurate weapon that fired a heavy bullet of an ounce for a distance of up to 400 yards, although there was no hope of accuracy above 100 yards. The pike was still the most decisive weapon of the battle – there had been many instances in the First Civil War when the fortune of the day had been decided 'at push-of-pike'. The infantryman (musketeer and pikeman) also carried a sword, which was not of much value to him. The pikemen usually formed the centre of the line, with the musketeers on their flanks. Their defensive armour consisted of an iron helmet and back and breast plates – known as a corslet – which greatly restricted mobility. Cromwell marched nine regiments of foot into Scotland, each up to establishment of 1,200 men and organized into ten companies.

The cavalryman of this time was usually armed with a pair of pistols and a sword. He rode a medium-weight horse of fifteen hands or just over, and wore a headpiece known as a pot, and back and breast plates over an apron-skirted buffcoat. Dragoons (mounted infantry, armed with sword and carbine) do not appear to have formed a part of either army in this campaign; Colonel Okey's 12th Dragoons were converted into a regiment of horse before Cromwell crossed the Border. The New Model Army was provided with a formidable train of artillery, and their principal cannon was a demi-culverin, a nine-pounder dual-purpose gun used as a siege gun, and in the field as heavy artillery; they also had the lighter calibre saker. It is true that in the battles of Dunbar and Worcester the artillery was to play an important although minor part, but on the whole at this time the chief value of cannon was in siege warfare, for their rate of fire (even with light cannon) was very slow, and sometimes as many as twenty horses were needed to drag the guns over bad roads, while teams of oxen were required to haul the heavy siege pieces.

When the English army crossed into Scotland on 22 July they found that Leslie had successfully denuded the land of everything that could be of use to the invaders. This meant that Cromwell had to hug the coast, for his only source of supply was from the sea, and Dunbar was the only good harbour between the Border and Edinburgh. Even so his army was constantly at risk, for supplies could not be landed when the weather was bad – and that summer was notorious for gales and rain. Moreover, inability to commandeer wagons to transport supplies added to the difficulties of campaigning inland.

Leslie had taken up a strong defensive position from Leith to the foot of Canongate, thus covering Edinburgh and blocking the coast road. When Cromwell arrived before it on 29 July he found the position too strong to assault. The army had no tents, and that night it poured with rain, making the ground very wet. The next morning, with his troops miserably wet and short of provisions, he had little choice but to withdraw to Musselburgh. The withdrawal soon had the aspect of a retreat, with the rearguard hard pressed, for Leslie had seized his chance and gave the struggling, moiling English no respite. On arrival at Musselburgh the army found no supplies, for landing had proved

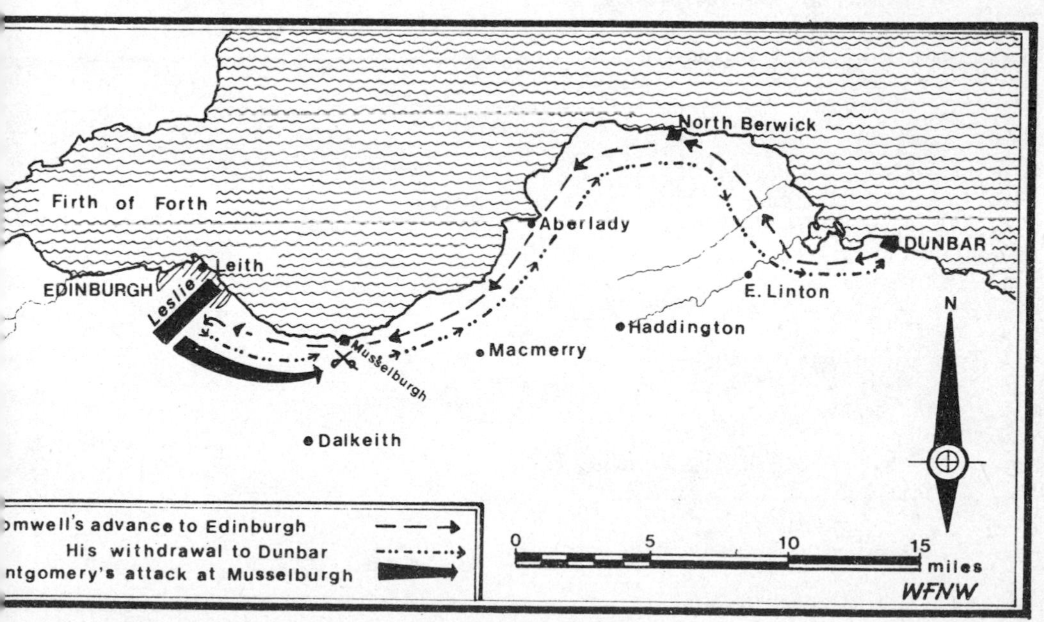

18 Cromwell's Scottish Campaign: his first advance

impossible, and at once the troops were called upon to stand and repulse a determined attack by General Montgomery commanding 1,000 horse. On 6 August the army was back at Dunbar, where supplies and – what must have seemed almost as important – tents had been landed.

Battle casualties had not been great, but already sickness caused by the terrible weather had considerably reduced Cromwell's fighting strength – nor could morale have been high after this disappointing beginning to a campaign against a numerically superior enemy fighting on their own ground. It could not have been easy for Cromwell to decide what course to follow. *He had perhaps three options open to him:*

(1) He could remain on the defensive at Dunbar for a little while in order to give his troops time to recover their strength, and in some cases their health. This might also give him a chance to improve his land communications. He would have to accept the fact that Leslie might advance his whole army against him before he was ready to go over to the attack.

(2) He could advance at once and, keeping his line of withdrawal along the coast open, assault the Leith–Edinburgh line.

(3) He could hazard a flank march round the Scottish right in an attempt to reach the Forth, and thereby cut Leslie's communications.

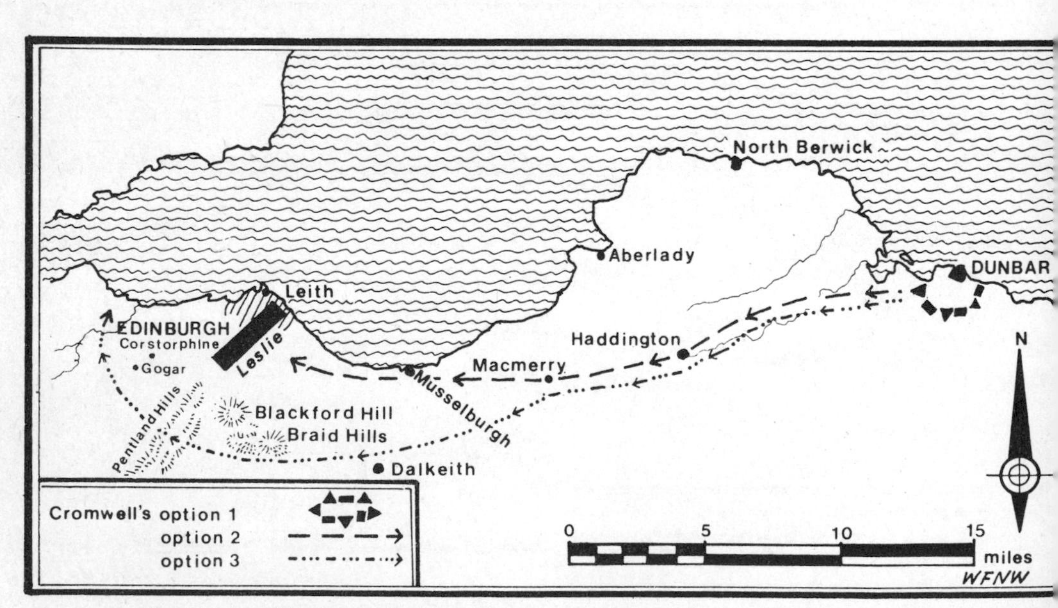

19 Cromwell's Scottish Campaign: his options for his second advance

Cromwell decided to march at once in an attempt to encircle Leslie's right and cut his communications (option 3). By the evening of 13 August he had occupied Braid Hills with outposts on Blackford Hill, south of Edinburgh. But once again he was defeated by lack of supplies and forced to return to Musselburgh. Nothing daunted he returned to the Pentlands on 18 August, but by now he was too late, for Leslie, realizing his intention, had deployed his army on the rising ground above Corstorphine, and as Cromwell moved west so Leslie moved parallel, and took up a new position near Gogar on ground that was too boggy to allow Cromwell to attack. The English army could do no more than march back in continuing appalling weather first to Musselburgh, and from there – under attack for some of the way – to the safety of Dunbar, which they reached on 1 September.

Was Cromwell's decision the right one, for it cannot be denied that in this first phase of the campaign he had been outgeneralled and achieved nothing? Moreover, through wandering about and exposing his unprotected troops to the atrocious weather he had reduced his fighting strength to little more than 11,000 men. The Council of State had ordered him to defeat the Scottish army, so that was his primary objective – although his animosity towards the Scots was nothing like that which he had shown towards the Irish Catholics, and there is some evidence that even at this late hour he had hopes of reaching an understanding. Obviously he wanted to bring Leslie to battle on ground of his own choosing, for although he knew his army to be of superior quality it was heavily outnumbered. Supposing he had been able to achieve the encircling operation, would he have been better off than awaiting Leslie at Dunbar? Probably not, for although it would have forced Leslie to give battle, he had that intention anyway. The chief merit in leaving Dunbar was one of morale. Troops that have suffered a setback, are a long way from home, and with uncertain communications, can become disheartened if kept hanging about under a command that appears uncertain of itself.

But it seems that Cromwell himself supplied the answer, for on his return to Dunbar for the second time (on 1 September) we find him writing to justify this move and his decision to fortify the town on four counts. First that it would provoke the Scots to engage, secondly 'that the having a garrison there would furnish us with accommodation for our sick men', thirdly that it would be 'a place for a good magazine' and was the only reliable harbour for the landing of supplies; and fourthly that Dunbar was a convenient place for the assembly of reinforcements expected from Berwick. Would not these excellent reasons have applied just as well on 6 August as they did on 1 September?

Leslie followed Cromwell back to Dunbar, marching parallel and a little to his south. On 2 September he took up a position above the sick and weary English on an eminence called Doon Hill at the very edge of the Lammermuir Hills, and sent a force to block the road to England at the Cockburnspath defile. Here he was virtually unassailable. The hill itself, which stands some two miles from the town, was too steep to be assaulted, and Leslie's left flank was guarded by the Brox Burn, which flows north down the hill, turns east and then north again before emptying into the sea just north of Broxmouth House, which stands on the west bank of the burn. The burn was fordable, but in its upper reaches, it flows through a ravine between steep banks, and here it was a formidable obstacle. Lower down, as it approaches the sea on ground that is comparatively flat (especially in the area round Broxmouth House), it could be crossed without great difficulty. At the east end of the Doon Hill spur the Dry Burn winds its way to the sea.

Cromwell could do nothing better than to draw his army up 'standing in battalia in the town fields, between the Scottish army and the town, ready to engage'. It was essential that he should secure Dunbar, for with the road to England blocked he needed the port for supplies and, if necessary (and it must have seemed increasingly necessary), to embark what was left of his army and sail home. *Uncertain of Leslie's intentions, he had but two options:*

(1) To make use of his fleet and withdraw at least the infantry, and let the horse try to force a passage through the defile.

(2) To stay where he was 'ready to engage', and perhaps get succour from England.

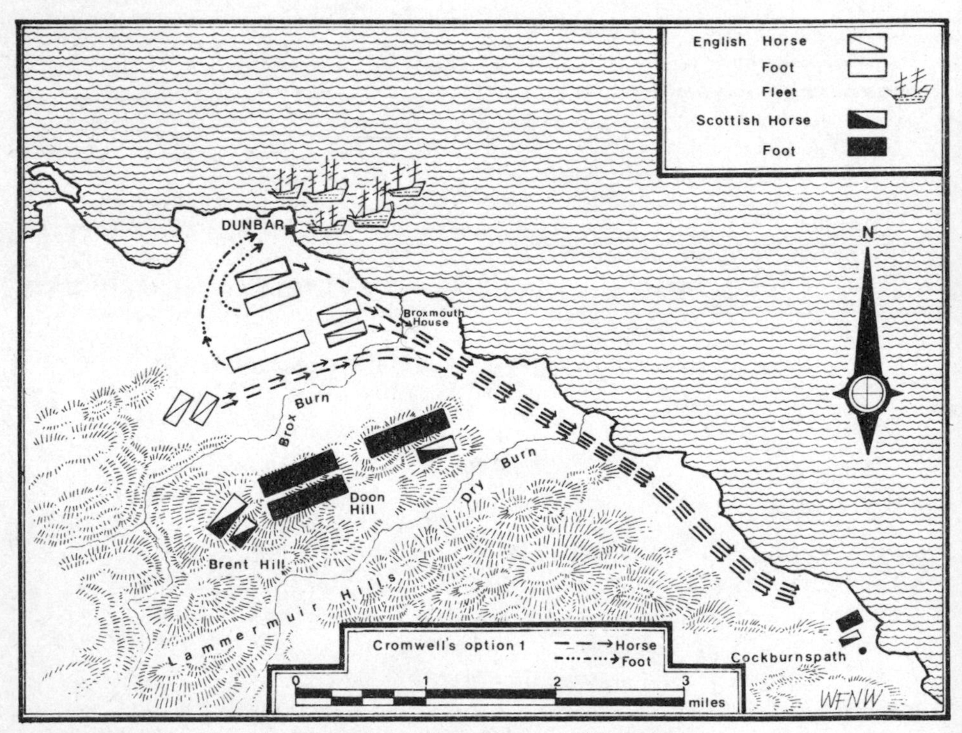

20 The Battle of Dunbar: first phase, Cromwell's options

Cromwell was well aware of the grave repercussions that would follow a withdrawal from Scotland after such a futile expedition. The damage to his own reputation would be enormous, and that to the cause of the Commonwealth almost as great. Moreover, he was a man of courage and steadfastness, not easily defeated. He therefore wrote to Arthur Heselrige at Newcastle requesting help: 'The enemy has blocked our way at the Pass at Copperspath through which we cannot get without almost a miracle . . . Our lying here daily consumeth our men, who fall sick beyond imagination.' This letter would have to go by sea, which underlines the seriousness of Cromwell's situation. Nevertheless, his decision to stay put – even supposing he could have made a successful evacuation – must have been the right one.

The initiative lay with Leslie; everything was in his favour. The English army, little more than half his numbers, had their backs to the sea and the road to England was barred. *He had two options:*

(1) He could march down the western side of the hill and give battle with his superior numbers.

(2) He could remain in his strong position and allow the English to embark, and yet be ready to sweep down on their flank or rear should they try to force the defile at Cockburnspath.

There has been much unresolved argument about whether the Scots' decision to abandon their strong position and come down to the lower ground during Monday 2 September was made by Leslie, or forced upon him by an ever-watchful Committee of Estates. Undoubtedly ministers of the Kirk exerted a powerful influence over military decisions, but it seems likely that Leslie at least agreed with them – if it was a Committee decision – for his interrogation of a prisoner, captured that afternoon when the Scots overran one of Cromwell's outposts east of the burn, shows that he thought the English were intending to embark and escape him. He was impatient of delay and eager for glory.

This is the only charitable interpretation one can put on what was a disastrous decision, because unless he thought that his chance of annihilating the English was slipping away fast there was no point in taking up a cramped position on the low ground some twelve hours before he intended to launch his attack. As will be seen, his dispositions for an attack were not unsuitable, but should he lose the initiative and have to fall back on the defensive he had left himself no possible room for manoeuvre.

By the afternoon of 2 September Leslie's descent of the hill was nearing completion. The horse and foot were followed by the artillery train and the baggage. He took up a position, roughly parallel with the burn, his army drawn up in the conventional way with the infantry in the centre and the cavalry on the wings. However, he transferred about two-thirds of the left wing cavalry to the right, where the ground was much more suitable, and the line, which extended for nearly two miles, had the punch in the area of the present railway line and road, a little west of East Barns and near to where the monument stands. By the evening the manoeuvre was completed, and Leslie had succeeded in getting much of his army nastily bottled up between the ravine and the hill. His intention was to cross the burn in its lower reaches and with weight of numbers envelop Cromwell's army with a powerful right hook.

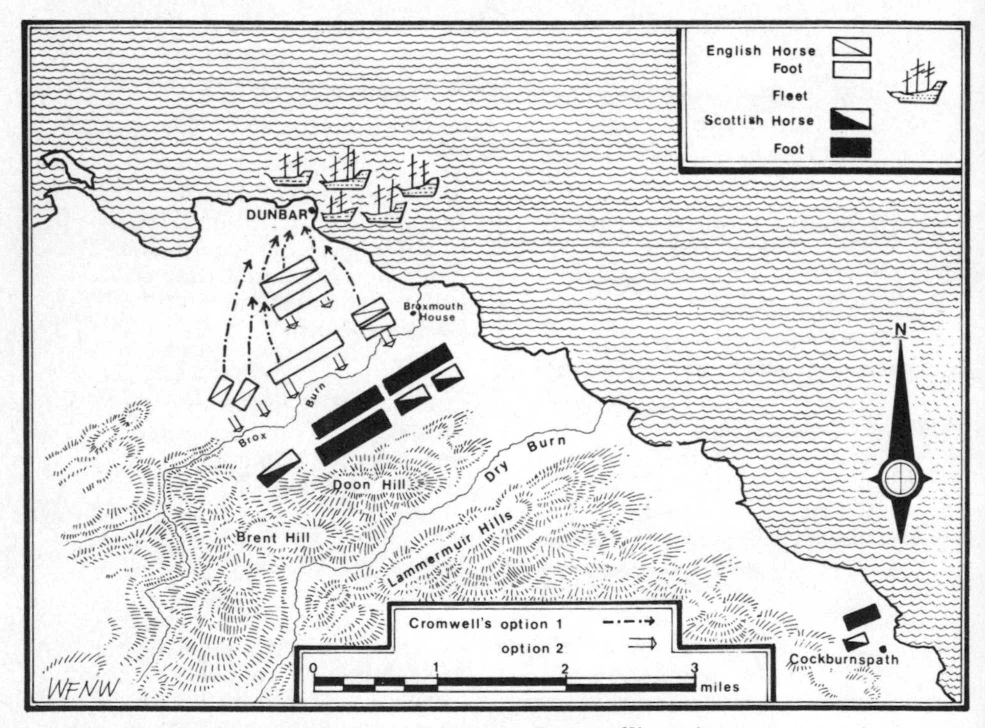

21 The Battle of Dunbar: second phase, Cromwell's options

All this was not lost on Cromwell who, having advanced his army as close to the line of the burn as was safe, was watching the Scottish army's awkward progress with Lambert from Broxmouth House, which he had garrisoned earlier in the day. In describing the events later he comments, 'We could not well imagine, but that the enemy intended to attempt upon us or to place themselves in a more exact condition of interposition.' At about nine that night Cromwell held a council of war; *the council had three possible options to consider:*

(1) To withdraw along the lines previously considered – if indeed there was still time.

(2) To take the initiative and attack. If this course was adopted a plan had to be formed.

(3) To stay on the defensive west of the burn, and oppose the crossing.

According to an officer of Lambert's regiment (Captain Hodgson) who fought at the battle – although he was too junior to have been present at the council – there was considerable support for the first option. This is surprising, for although some of the officers present may not have been consulted earlier, it should have been obvious to all that if ever this course of action had had a chance of success, by now it was too late. The third option had virtually no support, and the council eventually agreed (according to Captain Hodgson, at the instigation of Lambert) that the best hope lay in a dawn attack, which might have the advantage of surprise. It was a bold decision but clearly the right one; the plan of attack bears the hallmark of Cromwell's genius, with no doubt a touch of Lambert, and the blessing of Monck, who was called in to advise.

Like all good plans it was simple. At daybreak Lambert and Fleetwood, with six regiments of horse, would spearhead the attack on the enemy's right wing, for if that could be routed the infantry in the centre and left could be rolled up and pulverized between the upper and nether millstones of ravine and hill. Monck would be in close support of the cavalry thrust with three and a half regiments of foot. The remaining two brigades of infantry (Pride's and Overton's) should form the reserve and secure the artillery, whose task it was to pour shot into the centre and left of the Scottish line while the main business was taking place on their right. Cromwell positioned himself and his regiment of horse near to Pride's brigade with a view to assaulting the enemy's right flank at the decisive moment of the battle.

At Dunbar on 3 September 1650 Oliver Cromwell won what was undoubtedly his greatest victory. In spite of some temporary setbacks the plan adopted at the council of war on the previous evening worked very well. Complete surprise was gained. The night of 2–3 September followed the pattern of many of its predecessors that summer in being wet and windy; although the Scottish army stood to twice in the early part of the night, when Leslie found that these alarms were false, and thinking it unlikely that the hopelessly outnumbered English would venture an attack, he gave orders that only the file-leaders need keep their matches alight (and they must have found it a tricky business). Some of the officers are said to have sought shelter away from their commands; the horses were unsaddled, and the men huddled under cornstooks to get what sleep they could. Meanwhile, the English, groping in Stygian darkness – for it is no easy task to get regiments to an assembly point on a black night – formed up for a dawn attack.

The difficulties encountered caused the attack to begin a little later than planned, and it was about 5 a.m., as a waning moon was just piercing the clearing rain clouds, when Lambert's men crossed the burn virtually unopposed. The Scottish musketeers took time to get their matches lighted, and such was the stealth of the attack that the whole host was caught unprepared; nevertheless, no sooner had they recovered than they fought with great determination and courage. Lambert, who was outnumbered by almost two to one, was soon in difficulties, nor could Monck, who came in on his right, make much progress. But Pride's brigade swept into the gap between Lambert and

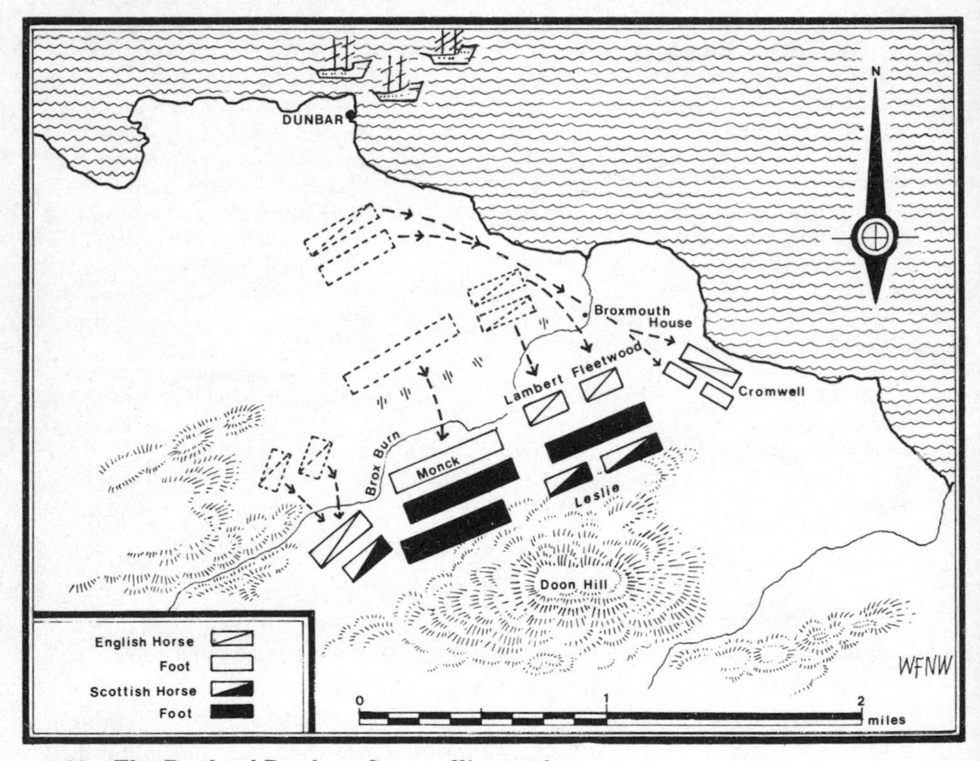

22 The Battle of Dunbar: Cromwell's attack

Monck, whose men soon renewed their charge, this time with greater success.

By now the sun was through, warming the darkly purple hills that formed a stern backdrop to the conflict. The fighting on Leslie's right was a desperate affair at push-of-pike, and the hard-pressed Scots bent and then broke before the storm. This exposed their centre, and Cromwell, with that coolness of judgment not given to every general in the heat of battle, saw his opportunity and swung his horse in a right hook against the Scottish flank. Soon their army was hopelessly wedged between the ravine and the steep hills; great numbers were now a hindrance and merely fodder for darting pikes and flailing swords. It is said that 3,000 perished and as many as 10,000 were made prisoner. With all the odds against him, Cromwell had won a magnificent victory; the Scots had suffered a disaster of the greatest magnitude.

Leslie's defeat was no disappointment to Charles and the Royalist faction, who saw that the Kirk must needs turn to them if a new and more efficient force was to take the field. Indeed it was not long before Charles's personal position had sensibly improved, and on 1 January 1651 he was crowned at Scone and made nominal Commander-in-Chief of the new army. Leslie, who was his second-in-command, retained operational control; Middleton was made Lieutenant-General of the Horse and Massey (a Parliamentarian in the First Civil War, and the hero of Gloucester) had command of Charles's English contingent. Such was the rate of recruitment that by June Leslie had some 15,000 foot and 6,000 horse under command.

Cromwell's army too had been increased, for the Council of State now realized that he was engaged in what could be a long war. Some new regiments were raised – including one of dragoons under Colonel Morgan – and others posted from England. By June Cromwell had fifteen regiments of foot, eleven of horse and one of dragoons, and so there was now not much difference in the strengths of the rival armies. However, for some months after Dunbar Cromwell had to weaken his army by leaving garrison and siege troops in Edinburgh, for although the city surrendered at once the governor of the castle continued defiant until surrendering at the end of December.

During the autumn and winter months, while the Scots were building up a strong defensive position in and around Stirling, Cromwell marched and countermarched between Edinburgh, Linlithgow and Glasgow without achieving very much – a similar pattern of events to those before Dunbar. Then in February 1651 he contracted a serious illness, which incapacitated him off and on until the end of May. By the end of June Leslie felt that his army was strong enough to give battle, if this could be achieved on favourable terms, and he took up a good defensive position in the Torwood Hills. Cromwell came after him, but this time Leslie was not to be drawn from his position, which Cromwell felt was too strong to assault.

The situation was rapidly approaching stalemate, which was what Cromwell could least afford. Leslie occupied an impregnable position safeguarding Stirling, drawing his supplies from Fife and in communication with Perth, which was the present seat of government. All Cromwell had so far accomplished was some fruitless skirmishing with outposts. There were perhaps three options open to him if we discount the thought of withdrawal into England, which he could not now have entertained. *These were his options:*

(1) He could march round the enemy position and place his army between Torwood and Stirling, hoping the threat would be sufficient to bring Leslie off his hillside.

(2) He could cross part of his force into Fife, endangering Leslie's supplies, and threaten the Torwood position with the bulk of his army.

(3) He could remove his whole army into Fife, cutting Leslie off from his supplies and from Perth, thus forcing him into battle, although leaving the door open for him to move into England.

Cromwell decided to cross part of his army into Fife, and confront Leslie with the remainder of his force (option 2). The plan was sound in conception, although it very nearly failed in execution through his misjudging Leslie's reaction to the threat to his supplies. And he had later to adopt the third option before he had Leslie where he wanted him.

Cromwell ordered Colonel Overton to cross the Forth at Queensferry with a regiment, four independent companies and four troops of horse, and establish a base on the peninsula off Inverkeithing Bay. Lambert appears to have shown more perspicacity than his chief on this occasion for, realizing that if Leslie reacted strongly to this threat Overton would be in grave danger, he took two regiments of horse and foot across on his own initiative. Cromwell later admitted that without these the Scots 'would probably have beaten our men from the place'. Leslie sent 4,500 men under Major-General Browne to confront Overton, and kept his horse to deal with any threat from Cromwell. In a short, sharp battle near Inverkeithing Lambert and Overton smashed the Scots, killing some say as many as 2,000 and taking General Browne prisoner.

Marching and counter-marching followed, all too familiar by now. On learning of Browne's defeat Leslie determined to march his whole army against Lambert, but Cromwell was too close to him. The Scots were already a few miles beyond Stirling before Leslie realized the danger; quickly swinging back on his tracks he just regained his former position in time. Still unable to force an issue with Leslie, Cromwell now transferred the bulk of his army across the Forth at Queensferry, leaving only a few regiments to keep watch on the enemy at Stirling. On 31 July Cromwell invested Perth and on 2 August it surrendered.

Leslie now found himself in a very unpleasant position, with most of Cromwell's army squarely astride his lines of communication. *He had but two options:*

(1) He could attempt to steal a march on Cromwell and make quickly for England, recruiting what Royalists he could as he went.

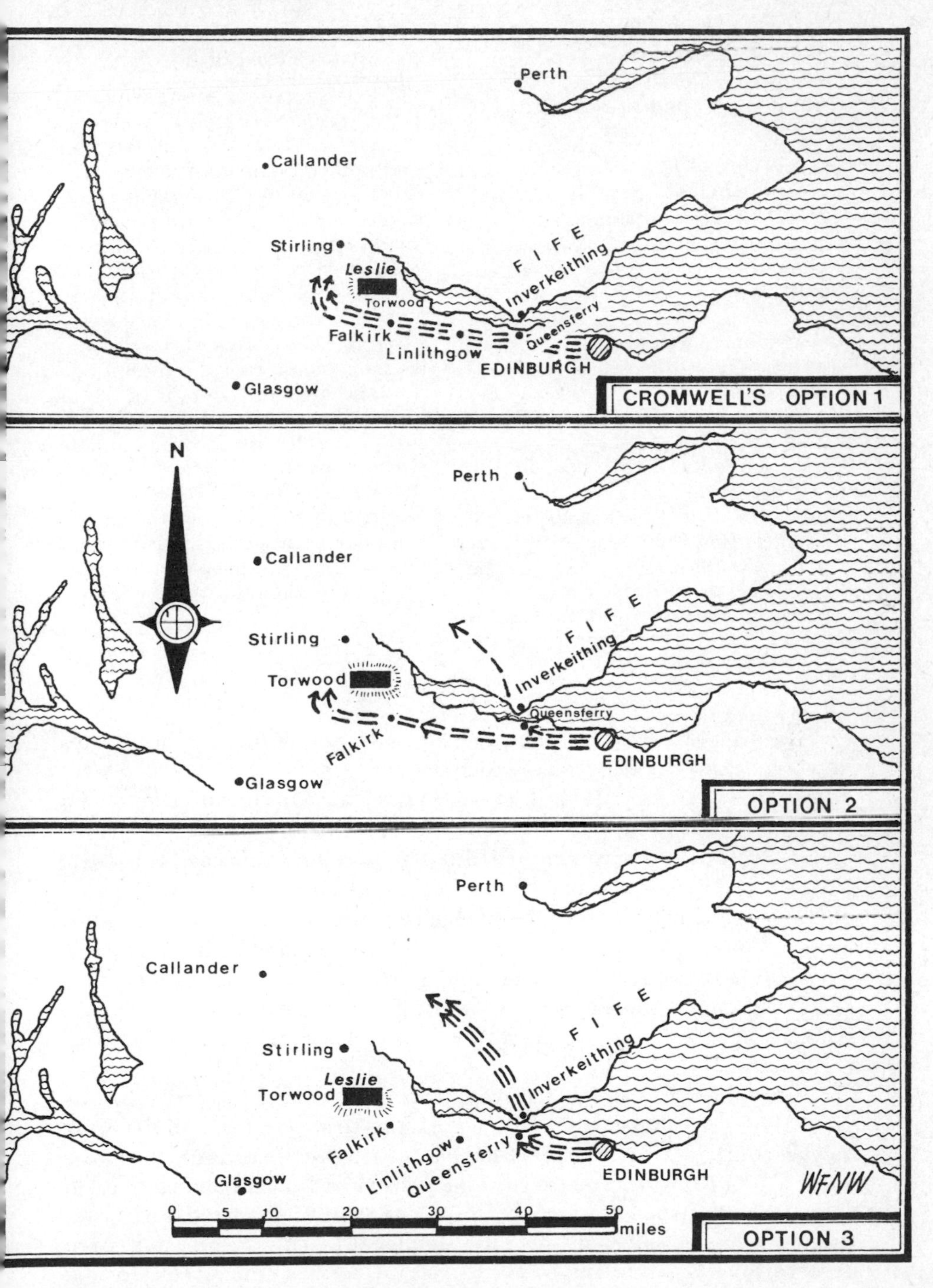

23 *Cromwell's options after Dunbar*

(2) He could interpose his army between Cromwell's troops, at Perth and Edinburgh, and in an attempt to reopen his supply lines seek battle on ground of his own choosing.

The Scots opted to march for England. It is quite possible that this was not Leslie's decision, for there are indications that he favoured giving battle in Perthshire. But Charles was rapidly becoming much more than a figurehead Commander-in-chief, and he was anxious to go into England, where he was convinced he would receive great support.

It is easy to say with hindsight that this was the wrong choice, but in fact it was probably the right one. On the march south Leslie said to Charles (who wisely kept it to himself) that he 'well knew that army, how well soever it look'd, would not fight'. And in the event, so far as he himself was concerned, this proved all too true. But there was no reason to assume that the Scots would not fight outside their country, and indeed most of them fought very well. Nor at this stage could it be known for certain that dislike of the Scots, fear of reprisals, Charles's enforced adherence to the Covenant and a general lassitude would combine to make recruiting in England a farce. Again, it could not be known that Cromwell looked upon such a march with favour, and had plans to defeat it. In short, Charles felt that he had a better chance of raising England by his presence than Leslie had of defeating the English army in Perthshire. Most people in his position would have thought the same.

The Royalist army started their march south on 1 August with Charles now the acknowledged Commander-in-Chief, and by the 15th they were at Wigan. On the south bank of the Mersey at Warrington they had a minor confrontation with a combined force under Generals Lambert and Harrison, but the English did not seriously contest the issue because Cromwell was anxious to avoid battle until his carefully laid plans were completed. Here the Earl of Derby joined the King with a few hundred men, and he was promptly sent off to Lancashire to recruit more. In this he failed, being defeated at Wigan by Colonel Lilburne. At the same time General Massey, himself a Presbyterian, was sent to Gloucester to try to gain recruits from his own sect, but he too met with no success.

Meanwhile, Charles and his weary Scots continued towards Shrewsbury. Here the Governor refused a summons to surrender, and the army moved south. At Worcester Charles was more fortunate: with the approach of his army the mayor and citizens of this consistently loyal Royalist city turned against the Parliamentary Committee, and on 23 August Charles entered the city unopposed. The march had been desperately disappointing so far as recruiting was concerned, nor did proclamations issued from Worcester have any great success. It is doubtful if as many as 2,000 had rallied to the King's standard since the army crossed the Border, and it cannot now have numbered any more than 16,000 men. Their morale was far from high and their commanders did little to inspire them. The Duke of Buckingham was sulking because he thought he ought to command the army; Leslie continued to brood over the fighting prospects of his troops, and, to make matters worse, was not on speaking terms with General Middleton.

Worcester was a good place for a temporary halt, and in any case for the moment the Scots, having covered 300 miles, were in no mood for further marching. Food and forage were fairly plentiful, and the city commanded the approaches to Wales and the south-west, from where Charles still hoped to draw reinforcements. The fortifications had fallen into a sad state of disrepair, but men were hastily pressed for what repair work was possible in the time. Particular attention was paid to the earthwork, known as Fort Royal, just outside the city wall at the south-east. By 28 August much had been done, and Charles probably believed himself to occupy a fairly strong defensive position.

It was now more than time for a number of important decisions to be made, and Charles held a council in the Commandery – a building (which still stands) just west of Fort Royal. *There were three possible options:*

(1) To go forward in an attempt to threaten London.

(2) To take up as strong a defensive position as possible.

(3) To march into Wales.

The first option, to march for London, might have seemed attractive, but in fact by 28 August it was too late. Of the other two courses open to the Royalists it is possible to argue that the most sensible one – which is said to have been proposed by the Duke of Hamilton – was to march out of the trap that was closing on them and into Wales. But this does not appear to have had much support, and Charles decided to take up as strong a defensive position as possible (option 2).

This could have been done in one of three ways:

(1) By safeguarding the right flank in depth, holding the bridge at Upton in strength, and then having another line on the Teme with outposts south of the river on the Powick ridge where the church stands; at the same time keeping about half the army to man the city defences.

(2) By concentrating the entire force within the city and its immediate environs.

(3) By manning the fortifications with the bulk of the army, but holding the line of the Teme with a strong delaying force, and being prepared to reinforce it should the enemy's main attack develop on that flank.

24 The advance south: Charles II's options

Charles decided on the first option. But, perhaps because by now he had become despondent – and who can blame him – and a little careless of detail, though he made what must have been the right decision he failed to implement it properly. The army was divided into two parts. The larger was to man the defences of the city, but General Montgomery with two brigades, and Colonel Pitscottie's Highlanders, were to hold the line of the Teme with outposts on the ridge south of the river, and General Massey was to take 300 men and destroy the bridges at Powick, Bransford and Upton. This left only one bridge across the Severn intact – that leading from the city to St John's suburb (known as The Severn Bridge), which allowed access for troops to the suburb itself and the Teme meadows in the area of the confluence of the Teme and the Severn. This was an admirable plan, but why give Montgomery a mere 300 men to defend an important crossing point of the Severn? At a later council Charles decided that in order to strengthen his defensive plan he would take the initiative to the limited extent that was open to him by mounting a strong raid on the night of the 29th with 1,500 men under General Middleton and Colonel William Keith against the batteries on Red Hill and Perry Wood, which had just started to open fire on the city.

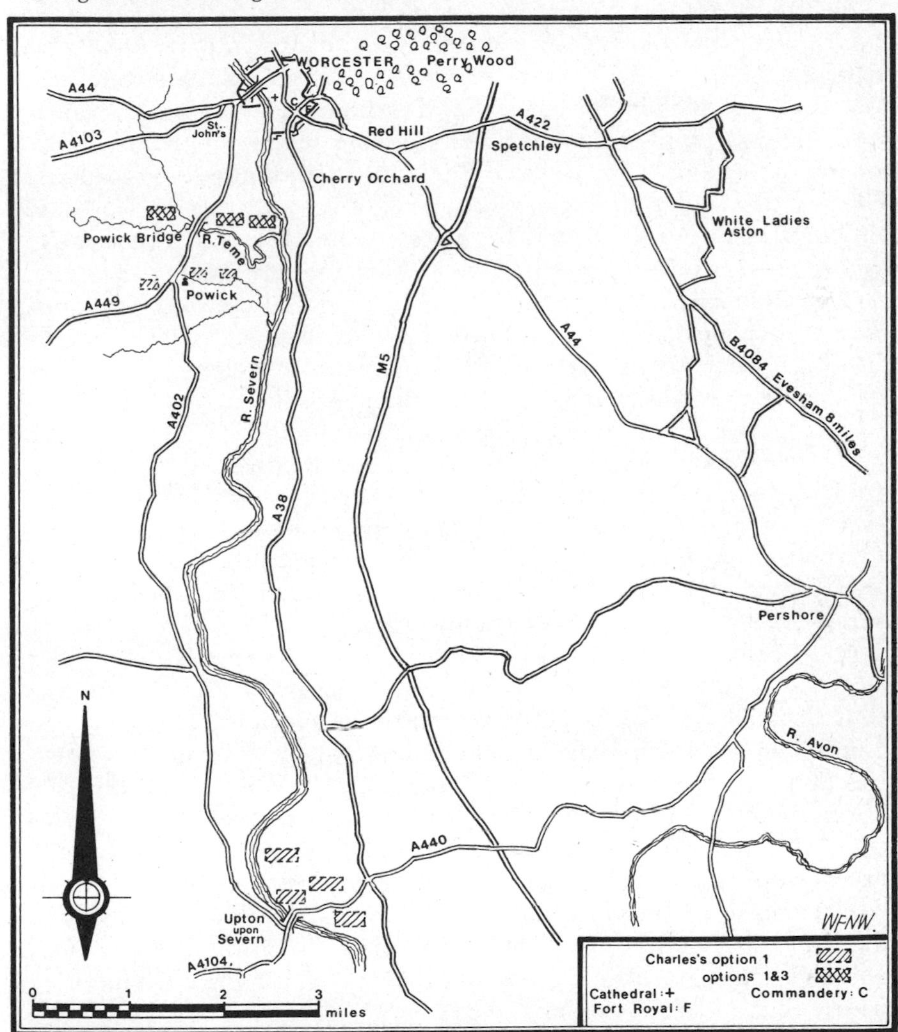

25 *The Battle of Worcester: the battleground and Charles's preliminary options*

When Cromwell marched into Scotland in July 1650 the Council of State were at pains to see that England was not denuded of troops, for Royalist risings could always be expected. New regiments were raised, and measures taken to activate the old trained bands of the First Civil

War. More new regiments were raised after Dunbar; some of these joined Cromwell in Scotland, and others were to strengthen the defence in England, now under General Fleetwood, who had been transferred from Scotland for the purpose. And so Cromwell had a good supply of troops, strategically placed, to call upon.

He was in no great hurry to pursue the Scots army, for he had foreseen their intention, and indeed welcomed the march into England as a necessary step to finish the business. He had laid his plans accordingly. A garrison was left in Perth; Monck with some 6,000 men was to besiege and take Stirling; Lambert was sent off independently with about 4,000 horse with orders to join General Harrison on the borders and to harass the Scottish army, but not to risk a serious engagement; the Northern and Midland County Committees were written to with instructions to raise troops and remove all animals from the advancing Scots; and Cromwell himself, with the bulk of the infantry, left Leith on 6 August. These troop movements were aimed at placing a ring of steel round the invaders, and recent arrests of prominent Royalists throughout England had further strengthened Parliament's hand. Nevertheless, if Charles could march rapidly and if his recruiting proved successful, there must have been every chance – despite these thorough precautions – that he could pose a serious threat to London. Cromwell either discounted the risk, or considered it justified.

Lambert, having broken off his engagement with the Scots at Warrington, withdrew to Knutsford and from there marched to Warwick, where he was joined by Cromwell on 24 August. That same day Colonel Lilburne had routed 1,500 Royalist reinforcements under Lord Derby at Wigan on their way to join Charles. Derby himself was wounded, but managed to get to Worcester with thirty men. Fleetwood had moved to Banbury, and joined Cromwell at Evesham on 27 August, together with General Desborough and Lord Grey of Groby with 1,100 horse, who were advancing from Reading. The net was drawing ever tighter round Worcester, and although Cromwell had taken the precaution of ordering siege tools, with an army numbering about 28,000 men to Charles's 16,000 he could not have allowed his mind to dwell too long on siege warfare. At Evesham Cromwell held a council and made his plan of attack. There were several options open to him, all of which must include blocking the route to Wales. *His five options were:*

(1) He could avoid making his main thrust on the east side of the Severn, which could be the most strongly defended; keep a light containing force there, including artillery on the dominating features of Red Hill and Perry Wood, and pack his punch on the west side after forcing the river at Upton.

(2) On the other hand, if he wished to avoid splitting his

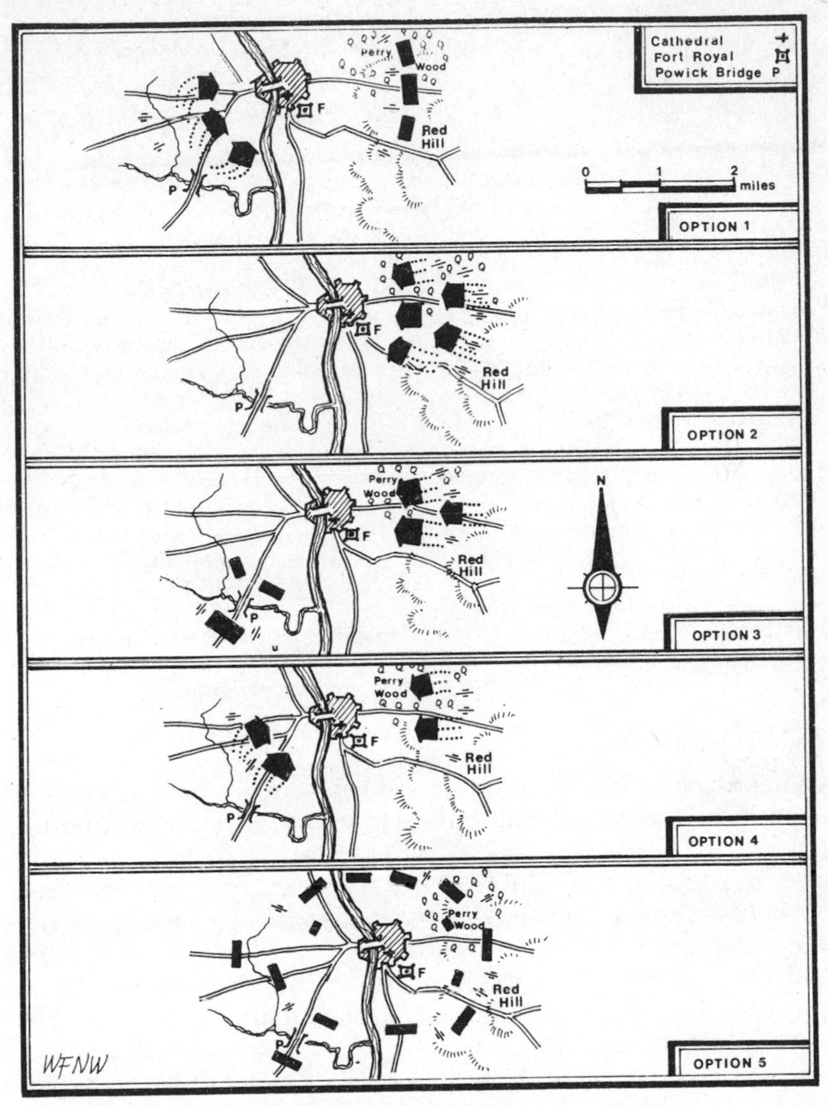

26 The Battle of Worcester: Cromwell's options

army and having two difficult river crossings, he might consider that he had a better chance of a quick victory if he concentrated his entire force on the east bank, and relied on his heavy guns and superior numbers to force the city defences, which he probably knew were in poor shape.

(3) He could send a force across the Severn of sufficient strength to bar the King from Wales, and still concentrate his main attack on the east side.

(4) He could attack in strength on both sides of the river, occupying the dominating

features of Red Hill and Perry Wood.

(5) He could encircle the city, cut off supplies and wait for the Scots to break out or surrender.

Provided Cromwell blocked the way to Wales and the south-west he could expect to defeat Charles by any of the options open to him. He had to decide which way would be the surest and would involve the fewest casualties. He chose to split his force and attack in strength on both sides of the river (option 4), thereby providing us with the only example of a major river crossing battle to be fought in Britain. General Fleetwood, with Generals Lambert and Deane under command, was to cross the Severn at Upton and with 11,000 men to march up the west bank and attack the Scottish right. In order to cross the Teme and Severn (for to maintain contact was essential) sufficient boats were to be collected and hauled upstream with the advancing army to form two bridges 'within pistol shot' of each other. Cromwell would take the rest of the army via White Ladies Aston and Spetchley to occupy the Perry Wood–Red Hill heights above Worcester, with his left on Bunds Hill above the confluence of the two rivers.

It was a plan simple in design, but bold in concept, for it involved the hazard of quite probably two opposed river crossings, and by dividing the force it broke a principle of war. Because it came off we are inclined to think it was the best plan, but it certainly was not the most economic in manpower; of all the choices open to Cromwell he probably selected the one most capable of going wrong.

In the early hours of 29 August General Lambert, who was spearheading Fleetwood's army, found that the bridge across the Severn at Upton had been destroyed, except for one plank which still remained more or less intact. He hand-picked eighteen intrepid dragoons to carry out the unpleasant crossing high above the swirling water; and while it was still dark these brave men straddled, for they dared not walk, the single plank. Once across, their presence was quickly discovered, and there followed a fierce fight, first in the churchyard and then in the church itself. To reinforce these dragoons proved quite impossible until in desperation, and by a lucky chance, Lambert found that the river was just fordable for horses a little lower down. By half swimming and half fording his men struggled over, and taking General Massey's 300 men in rear forced them back against their improvised entrenchments, and through weight of numbers drove them, after a desperate struggle, from the river and from further interference. The next day the bridge was repaired, and Fleetwood was able to bring his entire force across.

By 2 September boats and planks for the bridge had been collected, and everything was in readiness. But the Lord-General had decreed that the attack was not to go in until the 3rd, for he had become increasingly superstitious, and that day was the anniversary of his great victory at

Dunbar. Accordingly at about 5 a.m. on Wednesday 3 September Generals Fleetwood, Lambert and Deane began their slow advance (for the pace was regulated by the speed of the boats) up the west bank of the river. By the early afternoon this force was in contact with the enemy around Powick Church, and after a sharp skirmish the Scottish outposts withdrew behind the Teme. Fleetwood's plan was for Deane to attempt to force the bridge at Powick, while Lambert's men constructed and made use of the two boat bridges – the one across the Severn to be placed just above the confluence.

The fighting along the river bank was extremely heavy. The Powick bridge was found to be partially intact, but Deane could not force it, and his cavalry would anyway have to find a ford further upstream. A forlorn hope was somehow got across the Teme in boats near the confluence, to protect the bridge builders, and in spite of strong opposition from the Highlanders the pontoon bridges were constructed. It was a splendid feat, and it is a pity we know so little about its execution. The Highlanders, tough and desperate men, fought off superior numbers with great gallantry, until Cromwell, seeing that this was the vital sector of the battle, brought his men across the Severn

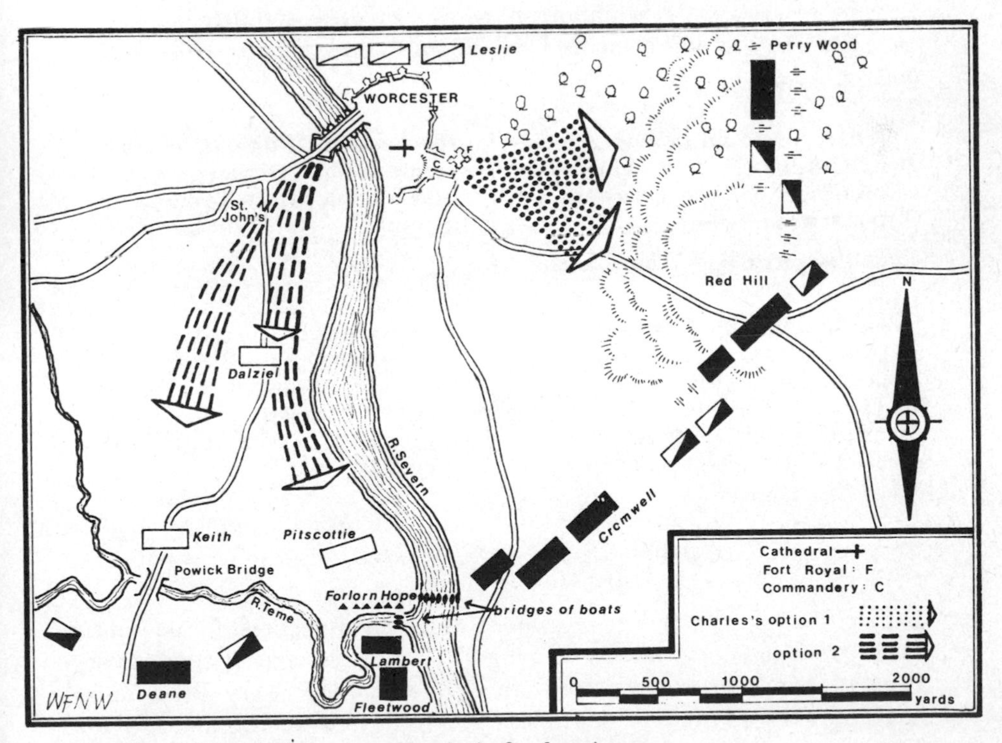

27 *The Battle of Worcester: Charles's final options*

pontoon and took the enemy in flank. Gradually the Highlanders gave ground, and when it came to in-fighting in the enclosed ground on the north side of the river the experience of the English in this type of warfare proved decisive. The eventual withdrawal of the Highlanders left Keith's brigade with its flank in the air, and with no option but to conform. Dalziel in reserve either could not or would not stem the rapidly accelerating retreat.

Charles's earlier attempt to gain the initiative with a two-pronged attack on the Perry Wood and Red Hill batteries had not met with the success that such a spirited venture deserved. The success of the operation depended largely upon secrecy, and a tailor in the city, who learned of the project, proved to be a Parliament spy. The raiders were repulsed with casualties, the tailor was hanged, and the batteries went on firing into the city. Apart from this there was very little action on the east bank of the Severn in the initial phase of the battle.

Charles established an observation post at the top of the cathedral tower, which gave him an excellent view of the whole battlefield. During the river battle, which lasted for a full two hours, he paid at least one visit of encouragement to the forward troops. When he saw Cromwell reinforce the west bank, and could appreciate the way the fighting was going in the important river sector, he had to make a quick decision. He was operating on interior lines, and *there were two options open to him:*

(1) He could launch an attack on Cromwell's now depleted – although still quite strong – right wing.

(2) By switching troops across the Severn Bridge, he could reinforce his brigades on the north bank of the Teme in an attempt to stabilize the position there.

In deciding to launch an attack on Cromwell's depleted right wing there is no doubt that Charles made the right decision. Cromwell had offered him an opportunity which if seized at once and executed with resolution might conceivably give him that chance of victory which until then had seemed hopelessly remote. Just as it is an excellent policy to reinforce success, it seldom pays to reinforce failure.

Before the days when wireless was sufficiently developed for use in the field, great importance was attached to orderlies, and at times everything depended on their courage and efficiency. Very little detail survives of how orders were transmitted in the field during medieval and later battles, but decisions such as Charles had just made would have had to be relayed quickly to subordinate commanders, who were probably unable to leave their posts for what became known as an O (Orders) group. Staff officers, as such, had not yet appeared on the establishment, and commanders, when unable to deliver orders personally, would rely on officers or troopers (the gallopers of later days), depending on the importance of the directive.

Charles ordered a similar two-pronged attack to the one that had failed four nights previously through lack of surprise. This time he personally led the right thrust against Red Hill, while the Duke of Hamilton was in command of the troops detailed for the attack on Perry Wood. The Duke of Buckingham and Lord Grandison were in support with cavalry. On this occasion there was no need for surprise; leadership and courage were called for and given. The initial thrust gained instant success. In a battle that lasted for three hours the enemy foot gave way all along the line, and cavalry from Whalley's and Harrison's commands, who came to the aid of the militia, could not stabilize the front. It needed only Leslie's large body of horse, who were drawn up outside the north wall of the city on the Pitchcroft, to reinforce this success and the whole of Cromwell's right wing might have been routed. But both now and in the last desperate phase of the battle, Leslie and his men remained supine, refusing to fight.

As it was Cromwell, realizing the extreme danger, hastily recrossed the Severn. Just at the time when the Royalist attack was beginning to lose its impetus (the Duke of Hamilton had already been seriously wounded) he took personal command, and with the help of the three brigades he led drove Charles and his men back into the city in considerable confusion.

The last phase of the battle was a shambles, and Charles narrowly escaped capture at the Sidbury Gate. There was no time to organize any sound defence of the city, and anyway by now there were few men left willing to become further involved in the bloody and bewildering business of street fighting. Amid many recorded and unrecorded acts of gallantry the whole army was pushed back through the city with heavy casualties. Dalziel's brigade on the west bank of the river laid down their arms having scarcely struck a blow, and the Parliamentarians on that flank found the Severn Bridge intact. Fort Royal was stormed and captured with great dash, and its guns turned on the milling turmoil in the tortured city. It was time for men to be thinking of how to save themselves, and soon Charles was starting on the most thrilling adventure of his life.

It was a total disaster for the hopes of the monarchy. Of the 16,000 men – most of them Scots – who started the battle, very few got home. Somewhere between 2,000 and 3,000 were slain in the fight, and some 10,000, including 640 officers, were taken either on the battlefield or as they were making their way north. Well might Cromwell write to the Speaker next day, 'The dimensions of this mercy are above my thoughts. It is, for aught I know, a crowning mercy.' For Cromwell it was certainly a mercy, and very nearly a crowning.

5 The Saratoga Campaign
Setting the Scene

THE causes of the American War of Independence were too many and too complex to be discussed here. A lot of ill-informed and muddled thinking for some twelve years from 1763 by successive First Lords of the Treasury, misplaced fiscal legislation, some good intentions by Grenville misconstrued by the colonists, and the bellicose attitude of King George III, were compounded by the crowning folly of attempting to put back the colonial clock by coercion. Such futile fulminations, ineptitude and the divided counsels of Parliament were the crucible from which the strength and splendour of the United States of America were soon to emerge.

During the latter part of 1774 and the early months of 1775 events moved inexorably towards war. The Cabinet was convinced that the presence of less than 10,000 British soldiers would be sufficient to bring the rebels – as they were beginning to be called – to heel, and, although expertly advised to the contrary and urged to rely on an economic stranglehold through blockade, they refused to listen.

Accordingly General Gage, commanding in Boston at the time, was sent limited reinforcements – indeed neglect of the armed forces since 1763 made it difficult to spare men or ships from home waters. In April 1775 he received orders to move at once against the rebellious elements in Massachusetts, Connecticut and Rhode Island. Gage chose to march on Concord, where he knew there was a military depot. Secrecy was impossible, the raid was a failure, and during the withdrawal to Lexington the rebels took very heavy toll of his force. The sword thus unsheathed was not to be returned to its scabbard for more than six years of bitter struggle and cruel losses.

In its initial stages the war was fought by the Americans for better conditions within the British Empire – a return to the *status quo ante* 1763 would have been acceptable – but by the summer of 1776 the colonists had despaired of achieving any form of home rule; on 4 July they declared their independence. The Continental Congress, sitting in Philadelphia, was quick to see the need for a Continental army, and the colonies were requested to raise specified numbers for this purpose.

These were to be Federal troops, who were to serve for two or three years, while the short service (month or two) militia remained under the control of the colonies. To command their army Congress selected George Washington, a greatly respected Virginian, whose military experience far exceeded that of any other American-born officer.

After Lexington, Boston was besieged, and it was not until June, by which time reinforcements had brought his army up to 6,000 men, that Gage felt able to take the offensive. On the 17th his troops attempted to take the Charleston peninsula. The battle of Bunker Hill was a bloody affair in which the British,who were the technical victors, achieved nothing and lost over 1,000 men, while the Americans, although beaten back to the mainland, lost less than five hundred. In March 1776, finding their position untenable, the British evacuated Boston.

Meanwhile, stirring events had been taking place in the north of the country. In May 1775 a handful of Americans under Ethan Allen and Benedict Arnold had surprised the small British garrison at Ticonderoga and taken the fort, with its useful store of heavy cannon, which were eventually to be used to bombard Boston. Then in June General Philip Schuyler, and his brilliant second-in-command General Montgomery, were ordered to take the offensive against General Carleton's troops in Canada. In some desperate fighting during the winter of 1775–76 Montgomery and Benedict Arnold came close to taking Quebec, but shortage of troops, the death of Montgomery and the wounding of Arnold enabled Carleton to hold out.

In May 1776 General Burgoyne arrived at Quebec with English and German reinforcements, to act as Carleton's second-in-command in implementing part of a plan, mainly of his own devising, from a paper he had submitted to the Cabinet entitled 'Reflections upon the War in America'. This recommended an advance down the Hudson from Canada in conjunction with one up river by Howe (who had succeeded Gage) so as to cut off the New England colonists from the south, and thereafter defeat them in detail. During the late summer and autumn Carleton advanced south, but got no farther than Crown Point on Lake Champlain. Arnold (now recovered from his wound) contested his march both on land and water until his small American fleet was mostly sunk in a battle off Valcour Island. Carleton did not feel strong enough to tackle Ticonderoga; although defeated, the valiant efforts of Arnold and Montgomery had prevented an advance down the Hudson at a time when it might well have succeeded.

In November 1775 Lord George Germain had joined the ministry as Secretary of State for the Colonies, and became one of the three men most closely involved in the Saratoga campaign. He had had considerable military experience, ending up as Commander-in-Chief of all the British forces in Germany. But in 1759 at Minden he had failed to obey an order from his superior, Prince Ferdinand of Brunswick, to advance and as a result of a court-martial was declared 'unfit to serve His Majesty in any military capacity whatever'. Nevertheless, he appears to have assumed complete responsibility for the war in America, to the exclusion of Lord Barrington, who was Secretary at War, Lord Amherst, who although only a lieutenant-general was nominally Commander-in-Chief and military adviser to the Cabinet on American affairs, and General Harvey, who was Adjutant-General.*

General William Howe had replaced Gage in command at Boston in October 1775, and had been knighted after his capture of Long Island and New York in September 1776. He was now in his forty-fifth year, and good living had begun to take its toll. He had considerable experience in command, had been a soldier since the age of seventeen and had seen service in America in 1758, and at Quebec under Wolfe. His tactics were usually sound, although he was far too cautious and often failed to follow up a victory; his strategical grasp was less sure.

The third Saratoga personage on the British side was General John Burgoyne. He had served in a subordinate capacity with Gage at Boston and, as already mentioned, with Carleton in the north; he was, therefore, no stranger to the scene in which he was now about to play the principal role. Much has been written about this very diverse character – playboy, playwright, politician and last, but by no means least, soldier. A debonair, amusing, and ambitious man, who was brave to a fault; he proved to be not a bad choice when eventually it was decided to give him command of the expedition from Canada, for he had served with distinction in comparatively high rank in Europe for seven years. It is often said of him derogatorily that he was the best that could be found; this may be so, but he had paid much attention to the proper training of officers and men, was popular with both, and was by no means unsound strategically or tactically.

These three men, together with King George, were responsible for framing the strategy that led to the Saratoga campaign. During the year 1776 Howe had been strongly reinforced; regiments had been raised and troops scraped up from various garrisons, and some 17,000 Germans (with more to come) had been hired from the Landgrave of Hesse-Kassel, and the rulers of Brunswick and Hesse-Hanau. The year had gone quite well for the British, although it ended on a slightly sour note when Washington gained two minor victories against Howe at Trenton and Princeton. However, time was not on Britain's side; the Americans were gaining military strength each month, and France was only waiting to see how matters went before openly and actively aiding the colonists. It

* Burgoyne did correspond from America occasionally with the Adjutant-General.

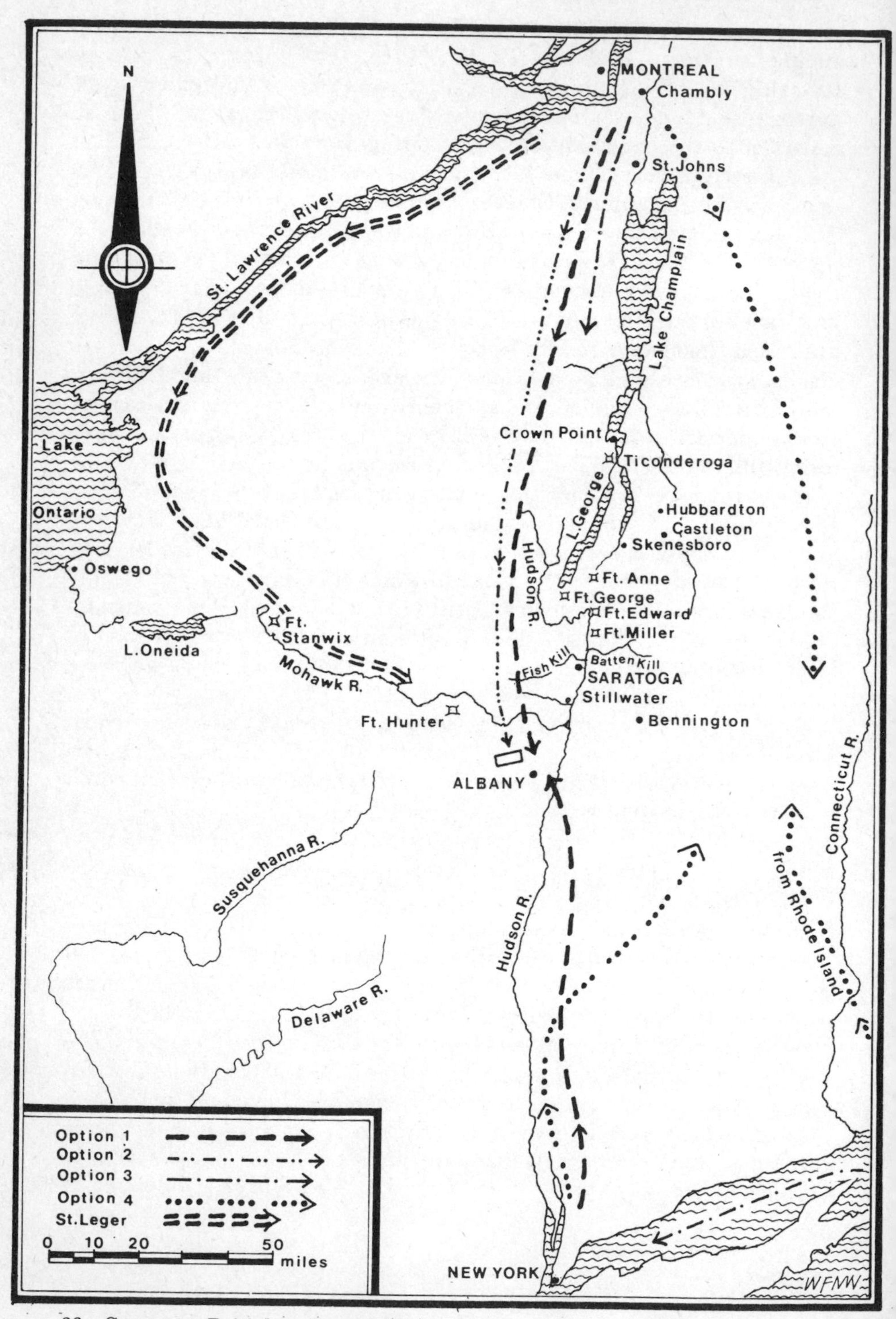

28 *Saratoga: British options, 1777*

was therefore necessary to conceive a plan that held prospects for ending the war in 1777, or early in 1778.

By this time the King's authority had been overthrown in every one of the thirteen colonies, but the heart of the rebellion was in New England; in particular it was the coastline with its ports that was so important to the Americans. Burgoyne had strongly disapproved of what he considered to be Carleton's pusillanimity in not attempting to take Fort Ticonderoga, and on his return to England in December 1776 he found Germain, who thoroughly disliked Carleton, a ready listener to his views. The following February Burgoyne submitted another paper, this time called 'Thoughts for Conducting the War from the Side of Canada'. But before Germain received this Howe, from 3,000 miles away, had also put forward plans for the 1777 campaign. Germain, who had no knowledge of the terrain – and despite his years in the army not much of strategy – was faced with having to advise the King (who had his own very pronounced ideas) on the various proposals submitted. These in substance amounted to four possible options for a strategy to end the war in 1777 or early 1778. In considering these options the planners in London had to bear in mind the length of time it took to communicate with the commander on the spot – a turn round of about twelve weeks – and therefore the need for concise orders. *The four options were:*

(1) To advance from Canada down the Hudson as far as Albany, in close co-operation with an army advancing north from New York to join forces at Albany. Thereby to control the river, isolate the New England colonies, and afterwards turn on the southern states. For this, and for option 2, it was estimated that at least 8,000 men would be needed for the advance south.

(2) The northern (or Canada) army to advance to Albany unaided, and thereafter await developments. Communications to be opened up between New York and Albany from the south – thus ensuring control of the Hudson – but no physical junction of the two armies at Albany. This would enable Howe's army to act independently in New Jersey and Pennsylvania, so long as sufficient troops were left to operate on the lower Hudson.

(3) To allow Carleton to make a threat from Canada with a limited force, but to reinforce Howe by sea, thereby enabling him to operate a strong unified command from his New Jersey base.

(4) To carry out a three-pronged thrust to isolate New England. The northern army to get control of the Connecticut river in conjunction with a force advancing north from Rhode Island, and another up the Hudson from New York.

The strategy of the Saratoga campaign has been closely studied by historians for over 200 years, and there has been a great diversity of opinion, firstly as to what was the original intention, and secondly as to why it went wrong. When the campaign was over all manner of excuses were put forward by those principally concerned, and often these contradicted what appeared to be the original thoughts and intentions. It seems probable that the second option was the one decided upon, with the addition of a force operating under Colonel Barry St Leger as a right hook down the St Lawrence to Lake Ontario, and from thence along the Mohawk valley to join forces with the northern army at Albany.

It has often been thought that Burgoyne envisaged joining hands with Howe's army at Albany (option 1); indeed, as the campaign developed some of his comments lend strength to this, but his subsequent statement before Parliament is clear enough. The object of the Canada army was 'to effect a junction with General Howe, or *after co-operating so far as to get possession of Albany and open communication to New York*, to remain upon the Hudson's River and thereby enable that general [Howe] to act with his whole force to the southwards' (my italics). This does not comply with the conditions of option 1.

The decision was probably not the correct one, for reasons that will be mentioned later, but it could conceivably have succeeded if Howe had taken sufficient interest in the northern army's progress, and ensured proper support for it on the lower Hudson. His determination to take Philadelphia (which had some prestigious, but no strategic, value), even though he knew that it meant having neither sufficient troops nor the time to offer any positive assistance to the northern army, condemned that army's chances. However, although Howe's seeming indifference may have made failure certain, in fact the campaign was virtually lost in London before it ever began.

Burgoyne, since his return to England, had been in close touch with Germain, although he did not submit his 'Thoughts' until the end of February. These basically contained most of the options already given, but during the previous three months Howe had put forward three different plans, and their contents must have been known to Burgoyne. At first Howe had called for 15,000 more troops with which to make a three-pronged thrust (option 4) from Canada, New York and Rhode Island, but, realizing that he would not get the troops (although it seems possible that he might have carried out the operation with those he already had), he soon changed to a plan for rendering what assistance he could to the northern army, but marching his own army on Philadelphia. This was later altered to taking his army by sea, and advancing on Philadelphia from the south.

The letter announcing the change from land to sea was not received in London until 8 May, by which time Burgoyne had arrived in Canada with his own plan approved. Howe's last plan was the worst possible choice, for it meant removing his army from between Washington and the lower Hudson. Nevertheless, Germain and the King approved it, merely expressing the hope that the operation would be 'executed in time for you to co-operate with the army ordered to proceed from Canada and put itself under your command'. Neither at this time nor at any other did Howe get definite orders to assist the northern army, and it soon became apparent that he thought – as indeed did Burgoyne – that it would be capable of looking after itself.

It is interesting to note that little attention (for it had been vetoed by the King) was given to sending troops by sea to reinforce Howe, leaving Carleton with some 3,000 men to take care of the northern border. This might have been the best solution, for there would then have been a unified command, and a large enough army to pack a powerful punch at the main American opposition. The plan adopted – provided Howe was able to co-operate with the northern army – looked quite attractive on paper. But at the planning stage, far too little consideration had been given to the type of country that the northern army had to operate in, and to the problems of supplying that army on the march and when it got to Albany. Furthermore, far too much faith had been put in loyalist recruitment.

The choice of commander for the northern army was not as straightforward as might appear from what has been written above. The grand strategy may have been largely based on Burgoyne's thinking, as developed in his 'Thoughts' (and indeed in his earlier 'Reflections upon the War in America'), and these were no doubt written with the intention that he should be given the command. But this was by no means a foregone conclusion. There were three contenders for the post – Carleton, Clinton and Burgoyne – and the selection was bedevilled by the likes and dislikes of those in high places.

When Gage was recalled from Boston at the end of 1775, Carleton was the senior general in North America and well suited to be given the overall command. But Germain, who harboured a long-standing grudge against Carleton, refused to entertain the idea. Thus was born the dichotomy which had such disastrous results. Carleton, for the same reason, was also passed over for command of the northern army – although the King, recognizing Germain's personal vendetta and appreciating Carleton's ability, insisted on his remaining Governor of Canada.

Clinton, who was senior to Burgoyne, and more experienced, had failed in an attack on Charleston, but had subsequently taken part with distinction at Long Island and New York. He suffered from consistently thinking that his services were undervalued, which was not the case, for he was recognized as a competent general. Why he did not get the northern army will always be something of a mystery; he could probably have had it for the asking – indeed, it seems that he was offered it and turned it down. In the end he received a knighthood and agreed to go back to America to serve under Howe.

This left Burgoyne, who never lacked for powers of persuasion, and whose suave manner and self-assurance found considerable favour with Germain. Added to this he had for some time basked in the royal grace. Two years earlier his endeavours to get the governorship of New York had been blocked by Lords North and Barrington, and he had had to be content (which he was not) with subordinate positions in Boston and Canada. Now came the opportunity so long awaited. Allured by the hope of glory and sustained by an overweening confidence, Burgoyne sailed for Quebec at the end of March 1777.

6 The Saratoga Campaign
Confidence Confounded

GENERAL BURGOYNE arrived at Quebec on 6 May 1777. It must have been an embarrassing meeting with his old commander, Sir Guy Carleton, for Burgoyne had now come to implement the plan that Carleton had failed to prosecute the previous year with troops hitherto under the latter's command. But Carleton was a big enough man not to bear Burgoyne any malice; indeed he was to give him unstinted and generous assistance. What resentment he felt – and it was plenty – he reserved for Lord George Germain.

Burgoyne remained only a few days in Quebec before proceeding to Montreal, where his army was assembling. He had under his command at the start of the campaign seven British regiments – the 9th, 20th, 21st, 24th, 47th, 53rd and 62nd – and eight German regiments commanded by Major-General Baron von Riedesel. The total British force (including 411 artillerymen) was 4,488, while the Germans numbered 4,699. At Skenesboro the army was joined by 148 Canadians, 500 Indians and 682 Tories (loyalists). There was also a large number of non-combatants.* As was usual at the time the grenadier and light companies were detached from their regiments to form an élite corps, and to this corps were added the light companies of those regiments – the 29th, 31st and 34th – which were to remain as garrison troops in Canada.

With the exception of the Indians, who were a constant source of trouble, it was a useful army of some 9,000 combatants. The British element comprised tough, well-trained and disciplined regular soldiers, while the Germans were hired troops who had spent the greater part of their lives fighting, to whom bloodshed had become a profession. In addition to this strong, well-equipped land force Burgoyne had a fleet of

*It is not easy to be absolutely accurate in regard to numbers for the campaign, because there has been much confusion between different methods of compiling returns; for instance Burgoyne included his entire strength when talking of rank and file in the returns at 1 July of his own army, but when using the same term for the German troops he excluded officers. I have relied on those figures compiled by the Park Historian, Charles W. Snell, who has researched the matter more thoroughly than anyone else, and arrived at what are almost certainly the most accurate tables possible for both armies at various stages of the campaign.

thirty powerfully armed boats. Transport overland was to be by carts, and for lake and river there were longboats and some bateaux.

The infantryman was armed with the famous Brown Bess musket, a smooth-bored flintlock firearm which fired a .75 calibre ball accurate only at a fairly short range. This did not greatly matter, because eighteenth-century tactics were based on advance in columns and deployment into line to give close-quarter volley fire, followed by the use of the socket-fixed bayonet. The soldier with his cartouche box, sword belt, knapsack, rations and water bottle humped about 60 lb., which of course made him a very cumbersome and unwieldy individual. Nor was the traditional uniform of red coat, waistcoat, close-fitting white breeches and black gaiters at all suitable for the steaming American jungles that he had to march and fight in.

It was on 18 June at Cumberland Head that Burgoyne issued a proclamation to his assembled troops. Unfortunately it was not the soldier that rose to the occasion but the dramatist, with the result that the troops got a piece of pompous rodomontade that left them gaping and the enemy laughing; two days later those Indians who had already joined the expedition were treated to a further display of high-sounding words. The expedition then got under way for Ticonderoga.

The army was loosely organized into three divisions or wings.* The Advanced Corps, commanded by General Simon Fraser, contained the grenadier and light infantry battalions, the 24th Regiment, sixty rangers (ten men selected from each of six regiments) under Captain Fraser, and some Canadian woodsmen; the right wing of two brigades was under General Phillips; the Germans formed the left wing, and they also were organized into two brigades. The artillery train of thirty-eight pieces of field artillery, ten howitzers and two eight-inch mortars was divided among the three divisions, and a central reserve artillery park. † The artillery was under the direct command of Major Griffith Williams, whose task in manhandling the guns and keeping them supplied along the treacherous forest tracks was among the most difficult in the army.

On 20 June, a few days after Colonel St Leger's force had set out for the Mohawk valley, Burgoyne's great fleet of gunboats and bateaux set sail up Lake Champlain. The fort at Ticonderoga had originally been built by the French in 1755, and called Fort Carillon. It was a stone star-shaped building with five bastions, and it stood (as it still does) at the end of a promontory jutting into Lake Champlain. It is overlooked to the

* See Appendix I for order of battle.

† This was only a fraction of what left Canada. After Ticonderoga most of the heavier pieces, and some of the light ones, were sent back or left at the fort, and some were later left at Fort George. As these weapons played little part in the main operations I have not included them.

south-west by Sugar-loaf Hill (subsequently renamed Mount Defiance), on which there were no defences, but across the lake to the south – and connected by a bridge of boats – Mount Independence had been fortified. About two miles to the south-west was the fortified Mount Hope, which posed very little threat to Ticonderoga but did command the portage route from the Lachute river to Fort George.

The troops allotted to General St Clair for the defence of the Ticonderoga complex were totally inadequate, being only about 1,576 Continental troops and three or four regiments of militia – whose tenure of service was unreliable – a total of no more than 2,500 men. Moreover, he was not well supplied with stores or ammunition.

In June General Schuyler (commanding the Northern Department) had come to the fort to discuss with St Clair the strategic plan, and it had been decided that if the fort became untenable St Clair was to fall back on to the higher Mount Independence and delay the British advance for as long as possible. Reinforcements of men and supplies would be sent if possible – and in fact 800 men and eighty head of cattle and sheep did arrive on 3 July. As soon as he learned that Burgoyne was advancing against him in considerably superior numbers *St Clair had three options in laying out his defence:*

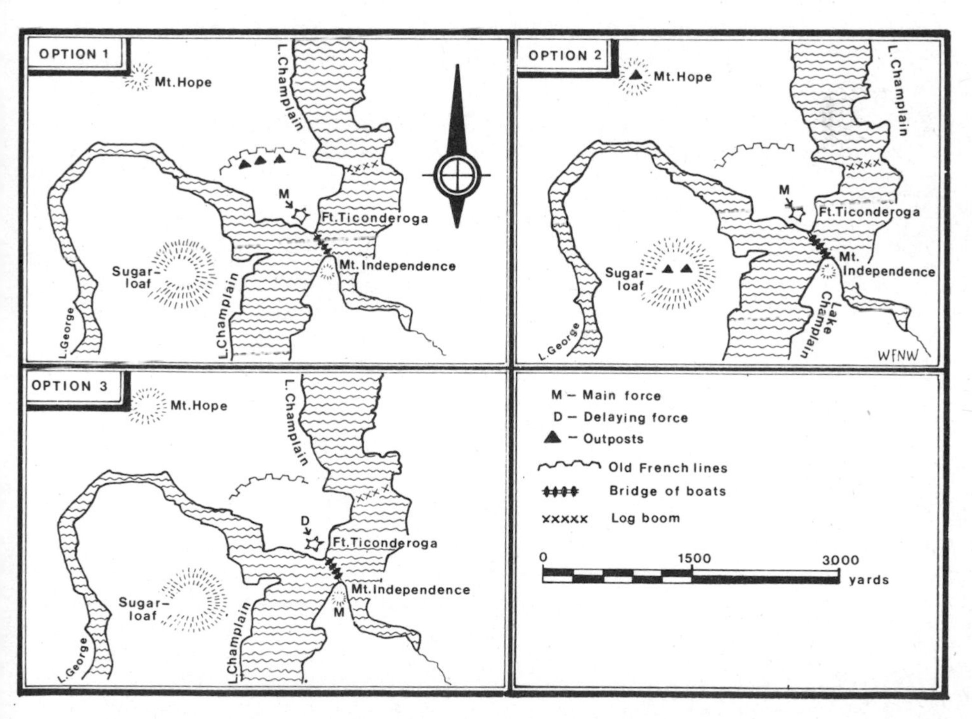

29 Ticonderoga: June, St Clair's options

(1) He could keep his troops collected manning only the fort, with an outpost holding the old French lines (constructed in 1755 across the promontory behind and to the west of the fort).

(2) He could keep in position the existing outpost on Mount Hope, and send another to the Sugar-loaf Hill, keeping his main force in the fort.

(3) He could anticipate an enveloping attack from both sides of the lake and use the fort for a delaying action, but concentrate his main force on Mount Independence.

General St Clair decided to leave a fairly strong outpost on Mount Hope, and concentrate the remainder of his force in the fort (a part of option 2). Even without the advantages of hindsight it does not seem to have been a very sound tactical disposition.

The importance of Ticonderoga has perhaps been somewhat exaggerated. Nevertheless, Burgoyne appreciated that his expedition could not proceed without its capture, and he expected to have a stern siege on his hands, which was one reason for his insistence on a large artillery train. His advance was cautious; a three-day halt was made at Crown Point, from where Fraser's Advanced Corps marched, while the remainder of the army continued by boat, disembarking three miles above Ticonderoga. A militia deserter had given Burgoyne a fairly accurate description of St Clair's dispositions, but the number of defenders had been exaggerated to about 5,000. From this he would have known that Mount Hope was occupied, but that Sugar-loaf Hill was not. *In planning his attack he had two options:*

(1) He could march with his whole army down the west shore of Lake Champlain, drive in – and possibly capture – the Mount Hope garrison, lay siege to the fort and recce the Sugar-loaf Hill with a view to occupying it with artillery if found possible.

(2) He could split his army, carrying out the above operation with one portion and sending a strong contingent down the east shore of the lake where the going was very marshy – in an attempt to cut off St Clair's retreat completely by this pincer movement.

Burgoyne was not content to reduce the fort; he intended to surround the area and cut off the garrison. For this purpose he ordered Fraser to seize Mount Hope and then swing in against the fort, while Riedesel's Germans marched down the east shore of the lake to block the isthmus behind Mount Independence. At the same time his chief engineer, William Twiss, was to reconnoitre the Sugar-loaf to see if guns could be got up it (option 2). It was a good plan deserving success.

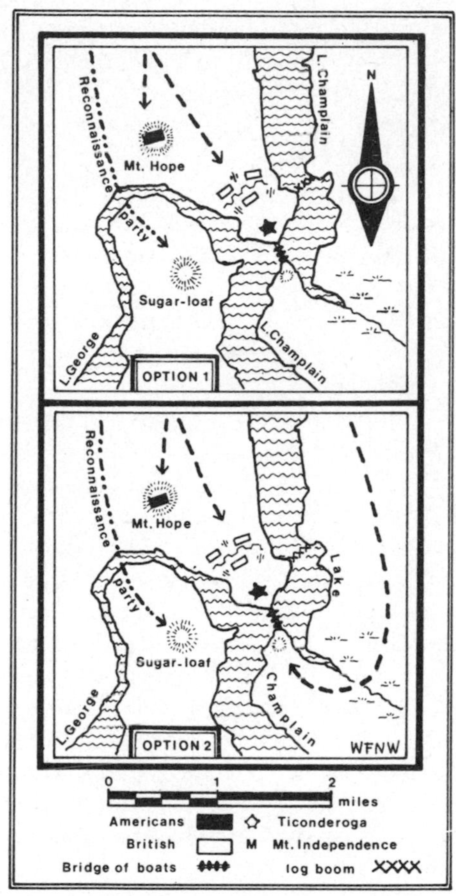

30 Ticonderoga: 1 July, Burgoyne's options

On 1 July Fraser's advance failed to catch the Mount Hope garrison, which St Clair had withdrawn just in time, but his men soon pushed them out of the old lines and into the fort. It is difficult to know why St Clair ever left them on Mount Hope in the first instance. By the next day Fraser had outflanked the fort from the west, and prevented any chance of evacuation along the portage route between Lakes Champlain and George. Riedesel, however, could not move so fast, the swampy ground bordering the East Creek proving a difficult obstacle to overcome. During 3 and 4 July there was some brisk interchange of cannon and small arms fire between the besieged and the besiegers, until on the 5th St Clair awoke to find that British artillery was threatening him from the top of the Sugar-loaf.

Some months earlier General Gates had refused the advice offered by the Polish engineer, Thaddeus Kosciuszko, to fortify this mountain, and now it had become a distinct embarrassment to the Ticonderoga garrison. However, the actual amount of damage a battery could do from this position and range (1,400 yards to the fort and 1,500 to Mount Independence) has often been exaggerated; undoubtedly it was a psychological threat, but St Clair's decision to evacuate the fort and withdraw to Mount Independence was probably due more to the danger of being surrounded than to the threat of being blasted out from Mount Defiance – as it was now called. Nor was he any longer safe on Mount Independence; on the night of 5–6 July, under cover of gunfire, the garrison stole away. The stores and wounded were ferried up the lake to Skenesboro (now Whitehall), while St Clair led the main force down the east side through Hubbardton.

It would have been a bold decision to have anticipated envelopment and concentrated the bulk of the troops on Mount Independence in the first instance (option 3), but it might have been the better plan. The truth is that St Clair was under-supplied with both troops and provisions; his defensive dispositions may not have made the best use of what he had at his disposal, but unless he had fought to the last man he could never have delayed Burgoyne for much longer. He was subsequently court-martialled for evacuating the fort, but completely exonerated.

So far everything had gone extremely well for Burgoyne. It was a pity that St Clair's men had extricated themselves from the Ticonderoga trap, but there was time perhaps to remedy that if the pursuit was rapid enough. Back in England King George, when he heard the news, was enormously elated, rushing into his Queen's room and exclaiming, 'I have beat them, beat all the Americans'; and to the army Albany by September must have seemed a target as easily attainable as it was desirable.

Burgoyne lost no time in ordering the pursuit. Fraser's Advanced Corps started their march at 5 a.m. on 6 July, and close behind – but moving at a slower pace than Fraser's light companies – came the Germans. As a temporary measure the 62nd (British) Regiment and the Brunswick Grenadier Regiment were left behind – the former to garrison Fort Ticonderoga, and the latter Mount Independence; the remainder of the force embarked with Burgoyne and made up the lake for Skenesboro. The Americans had built a fairly solid boom connecting Ticonderoga with Mount Independence, which they felt sure would defy the British fleet for some while, but it was blasted quite quickly, and Burgoyne caught the American boats at anchor near Skenesboro and destroyed many of them. However, a detachment of the 9th Regiment under Colonel Hill, which was landed and ordered to pursue the enemy towards Fort Ann, did not do so well and at one point was in danger of being annihilated.

Meanwhile St Clair, marching his men at a great pace, skirted round the top of Lake Bomoseen and after twenty-four gruelling miles reached Hubbardton. Here he halted briefly and made a rather curious decision. His rearguard was commanded by Colonel Francis's 11th Massachusetts Regiment, which the main body had outpaced, and *he now had three options:*

(1) He could allow the rearguard to close up and stand and fight at Hubbardton with his whole force.

(2) He could leave the rearguard to continue as such, trusting that it would not be overtaken and annihilated, while he continued the march to Skenesboro.

(3) He could detach a part of his force to stand and wait for the rearguard, and then to rejoin the main body.

St Clair decided to detach Colonels Seth Warner and Nathan Hale, commanding the Green Mountain Boys and the 2nd New Hampshires respectively, with orders to await Francis; as soon as the rearguard came in the three regiments were to march at once to rejoin the main body at Castleton (option 3). It is difficult to know why he decided to split his force in this way or, if he had to do so, why they were not given positive orders to check the pursuit. Colonel Warner had 900 men under his command, which, as it happened – although St Clair was not to know this – was about equal to Fraser's force. Either stand and fight, or get away as much of your force as possible. St Clair's decision was a bad compromise, unless the two regiments he left at Hubbardton were in no condition to continue the march; but there is no evidence that this was so.

General Fraser drove his men relentlessly from 5 a.m. to 4 p.m. over appallingly rough tracks, through open forest pierced by the flaming sun. It was a march that was to surprise the Americans, and it totally exhausted the poor Germans who trudged behind. A panting von Riedesel eventually caught up with Fraser at the evening halt, however, and the two men agreed upon a plan of mutual support if the enemy were to be met with the next day. In the early hours of 7 July Fraser's men continued the pursuit, and at about 5 a.m. his leading troops came upon the Americans lighting their camp fires for breakfast. Surprise was complete – an American prisoner euphemistically described his regiment as being 'in a very unfit posture for battle'.

The three American regimental commanders had agreed (contrary to definite orders given to Warner) to stand and fight if they were attacked, and on the previous evening had made a plan. This did not, somewhat naturally, allow for the forward regiment (2nd New Hampshires) to be completely surprised and routed in the first charge. However, the other two regiments resisted with spirit; taking cover behind fallen trees, and making use of the higher ground, they were soon doing great damage to the advancing British through their accurate snap-shooting. Both sides tried to outflank the other; the British were repulsed on the left, but on the right they managed to cut off the American retreat to Castleton, but not to dislodge the defence on what became known as Monument Hill. Fraser attacked the American position on this hill three times; twice the light companies were thrown back, and had it not been for the timely arrival of von Riedesel's advanced troops – who had left their camp at 3 a.m. – a third repulse would have been certain. The action had lasted forty-five minutes during which time the Americans, tough and desperate, had foiled every attempt by Fraser's men to break them; only the arrival of the Germans overwhelmed and dislodged these sturdy soldiers.

Although short it had been a very savage engagement. The British and Germans suffered 174 casualties, including Major Grant who was killed leading his 24th Regiment; the Americans, whose total casualties were twelve officers and 382 men (of whom 228 were prisoners), also lost a valuable officer in Colonel Francis, who continued the fight with a shattered arm but was killed in the last few minutes of the battle. It had been a victory for Burgoyne's men, but it should surely have been an eye-opener to them as to what lay ahead. Hubbardton had been a failure for the Americans, and probably it would have been better had it never been fought, but the American soldiers had shown themselves capable of some very hard hitting.

St Clair's force was in disarray. Two militia regiments ordered to turn back to reinforce Warner had virtually refused to go into battle; and when news was received of the disaster at Skenesboro there was little he could do but take what men he could collect on a wide circuit to the east of more than a hundred miles, and join General Schuyler's army at Fort Edward. His arrival there on 13 July brought Schuyler's force up to over 3,000 men, but the troops were mostly tired, dispirited, short of provisions and with little or no protection against the heavy rains that had recently developed.

The campaign had now reached a crucial stage. Schuyler begged Washington for reinforcements, but Washington had Howe to contend with; he was not sure which way that general would go – up the Hudson or to Philadelphia – and had to remain strong in New Jersey. Burgoyne, on the other hand, with the pleasure of victory adding lustre and joy to the sumptuous entertainment his friend Colonel Skene was giving him at his house at Skenesboro, had written confidently to Howe on his prospects for reaching Albany. It would seem that he had Schuyler's demoralized troops, now concentrated twenty-three miles away at Fort Edward, at his mercy. It was important that they should not be given time to rally, re-equip or be reinforced. *Burgoyne had five options when considering his next move:*

(1) He could take the direct overland route of twenty-three miles from Skenesboro to Fort Edward. It was the way that Skene advocated, and Burgoyne's engineer, Twiss, had reported after a reconnaissance that although it contained many problems they were not insuperable. But it was at best a narrow, low-lying forest track, often swampy and always very difficult for transport; moreover, there were many ravines and the bridges would

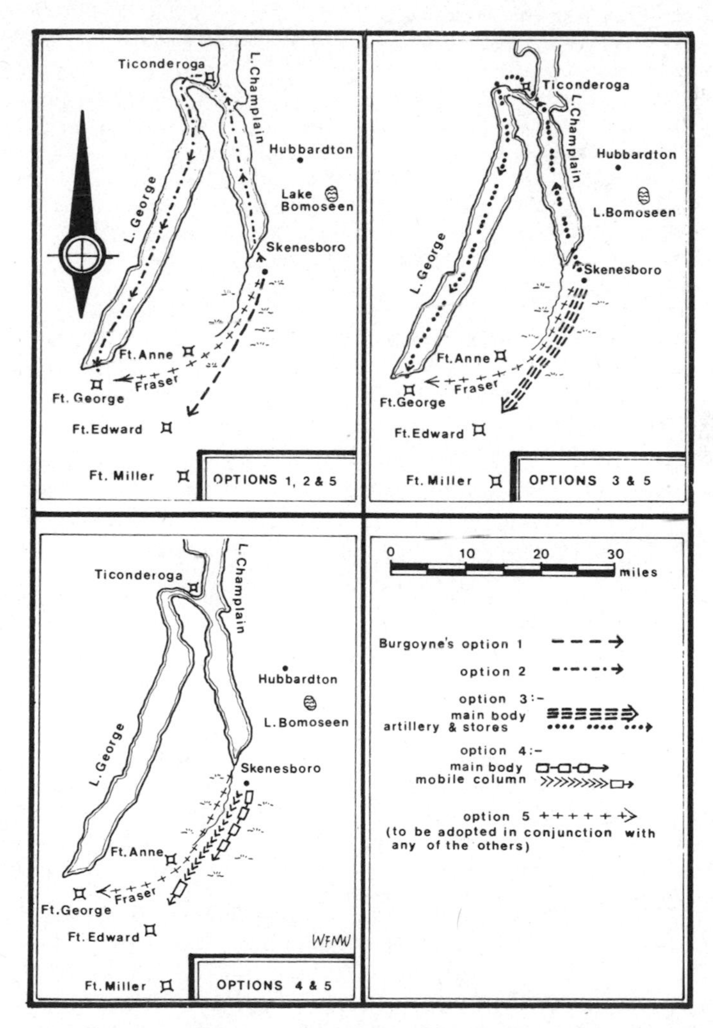

31 Saratoga: Burgoyne's options after Hubbardton

very likely have been destroyed by the retreating Americans.

(2) He could return by boat to Ticonderoga, carry his boats across to Lake George and sail for Fort George at the end of that lake. This was a longer route, but could well have been quicker, and it was the route that Burgoyne had considered the best when first planning the campaign. The short portage between the two lakes was over a reasonable track, and the distance between Fort George and Fort Edward was ten miles also over a good track. But the enemy was known to be holding Fort George, and there were two psychological disadvantages. Any backward movement, however advantageous strategically, could lower the morale of both his troops and the loyalists, and it would have prevented a deception movement on his left flank as a threat towards the Connecticut river.

(3) He could use the land route for the main column, and send the guns and some stores by boat with a small infantry escort.

(4) He could use the land route but send a strong, mobile column forward in advance of the main body to keep the pressure on the enemy and prevent them from sabotaging the track.

(5) Irrespective of which route he took, he could send General Fraser's troops to take Fort George, where (according to Lieutenant Digby, who marched with the 53rd Regiment*) there was a large quantity of wagons, horses and stores. This would greatly reduce the valuable time wasted in waiting for supplies from Ticonderoga.

Burgoyne decided to march the army overland, less the German contingent, who were to remain at Castleton 'giving a jealousy' (as he termed it) towards the Connecticut river; the heavy artillery and stores were to take the lake route (option 3). It is a decision that has been frequently criticized; as might have been expected, the enemy had destroyed the bridges and blocked the difficult forest track in many places. The need to send a pioneer force to clear the route, and also the time required to collect supplies, meant that Fort Edward was not reached until 29 July – three weeks after Burgoyne's arrival at Skenesboro. Although Schuyler did not feel he was strong enough to defend Fort Edward he had had time to build up supplies, equipment and numbers, and withdraw to a defensive position laid out by Kosciuszko at Stillwater.

In defending this decision later † Burgoyne emphasized the need to maintain morale by going forward rather than backwards; he also stressed that in this case the enemy would certainly have remained in Fort George, and compelled him to lose much valuable time in laying siege to that fort. Furthermore, when the garrison withdrew they would probably have been able to destroy the road.

Whatever the reason, the decision was surely the right one. To break off contact with the enemy completely must have been wrong; the

*Journal of Lieutenant William Digby, p. 227.

† *A State of the Expedition from Canada as Laid Before the House of Commons*, p. 12.

embarkation, disembarkation and portage between the lakes would have been a slow business, and Burgoyne might well have been right about the garrison in Fort George, which, as it was, fearing to be cut off, withdrew without a fight. The criticism within the army (Lieutenant Digby writes that doubtless Burgoyne 'had his proper reason for so acting, though contrary to the opinion of many') may have arisen because Burgoyne failed to send a flying column to push the Americans as far as possible while they were off balance, and perhaps another to seize the supplies at Fort George. It seems that Burgoyne was at fault here, but we know too little of the conditions at the time to be certain.

After their gruelling march the British soldiers were overjoyed to reach the Hudson at last. The trail had been testing and wearisome, but the enforced rest at Fort Edward in order to replenish supplies would soon revive men whose morale was still very high. But Burgoyne himself cannot have been entirely happy. The farther he went the larger the drain on his resources. Forts Ticonderoga, George, Ann and Edward all had to be garrisoned, he was short of animals, there was the usual wastage from sickness and desertion, and by early August he knew for certain that Howe had no intention of coming up the Hudson. General Clinton, who had arrived in New York from England on 5 July, although sympathetic – and disapproving of Howe's determination to take Philadelphia, which could not be achieved in time for him to help Burgoyne – was at present not strong enough to leave New York. There also occurred at this time an incident which, although its importance has been exaggerated, had some local repercussions.

A girl called Jane McCrae was engaged to marry a loyalist who marched with Burgoyne. At the time the British troops arrived at Fort Edward she was murdered by one, or more, of Burgoyne's Indians. Her exact fate is uncertain – she was probably waylaid and scalped by a redskin with the splendid name of Wyandot Panther – nor was it important to anyone but the unfortunate girl and her fiancé. This tragic occurrence gained wide publicity later as a result of correspondence on the subject between Gates and Burgoyne, but the story that Schuyler circulated it throughout the district to undermine loyalists is almost certainly apocryphal. Burgoyne was as horrified by the incident as anyone, and with difficulty was restrained from taking very severe disciplinary action, which would only have increased the risk of further mischief. As it was many of the Indians deserted; although the best of them were valuable scouts, the contingent on the whole was very troublesome.

While Burgoyne's numbers were diminishing, Schuyler was receiving reinforcements of men and senior officers. By the time he left Fort Edward he had at least 3,000 Continental soldiers and about 1,600 militia, and Washington had sent him General Lincoln and the colourful, unpredictable but valuable Benedict Arnold. However, numbers count for little if morale is low, and the American soldiers had had just about enough withdrawing when Schuyler took them another twelve miles down the river from Stillwater. The militia were deserting at an alarming rate, and for some time there had been considerable dissatisfaction among the New Englanders, and indeed among the members of the Continental Congress, at Schuyler's handling of the Northern Department's army. He was being blamed – most unfairly – for the loss of Ticonderoga, and even in some quarters accused of treachery. Burgoyne still held the advantage both in numbers and morale, but time was running out, logistics were an increasing worry, and he was very short of animals.

The last deficiency weighed heavily with von Riedesel, for his dragoons could be seen at times floundering about in their jackboots. Shortly after the fight at Hubbardton, when his division was at Castleton, von Riedesel had tried to persuade Burgoyne to let him make a tip-and-run raid towards the Connecticut river to snatch horses for his cavalrymen, but at the time Burgoyne was too preoccupied with getting his troops to Fort Edward to give the request much attention. Now the proposition came up again and Burgoyne was faced with having to make the decision. *His options were:*

(1) To mount a fairly large-scale raid at right angles to his line of advance, which would cause delay but bring in supplies.

(2) To press on down the river while he still had Schuyler on the run.

Burgoyne decided to mount the raid into Vermont. With the benefit of hindsight the decision has come in for a certain amount of criticism, but he was almost certainly right. Although it was known that there would be opposition it was thought (with some justification) that this would be negligible, and supplies were badly needed if the advance was not to be a continual stop-go affair. His 'political adviser', Philip Skene, assured him that the country was rich in horses, and that there were a great many loyalists who could be recruited. Added to this Burgoyne had always favoured a diversion towards the Connecticut river.

Having decided to mount what was called – with good reason – 'a secret expedition', the next question was the composition of the force. Von Riedesel had conceived the idea and done much of the planning – although Burgoyne had altered some of the details – so it was understandable to give the command to a German, nor did it matter that Lieutenant-Colonel Baum spoke no English, because Skene was to accompany the expedition as interpreter and loyalist recruiting officer. But bearing in mind that the operation, to be successful, had to be carried out by a hard-hitting, mobile force its composition was in some respects most unsuitable.

Baum was to command 170 of his own dismounted Brunswick Dragoons, 150 grenadiers and light infantry under Major von Bärner, a somewhat motley collection of Tories, Canadians and Indians under Colonel Peters, fifty of Captain Fraser's marksmen, and two three-pounders from the Hesse-Hanau battery; presumably to satisfy the German craving for martial music, although it was scarcely conducive to secrecy, some musicians were thrown in for good measure. There were also a number of women camp-followers. Colonel Baum's dismounted dragoons were at a grave disadvantage in such a venture, there was little point in having so many Tories and Canadian woodsmen, and none at all in allowing women to accompany this raid. It would surely have been better to have sent a composite force of the British and German grenadiers and light companies under the overall command of Colonel Breymann (commander of the German reserve).

The original plan was that the force should raid in the area of Manchester, Arlington and the Connecticut river valley, but just before Baum was ready to start information was received concerning a large supply of horses at Bennington guarded by only 400 militia. Baum was therefore to march straight for that place, and part of his instructions were that his corps was 'too valuable to let any considerable loss be hazarded'. This change of direction meant that his right flank was exposed to an attack from Schuyler, whose force was concentrated just across the river. This worried von Riedesel, but Burgoyne discounted it and believed that there would be little or no opposition. Nor was there from Schuyler, but Burgoyne's general optimism was to be sadly shattered.

John Stark, who was responsible for this reversal of fortune, was a figure almost as controversial and contrary as Benedict Arnold; like the latter he was possessed of indomitable grit and a flair for leadership. He had fought in the French war, and, quick to take up arms for the cause of independence, he had raised and led a regiment at Bunker Hill. But he was a proud man, and quick to take offence; on being passed over for promotion he resigned his commission. Now he was again willing to fight for freedom, provided he was given freedom to choose where to fight. He insisted that his New Hampshire brigade must be independent of Congress and the Continental army. Such was the need of the hour that his demand was accepted, and such was the reputation of the man that within a week he had enlisted 1,492 all ranks; by 8 August he was marching them to Bennington, where a military depot would take care of their needs.

Baum left Fort Miller on 12 August, and his troops made good progress marching through Cambridge to the Hoosick river and along the Bennington road. Somewhere in the vicinity of Cambridge he clashed with a small force and took some prisoners, from whom he learned that Bennington was held not by 400 militia, but by more than 1,500 well-armed troops. He at once sent back a report to Burgoyne, which included the information that his Indians were quite out of control, looting and rioting. At Van Schaick's mill, where the Owl Kill joins the Hoosick river, a small American detachment offered little resistance, but delayed progress by destroying the bridge. Later that day (14 August) Baum had a view of Stark's army, which had advanced in support of the detachment at Owl Kill. Baum was not prepared to offer battle and Stark withdrew his men.

The fifteenth of August was very wet, and Stark – who had received reinforcements that day and now, with about 2,000 men, had a three-to-one superiority – was unable to make use of his commanding firepower. Baum used the respite to make what defensive preparations he could. The Bennington road ran east from the Hoosick river and parallel to the Waloomsac river, until the latter curved round westwards at the foot of

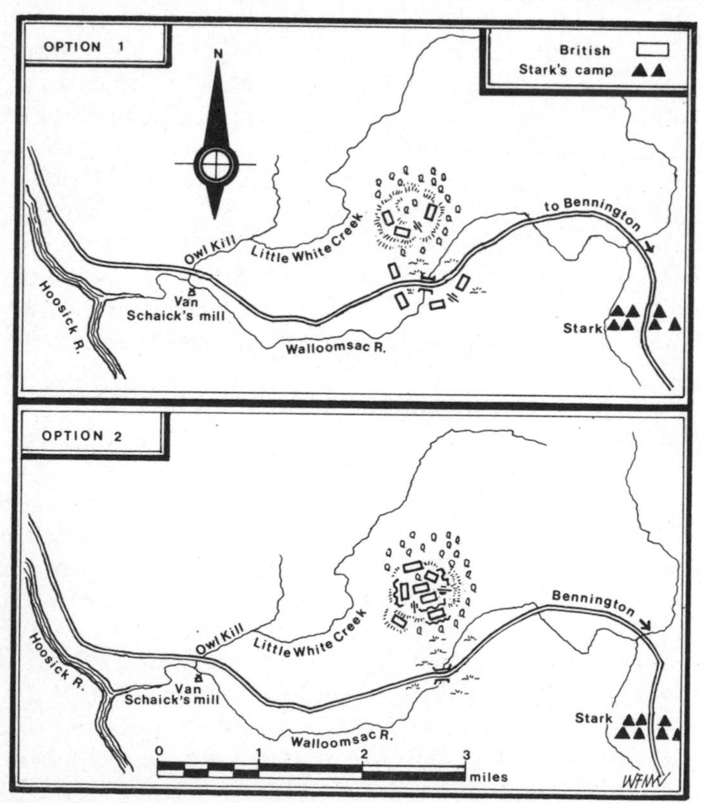

32 The Battle of Bennington: 16 August, Baum's options

a prominent eminence that rose some 300 feet above the swampy river valley. This hill afforded a good defensive position, although the contours and vegetation were such that the river (and the second of two bridges that crossed it) could not be commanded from the summit. In the area of the hill, and to the north of it, the country was densely wooded, but east and south the land was marshy and comparatively open. Stark had his camp a little over a mile to the east of the hill. Colonel Baum had 600 soldiers and two cannon. He had received word that reinforcements under Lieutenant-Colonel Breymann were on the way, but although he had sent Skene back to hurry them on, he had to be prepared to stand unaided if attacked on the morrow. He was in a very difficult and dangerous situation, for somehow he had to hold out until Breymann arrived. *In laying out his defence he had two options:*

(1) To dispose his small force so as to hold strategic points, and give so far as possible all-round protection.

(2) To hold the hill with his entire force, throwing up the maximum defences that time would permit.

Baum decided to disperse his force (option 1). One of the three-pounders, twenty of Fraser's rangers and the dragoons held the hill, where they spent the whole of the 15th entrenching and throwing up breastworks. The Indian contingent was also on the hill, manning a plateau behind the main position, and there were fifty German chasseurs placed forward on the eastern slope. On the low ground Baum had four groups. On the further side of the road and 250 yards below the bridge were 150 Tories; protecting the bridge to the south of the road were the Canadians; immediately north of them across the road were fifty German jaegers, twenty of Fraser's marksmen and the other 3-pounder; and in a meadow south-west of the main position Baum had placed fifty Germans and the remainder of the Tories with the task of guarding the road and the bridge from the rear.

He had scattered his small contingent right across the landscape, and subjected them to being overrun in penny packets. He seems to have placed much importance on guarding the bridge, but the Waloomsac at that time of year was scarcely an obstacle. It is true that the time was too short to make a proper redoubt on the hill, but as it was known that a relief force was on its way it would have been wiser to defend the hill in depth with what abatis could be constructed in the time available.

Stark, when he first got news of the raid, had sent urgently for Warner and those of his men that had survived Hubbardton and who were now at Manchester. The troops did not arrive in time for the fight on 16 August, but Warner had come ahead of them, and together with Stark formulated the plan of attack. Time was short if Baum was to be defeated before Breymann arrived, *and Stark had perhaps three options:*

(1) He could carry out a frontal attack on the main position with the majority of his force (say 1,300 men), and send detachments to isolate the two furthest outlying groups.

(2) He could split his force into three principal components, and carry out a holding operation in the front of Baum's position, while sending two wide-flanking detachments to form a pincer movement, co-ordinating his frontal attack with their flank attacks.

(3) He could carry out a less elaborate flank attack on one of the flanks, and combine it with a frontal assault.

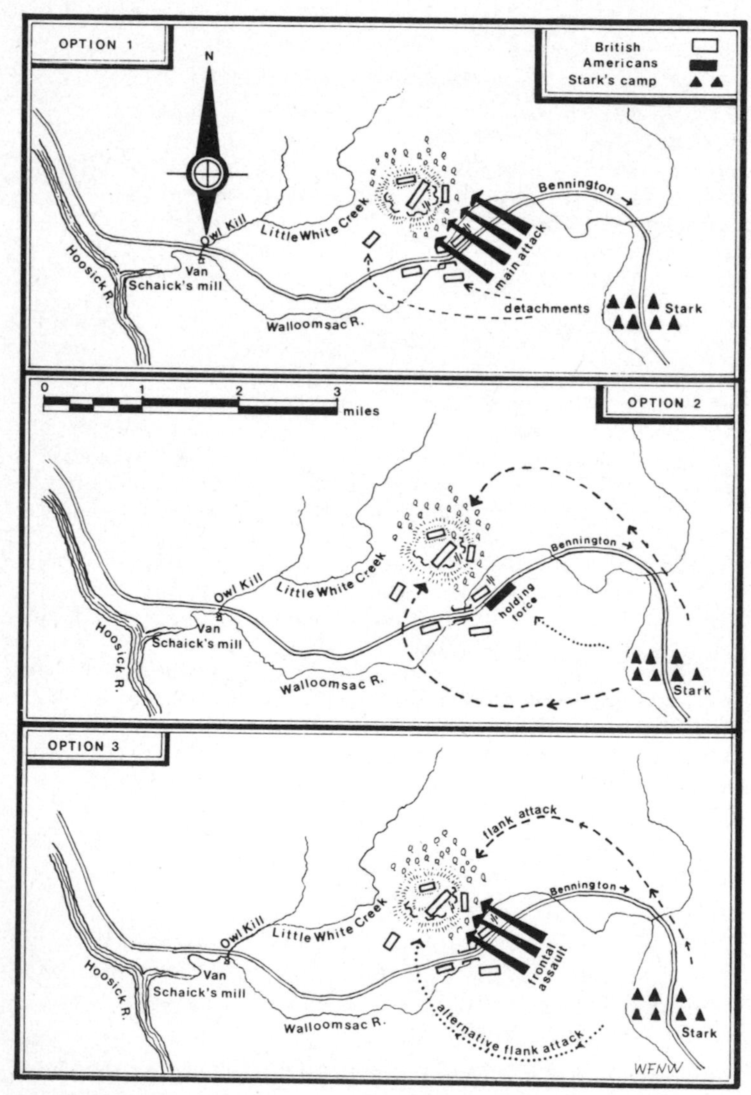

33 The Battle of Bennington: 16 August, Stark's options

Colonel Stark chose the pincer movement (option 2). It was a very bold plan, needing extremely careful co-ordination, and, on account of the wide detours over difficult country – especially for the northern pincer – it would need time. Because it was brilliantly executed it has been hailed as the best plan. It would be churlish to diminish the glory that Stark gained that day, but it was a plan that could very easily have gone wrong, and with his overwhelming superiority a lesser man might have chosen to pound his way to victory, or at least to try something less elaborate. And that lesser man might have been right.

The frontal diversion could not develop into a full attack until the two wide encircling movements were in a position to open the battle. We are told that they made the detours through this difficult wooded country in not much more than three hours; it was a remarkable feat.

The sixteenth of August dawned wet, but patches of blue between billows of white cloud told Stark that soon he could fight with dry powder. Nevertheless, it was after 11 a.m. before Colonel Nichols led his 200 New Hampshire men out on the long trail round the north of Baum's position, while Colonel Herrick with his 300 Vermonters and Bennington militia performed a similar operation to the south. Colonels Hobart and Stickney were sent to the rear right against the Tories in their exposed position south and east of the river, a force of about 100 men demonstrated against the front to keep Baum guessing, and Stark held some 800 men in reserve for the main frontal attack.

Shortly before 3 p.m. Baum's soldiers saw parties of armed, shirt-sleeved farmers infiltrating their position; taking them to be some of the loyalists that Skene had promised Burgoyne would rally to King George, they rashly withdrew their picquets so as not to intimidate these strangers. No one knows whether they were there by design or accident, but when Nichols's men, having at last reached their position on the northern flank, opened fire these simulated Tories joined in the shooting – at the Germans. Nichols's fire was the preconceived signal for the whole attack. The outposts were overrun very quickly, but on the hill the fighting was some of the fiercest of the war. Stark's men stormed the steep, rock-strewn slope, and marksmen had soon picked off the German gun team. Then for two hours it was hand-to-hand grapple, with ammunition beginning to run low. When most of Baum's men had found the pace too hot his dragoons stood by him; unsheathing their sabres bright and long, they carved their way through the less heavily armed Americans. But the numbers against them were too great, and when their gallant leader fell mortally wounded they, like the others, melted away into the woods.

It was a great victory for Stark, but the spoils were nearly snatched from him. Breymann, advancing slowly and dressing his ranks with misplaced Teutonic thoroughness, was at last close at hand. His force of about 640 men and two six-pounders (carried with their ammunition on carts), had set off at 9 a.m. on 15 August. The road was deep in mud, it was raining hard, and he had twenty-five miles to cover; even so, in view of the urgency of his task, he could have improved on the first day's march of half a mile an hour. He eventually arrived near to the battlefield, having had to contest a few minor skirmishes, a short while after the last of Baum's dragoons had disappeared into the wilderness.

Stark's men were scattered, enjoying the fruits of victory, but Warner's troops (some 330 men), who had been marching at the same sort of leisurely pace as Breymann's, had now arrived on the battlefield. Stark collected what men he could, but without Warner's reinforcements he could not have dealt with Breymann. As it was the fight was fiercely contested, with Breymann's cannon doing considerable execution, but eventually, with his flank turned and his ammunition running short, he broke off the engagement at about sunset. His withdrawal soon took the shape of a near rout; Stark was to say later, 'Had day lasted an hour longer, we should have taken the whole body of them.' As it was Breymann lost his guns and about two-thirds of his men.

Altogether Bennington had cost Burgoyne 527 of his German troops (twenty-eight of whom were officers) and thirty-seven rangers. The Americans, whose casualties were no more than thirty killed and forty wounded, had also taken four guns and a good number of muskets, sabres and ammunition. Moreover, American morale had received a tremendous fillip; it could be said that this was the turning point of the campaign.

Burgoyne had scarcely recovered from this serious blow when news came of St Leger's failure some sixty miles to the west in the Mohawk valley. It will be remembered that this was to be the right hook of the invasion, and that the two armies were to join at Albany. More than half of St Leger's force of 1,800 were Indians, and at Oriskaney they became involved in the wrong sort of fight, for they were not suited to a bloody hand-to-hand battle lasting some eight hours. Moreover, on returning to their camp in low spirits they were further demoralized by finding that it had been looted.

St Leger had to take Fort Stanwix, which was defended by Colonel Gansevoort with men from New York and Massachusetts regiments, before he could advance up the valley. The battle of Oriskaney on 6 August was technically a victory for St Leger, because Colonel Herkimer's militia had not been able to relieve the fort; on the other hand, St Leger had not gained it, and before he could starve the garrison out Benedict Arnold had marched to its relief. Arnold succeeded in frightening the Indians with a clever ruse as to his numbers, and they refused to fight any longer; this so depleted St Leger's force that on 23 August he had no alternative but to re-embark on Lake Ontario and retrace his steps to Canada.

By the beginning of September Burgoyne was in a position that can only command our sympathy. The hard-won triumphs of a difficult and dangerous advance of over 100 miles into enemy territory had suddenly turned sour on him; he now found himself alone and unsupported with an agonizing decision to make. His army had been reduced to 6,074 men fit for duty (this included 300 reinforcements just arrived from Canada, 680 Tories, 148 Canadians and 90 Indians), and there was little hope of any further local recruitment. On the other hand General Horatio Gates, whose replacement of the much aspersed Schuyler had considerably helped local recruiting, had also received substantial reinforcements from Washington, and with Arnold back from Fort Stanwix his army now outnumbered Burgoyne's by more than two to one.

Burgoyne had probably never envisaged a physical junction with Howe, but he must have hoped for pressure to be applied on the lower Hudson sufficient to reduce the weight of troops against him; he now knew that this was most unlikely to be the case, for General Clinton in New York appeared to be too weak even to attack the Highlands. The supply problem had become acute, the country assumed a more difficult aspect every day, and Hubbardton and Bennington had shown Burgoyne very clearly the tough resolution of his enemy.* *There were three options open to him:*

(1) To continue the advance in accordance with the plan, even though he knew that an important part of that plan might never materialize. This would mean abandoning his line of communications with Ticonderoga, for it would have become far too long, and the Americans were sure to threaten his left flank.

(2) To withdraw his army back to Ticonderoga while there was still time.

(3) To remain in his present position at Fort Miller, or withdraw to Fort Edward, there to await developments and hope that Clinton in New York might receive orders and reinforcements sufficient to make a diversion on the lower Hudson and draw off some of Gates's troops.

Burgoyne decided to proceed with the original plan (option 1). That he knew it was a fairly forlorn hope we learn from his letter to Germain of 20 August. His principal reason for going forward (expressed in this same letter) seems to be that he considered his orders gave him no latitude; but he had either forgotten or misunderstood that sentence, until they [Burgoyne and St Leger] shall have received orders from Sir William Howe, it is His Majesty's Pleasure that they Act as Exigencies may require . . .' No orders had been received from Howe, and in any event a general officer in Burgoyne's position, more or less incommunicado and fully aware of the dangers, could be expected to use his own initiative and take what action he considered best to safeguard his army.

*In his evidence before Parliament Burgoyne made it appear that circumstances were much more propitious than he knew to be the case at the time.

No one could have blamed Burgoyne for going back before it was too late; it seems as though the pursuit of glory – and perhaps he was also mindful of his general order issued back in June at Crown Point, which ended, 'This Army must not Retreat' – overcame sound military sense. It is true that the Earl of Harrington, a captain in the 29th Regiment and later A.D.C. to Burgoyne, stated in his evidence before Parliament that he thought 'General Burgoyne's character would not have stood very high either with the army, this country, or the enemy, had he halted at Fort William'. But one has to remember that throughout this enquiry Burgoyne's officers, very properly, remained absolutely loyal to their chief.

And so with the catalyst of threatening disaster at hand the opportunity for honourable withdrawal was allowed to pass. On 13 September Burgoyne's soldiers resumed their march to Albany. Their mood was valiant, but the perils were many.

7 The Saratoga Campaign
Defeat and Surrender

THE die having been cast Burgoyne had a choice of two routes to Albany – and he knew very well that if his army was to survive through the coming winter it would need the shelter and provisions that that town could provide. Albany is on the west bank of the Hudson, and so was Gates's army. Burgoyne was at Fort Miller on the east bank with food for his army for about thirty days and precious little chance of any further supplies getting through from Canada. *There were two options open to him:*

(1) To march as quickly as possible down the east bank, and hope to gain Albany unopposed. The crossing at Albany would be more difficult, however, for the Hudson is swollen there by the incoming Mohawk.

(2) To make the easier crossing, and fight his way through a numerically superior army that had taken up a defensive position to bar his route.

Burgoyne had to make his decision quickly, and he was without any definite knowledge of General Clinton's position in New York. He decided to cross the river and fight his way through. It has been heralded as a very bold decision, but it was the best if not the only one that he could have taken.

The crossing at Albany would have been very difficult, and Gates would surely have shadowed him with at least a part of his force, and arrived in time either to oppose the crossing, or to attack him immediately afterwards when off balance.

On 13 September the British troops, led by Simon Fraser, marched past their commanding general, and crossed the Hudson by a bridge of boats placed just above Saratoga near where the Fish Kill creek enters the river. On the next day the Germans followed them, and the supplies and baggage – guarded by six companies of the 47th Regiment – went by

river in the bateaux. On 15 September the bridge was dismantled; perhaps Burgoyne the dramatist recalled the words spoken by a greater general some 1,800 years before when about to cross another river – the Fiumicino, known to history as the Rubicon – 'My friends, stopping here will be the beginning of sorrow for me; crossing over will be such for all mankind. Let the die be cast.'

The army marched in three columns, the right wing, or Advanced Corps, was still under the overall command of General Fraser, with the Earl of Balcarres commanding the light companies and Major Dyke Acland the grenadier companies. This wing included the Brunswick Grenadiers, about 150 Tories and Canadians, some artillery, and what was left of the Indians – about 2,000 men in all. The centre was commanded by General Hamilton, and the German left wing by von Riedesel. Burgoyne marched with the centre and General Phillips with von Riedesel. Some recruits had joined the army just before the river crossing, which brought Burgoyne's strength up to 7,702 (3,818 British, 2,751 Germans and 1,133 auxiliaries).

The army had marched for six miles along a difficult track, crossed by many streams with broken bridges, which ran between the river and the steep wooded heights to the immediate west, before a foraging party being surprised near Sword's house by a strong American patrol gave Burgoyne his first indication of the enemy's whereabouts. This lack of information was understandable, for the country was very dense, Burgoyne had few Indian scouts left, and Gates had changed his mind more than once before selecting his defensive position.

Gates was a professional soldier who had served in the British army as an infantry officer, reaching the rank of major. Lacking sufficient money to purchase promotion he had retired in disgust, and in 1773 returned to America; there he had served as a young officer, and in 1775 threw in his lot with the colonists. He had seen plenty of action, and knew well the faults and virtues of the British soldier; he was not a front-line general, usually fighting his battles from the rear, and was always extremely cautious and unimaginative. It was not surprising therefore that he took his time in deciding where best to halt Burgoyne.

The land bordering the Hudson over which the two armies were operating had been settled and under cultivation for many years, mostly by industrious people of Dutch extraction who had given the flat alluvial valley an appearance of prosperity. In the wooded hills, which rose steeply from the comparatively narrow riparian strip, cultivation was less intense and the dwellings were primitive, being for the most part log cabins. When Gates took over command in mid-August his main army was at the mouth of the Mohawk, which was a totally unsuitable place at which to halt the British; he then advanced to Stillwater, but here too he found the ground to be more favourable to British tactics than American. He eventually settled on a place a mile or two upstream from Stillwater, where the Albany road passes through a defile between the river and the steeply rising wooded ground immediately to the west.

Here he found a naturally strong defensive position, which was to be made stronger through the prowess of Kosciuszko. The high ground was mainly clad with large oaks, hickory and a few pines, which shaded out much of the undergrowth and permitted fair visibility, but offered poor fields of fire. There were a few small clearings where farmers such as Freeman, Neilson and Barber had hewed a living out of the forest, and built cabins and tracks from which to work their smallholdings. What passed for the main road hugged the river, and bravely sought to overcome deep gullies; minor tracks branched up in to the hills. Near to where the road was forced a little way from the river, to circumvent a swamp, there stood a tavern belonging to Jotham Bemis, whose name had been given to the heights above the road.

Gates's army comprised twenty-eight regiments of foot, 200 light cavalry and an artillery component of twenty-two cannon. He took charge of the right wing in which there were three brigades (Glover's, Nixon's and Paterson's), while the left wing was commanded by General Arnold and had two brigades (Poor's and Learned's), Colonel Daniel Morgan's riflemen and Major Henry Dearborn's light infantry. It is not possible to give his exact total, but contemporary estimates, which seem reliable, put it at around 9,000 combatants. Stark was not prepared to join forces with Gates – indeed many of his levies would soon be time-expired – but he and General Lincoln held a watching brief across the river in the event of Burgoyne deciding to retreat.

Gates, with the help of Kosciuszko, was not long in laying out an almost impregnable position, with earthworks and breastworks stretching from the high ground overlooking the river to the west of Neilson's house, which formed the apex of his defensive position, the line being further strengthened by a deep ravine to its immediate front. On the right of the plateau, which was manned principally by Continental troops, artillery dominated the road and river; in the centre – held by Brigadier-General Learned's Continentals and troops from Massachusetts and New York – the fortifications ran north-westward to Neilson's barn, and here another battery was in position. To the west of the barn was another hill that had been only partly fortified, but the main position ran south-west to Gates's headquarters.

Contrary to what is often assumed, when Burgoyne first learned that the Americans were close to him it is impossible that he could have known the full extent, and strength of their defensive line, for he was without proper scouts and although operating in open forest the distance between the lines was about two miles. On the morning of 19 September, which incidentally dawned wet and misty, Burgoyne's army was encamped at Sword's house, about three miles from the American position. *Faced with a very difficult decision to make, he had four possible options:*

(1) To continue his march in column down the river road, and try to force the defile.

(2) To advance the army a short way, and then send out a reconnaissance in force to probe the extent of the American line.

(3) To launch an immediate frontal attack and hope to blast the position with his superiority in cannon.

(4) To feel for the left of the American position and try to turn it and roll the line off the bluff towards the river.

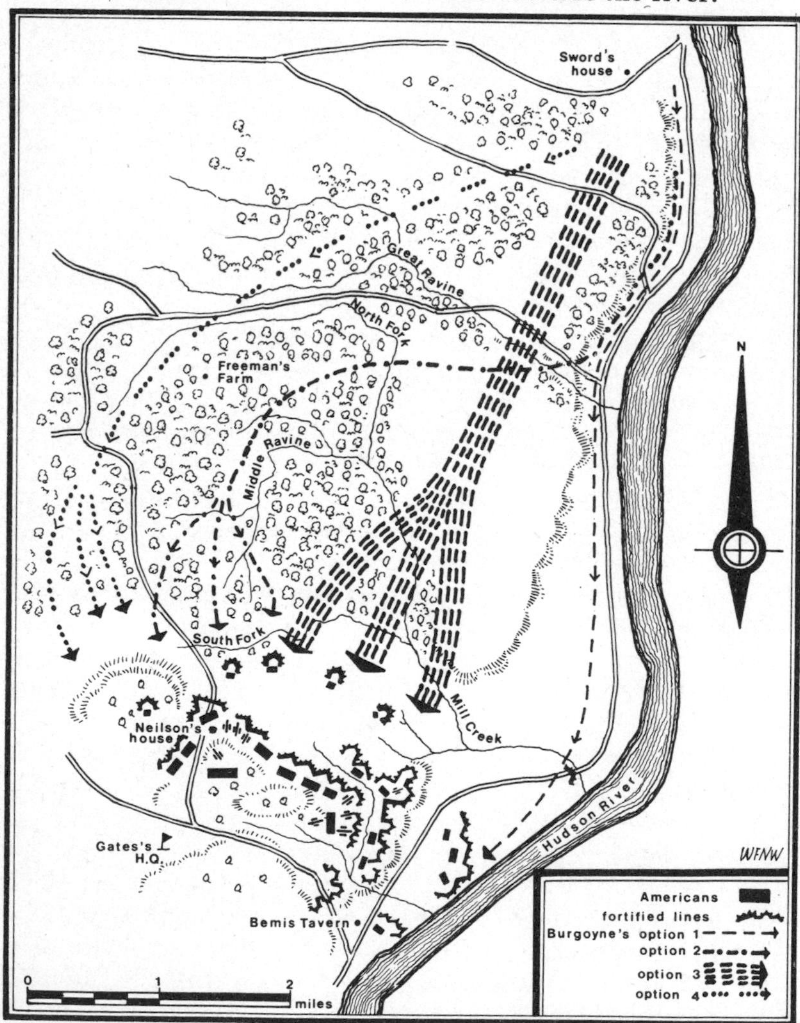

34 The Battle of Freeman's Farm: 19 September, Burgoyne's options

Burgoyne would very quickly have dismissed any idea of trying to force the defile, for even without knowledge of the exact American position he must assume that they would block the track in strength at that point. A reconnaissance in force would take time, and in that thickly wooded country might become engaged by superior numbers and be cut off from the main body. A frontal attack would very possibly have been successful if Gates had remained on the defensive, but experience had taught the British that in this type of country American marksmen had a good record against gun teams, and the artillery would have been the chief component of success. And so Burgoyne adopted option 4, and prepared a plan whereby Fraser's right wing, which would include Breymann's reserve corps, would sweep round to locate the American left and try to roll it up, while Hamilton, with four British regiments (the 9th, 20th, 21st and 62nd) and a brigade of artillery, would attack in the centre, leaving Phillips and von Riedesel with the left wing on the river road.

Burgoyne's army left Sword's house at 8 a.m. While Fraser's column took a wide sweep to the west through the thick woods, the centre column followed a wagon track which crossed the Great Ravine and then continued westwards to just north of Freeman's Farm. Here Mr Freeman, a loyalist who was to give his name to the forthcoming battle, had made a small clearing some 600 yards south of the Great Ravine. The clearing was on rising ground and extended for about 400 yards, sloping from east to west; at this point the centre column halted in order to enable Fraser to reach his starting position, which was some way further to the right. The Germans advanced about two miles along the river road, and halted by a track leading westward to the hills. It can be seen, therefore, that the three columns (particularly the centre and left ones) were well separated, and the dense nature of the country was to make any form of co-ordination difficult. When Fraser was adjudged to have reached his position a prearranged minute-gun signal would indicate the time for the columns to advance.

Whereas the British were feeling their way blindly the Americans had no such difficulty in quickly learning what their opponents were about. In the absence of Indians their scouts could get quite close, and anyway in the sunshine – which by 9 a.m. had displaced the rain and mist – the red coats and gleaming bayonets were easily picked out by treetop observers. Gates held nearly all the cards. He was inferior in weight of cannon, but he was numerically superior in troops, the ground was favourable to his riflemen, he was soundly barricaded, and time was totally on his side. *He had three options:*

(1) To sit where he was and let the British batter themselves in trying to break through his strong defensive position.

(2) To await events behind his defences, but send out riflemen and light companies to give strength to the left flank, which being only partly fortified was fairly vulnerable.

(3) To advance in strength and engage the enemy on ground that was unfavourable to the British tactics of volley firing and bayoneting.

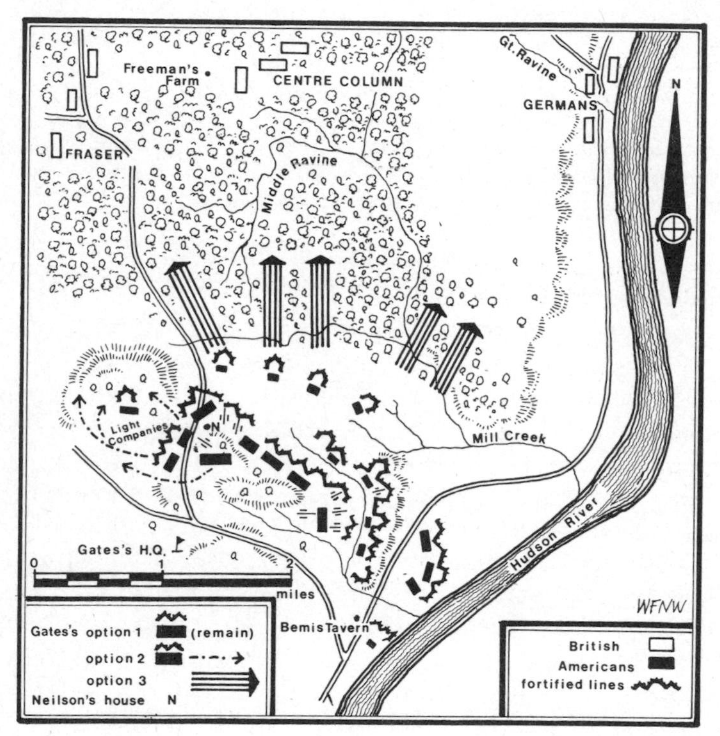

35 The Battle of Freeman's Farm: 19 September, Gates's options

Gates, as we have noted, was a very cautious general, and it is said – although chiefly by those who disliked him – that it took Arnold's cajolery to get him to agree to allow any man to leave the defensive position. Whether this was so cannot be said for certain, but Gates did decide to send Dearborn's light infantry and Morgan's riflemen to cripple the advance with sniping and skirmishing tactics. This was a mixture of options 2 and 3, although the action was to develop into a full attack (option 3). It was at best a half-hearted decision. Gates could have played his hand almost any way and taken all the tricks, although he might have been in trouble had he allowed Burgoyne, with his heavy guns, to close on the position and develop siege tactics. A swift and powerful attack on von Riedesel's Germans in their dangerously enclosed position, or a similar operation launched to hit Burgoyne between his right and centre columns, might have ended the campaign at a stroke. A general of Arnold's or Stark's calibre would have opted for this.

It was a little after 1 p.m. when Burgoyne judged the time right to give the signal for the general advance; at about the same time Morgan's men, advancing in line through the trees, came in contact with an advance picquet of General Hamilton's column commanded by Major Forbes. The picquet was virtually wiped out, but Morgan's riflemen, elated by this early success, dashed forward only to be thrown into complete confusion when they ran against the main body. Thus was the first bloodshed in what was to be a long and savage pounding match, lasting until darkness cast a merciful veil over soldiers who for the most part had fought themselves to a standstill.

The brunt of the fighting was borne by Hamilton's centre column, where Burgoyne had positioned himself. Of the four British regiments that held the line here – from right to left the 21st, 62nd and 20th, with the 9th in reserve – the 62nd took the greatest punishment, ending the day with only some seventy all ranks out of 250 who had gone into action. On the American side it was Arnold's fight; acting virtually independently he had ordered out Continental troops from his division. Having failed to make headway against Fraser's light companies on the left, these were then switched by their intrepid general to thrust through the gap between Hamilton's right regiment – the 21st – and Fraser's column.

Fraser gave what help he could, but Morgan's men had rallied and were keeping his troops occupied; the 21st were forced to face to their right, and this exposed the 62nd to cross-fire. The whole of Hamilton's front came into close intense fire action. American soldiers sank to the ground, writhing under the effects of the volleys, but more came on, for now the brigades of Learned and Poor from Arnold's wing had joined the fight. Moreover, American sharpshooters were making good use of the terrain, and doing great damage to key personnel such as officers and gun teams. Time and again Burgoyne's guns changed hands, but the enemy could never use them, for the gunners always removed the linstocks. The fight round Freeman's small clearing raged for three hours with scarcely a break; in the frenzy and turmoil of this furious battle the British found firearms a liability, and resorted to their favourite weapon – the bayonet.

By the late afternoon the British centre was near to breaking, but Arnold's men were incapable of that last bloody punch which would have pierced the line. Both army commanders now come in for some criticism – Gates to a greater extent than Burgoyne.

What was happening on Burgoyne's left? At 3.30 p.m., by which time the fight in the centre had become really hot, von Riedesel and his Germans were still inactive on the river road. Burgoyne, who knew very well how critical the position was, had inexplicably made no move to send for these troops. The initiative came first from General Phillips and then from von Riedesel – and it was nearly too late. Phillips, riding up from the river and seeing how desperate was the situation, ordered up four of von Riedesel's guns and personally led the 20th in a charge to reduce the pressure on the 62nd, while von Riedesel, acting on orders received from Burgoyne as late as 5 p.m., led two companies of the Brunswick von Rhetz Regiment, followed by his own regiment, to attack the enemy's right flank. The remainder of his force was left to guard the supply train and line of withdrawal.

It was about 6 p.m. when the two companies of the von Rhetz Regiment poured their first withering volleys into the Americans, and soon they were joined by the von Riedesel Regiment and two six-pounders of the Hesse-Hanau artillery which opened up with grapeshot. The American soldiers had fought bravely for seven hours, but now, dismayed by the carnage of the field, and conscious that victory had eluded them, they broke off the fight and withdrew into the gathering darkness, leaving those of their conquerors who had survived to bury the dead and succour the wounded.

To observe General Gates's conduct of the battle it is necessary to go back in time to around 4 p.m., when the whole of Arnold's wing was committed and it was clear that only a small extra shove would topple the British line. It is alleged – although there appears to be no positive evidence – that Arnold sent urgently to Gates for further reinforcements, but Gates 'deemed it prudent not to weaken his defensive position any further'. Whether or not Arnold sent this request, had Gates been further forward than his rear headquarters he would have seen for himself what the position was instead of having to rely on Arnold, whose impetuosity he mistrusted. By such poltroonery are battles lost.

Burgoyne's men were the victors, but it was a pyrrhic victory, for with almost 600 casualties (including thirty-five officers) the army had lost about a third of the troops actually engaged. The Americans lost only 283 officers and men, with a further thirty-three reported missing. That night the exhausted British and German soldiers bivouacked on the field they had so dearly won, while their commander tried to decide his next move.

In making his decision he would have to balance the respective condition of the two armies. Four regiments of his own had suffered heavy casualties. The hospital tents (where those gallant wives – Baroness von Riedesel, Lady Harriet Acland and others – whose chivalrous romanticism had led them to the front line, were doing splendid work) were overcrowded, and his ammunition needed replenishing. However, his two wings were comparatively unscathed, and he knew that although the American casualties might be less than his own they had taken a fearful hammering, and might be caught off balance. *He had three options:*

(1) To fall back to Ticonderoga while there was still time before the lakes iced up.

(2) To order an immediate attack on the morning of 20 September.

(3) To delay his attack until the 21st, in order to rest his troops, and allow more time for the replenishment of ammunition and the evacuation of the wounded.

In spite of Burgoyne's previous decision on the matter of withdrawal we know from a later communication to Clinton that he did not entirely dismiss the idea at this time; indeed, it seemed a viable option. His army was losing men at an alarming rate, the Americans could expect substantial reinforcements, there was no positive news – and apparently little hope – of help coming from the south, and even if he made Albany his winter supplies must come from New York. And it was absolutely the last chance of extricating his army. As it happened – although Burgoyne was not to know this – enemy troops, operating east of the Hudson, had now pretty well closed the back door. However, as previously, Burgoyne decided against withdrawing.

General Phillips, when a prisoner at Boston, told the Americans that Burgoyne had decided to attack their position at dawn on the 20th, and this was confirmed by a deserter from the 62nd who had entered the American lines that morning. Phillips said that it was Fraser who persuaded the General to postpone the attack until the 21st, because his men were not fully recovered from the exertions of the battle. But this story is suspect. Fraser's men were only lightly engaged on the 19th, and Lieutenant Digby in his journal for 20 September writes. 'It was Gen Phillips and Fraziers opinion we should follow the strike by attacking their camp that morning . . .' i.e. the morning of the 20th. Whatever the reason, the attack was postponed until the 21st (third option). Had it gone in as planned with a body punch on the American left flank and a holding operation against the centre and right, in the misty conditions of the morning of the 20th, it would have found the Americans totally disorganized and could well have succeeded. The decision was, however, probably the right one, for although by the 21st success was not so certain, the army was not ready on the 20th for another major engagement.

On the night of the 20th, or early on the 21st, Burgoyne received a message from Sir Henry Clinton sent off from New York on 11 September. In this he said that if it was Burgoyne's wish he could probably spare 2,000 men to make a push at Fort Montgomery, which guarded the Highlands on the lower Hudson. Burgoyne was to let him know. It did not take Burgoyne long to reply that such a move would be of the utmost help in probably drawing off some of Gates's men. Accordingly that night a messenger left on his perilous journey to Clinton in New York with a despatch outlining Burgoyne's plight. The messenger got through, and Clinton for the first time learned of the alarming situation at Saratoga. Contact thus opened was continued at great risk, and soon Burgoyne was found making the surprising request for 'explicit orders, either to attack the enemy or to retreat across the lakes while they were still clear of ice'. But meanwhile he was faced with the need for another immediate decision:

(1) Should he proceed with the planned attack while he still held some advantage?

(2) Should he dig in and remain inactive until hearing further from Clinton?

Burgoyne decided to take no action until news was received from Clinton. If the decision to postpone the attack from the 20th to the 21st was a right one, the decision to postpone it indefinitely was almost certainly a wrong one. In his statement to Parliament Burgoyne made much play on the time factor. Many of his wounded were recovering rapidly, there was some hope that St Leger and his force would soon join him from Ticonderoga, and Clinton's attack might open the way to Albany. But Burgoyne knew very well that what Clinton had offered was no more than a diversion; in any case with the communication delays he could not be expected to mount his attack before the end of the first week in October. In fact time was what Burgoyne could least afford. On 21 September he had a slender chance of fighting his way through, but Gates was gathering strength daily (by 7 October he had an effective force of 6,444 Continentals and 6,621 militia), and soon not only would the road be barred for ever, but the army, if not annihilated, must perish from starvation.

Why Burgoyne should have solicited orders from Clinton is not easily explained. Sir Henry – an experienced, cautious, but thoroughly reliable soldier – was, like Burgoyne, subordinate to Howe. He strongly disapproved of Howe's philandering in Philadelphia, but he was in no position to give Burgoyne orders; it looks rather as though the latter was beginning to apprehend disaster and was anxiously seeking to shift responsibility. In any case, a general in his position cannot wait sixteen days for orders, and if he was not going to fight he had to barricade and dig in.

By now every soldier in Burgoyne's army knew very well that the success of their mission was doubtful and the way ahead dark and hazardous in the extreme. This knowledge lent strength to their backs as they bent them to dig and build the best line of defence that the country and time would permit. The principal pivot points were three redoubts protected by log palisades and earthworks. The strongest of the three was called the Balcarres Redoubt, after the commander of the light and grenadier troops who held the position. Echeloned back from this, in order to prevent the right from being turned, was the Breymann Redoubt – a small square earthwork with sally ports, which housed his jaegers – and between these two redoubts were two stockaded log cabins held by the Canadian troops. The third redoubt, known as the Great Redoubt, occupied three hills overlooking the Hudson and the river

1 & 2 Two details from the Bayeux Tapestry. Above: *Duke William of Normandy extracting an oath from Earl Harold that he will not oppose William's accession to the English throne.* Below: *King Harold's fall at the Battle of Hastings. Harold, protected by his hard-pressed house-carls, is wounded by an arrow in his eye, while the Norman cavalry closes in remorselessly*

3 Edward III of England, victor at Crécy, from his gilt-bronze effigy in Westminster Abbey

4 An illustration from Froissart's Chronicles, showing Edward III, his standard quartering the arms of England and France, taking a town

5 Charles VI of France receiving English envoys, a miniature from Froissart's Chronicles

6 Henry V of England, portrait by an unknown artist

7 Marshal Boucicaut, one of the French commanders at the Battle of Agincourt. This miniature shows the Marshal with his wife, from the Boucicaut Book of Hours

8 A seventeenth-century satirical print showing the Scots holding Charles II's nose to the grindstone. This refers to the agreement signed at Breda on 1 May 1650, whereby the young King took the Covenant to embrace Presbytery and to enforce it upon his English subjects

9 General David Leslie, commander of Charles II's Army in Scotland during the Third Civil War

10 Oliver Cromwell, commander of the English army in the Third Civil War, from a portrait by Robert Walker

11 General John Burgoyne, commander of the British army at Saratoga, from a portrait painted in 1755

12 General Horatio Gates, commander of the American forces, from a portrait by Chappel

13 Left: *The Duke of Wellington during the Battle of Waterloo*

14 Left below: *Napoleon and the Old Guard before the Battle of Waterloo*

15 Below: *Field Marshal Prince Blücher, commander of the Prussian forces at Ligny and Waterloo. Portrait by Schadow*

16 General Robert E. Lee, commander of the Confederate armies in the American Civil War. This photograph taken by Matthew Brady in April 1865 shows Lee, seated, with his son G.W. Lee to his left, and Lt-Colonel Walter H. Taylor of his staff to his right

17 General Joseph Hooker, commander of the Union army at Chancellorsville, from a photograph by Brady

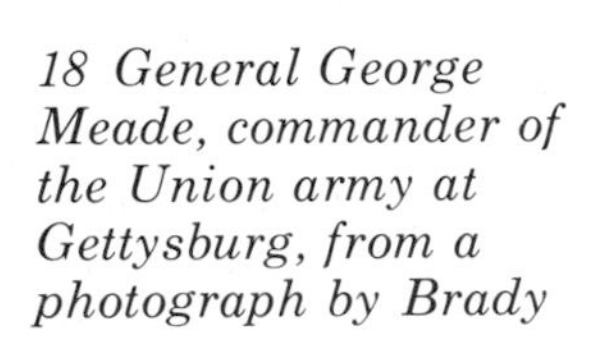

18 General George Meade, commander of the Union army at Gettysburg, from a photograph by Brady

19 General Sir Edmund Allenby, commander of the Egyptian Expeditionary Force in Palestine in 1917

20 General Erich von Falkenhayn, commander of the Central Powers' Seventh and Eighth Armies at the Third Battle of Gaza

21 Colonel Kress von Kressenstein (the tallest figure, with a stick) with his staff in 1918

22 Field-Marshal Kesselring, commander of the German armies in Italy in 1944

23 General von Mackensen, commander of the Fourteenth Army in Italy, looking at a map with his staff

24 *General Alexander, commander of Operation Shingle at Anzio*

25 *General Mark W. Clark, commander of the American Fifth Army at Anzio*

road. These hills were separately fortified and manned by von Riedesel's Germans facing south and east. The hospital, supply wagons and ammunition park remained near the river, guarded by the 47th Regiment and some detachments.

Along the whole line (which was held in much the same order as for the Freeman's Farm battle, except that von Riedesel's Germans were now on the high ground) as much use as possible was made of natural obstacles. It was a strong position, but the ground was so densely wooded and uneven that the field of fire was bad, and without reliable scouts (almost all the Indians had now disappeared) the defenders would have little warning as to where and when an attack might be made.

On 3 October Clinton, who had recently been reinforced by a strong draft from England, embarked 3,000 troops in sixty vessels and, escorted by a naval squadron, sent them up-river to carry out what proved to be the only really successful military operation of the campaign – the capture of Forts Montgomery and Clinton that barred the way up the Hudson at a narrow rocky defile just above Peekshill. But it came too late to be of assistance to Burgoyne; anyway, Clinton had made it clear to him by one of the messengers who plied their dangerous journey between the two commanders that he had not sufficient troops to fight his way to Albany – although, having taken the forts, and inflicted heavy casualties on the Americans, he did attempt to do so, and got within forty-four miles of Albany before the river pilots refused to take the vessels farther.

On 4 October Burgoyne called his senior generals to a council of war. Understandably, although he refused (much to Baroness von Riedesel's disgust) to dispense with the high standards of living to which he had adhered throughout the whole campaign, and continued to keep an outwardly calm and cheerful demeanour, he was getting extremely anxious by now. Food would not last beyond the end of the month, the cold weather was approaching and his troops were not kitted for it, and casualties, disease and desertion had torn the entrails out of the army, which now numbered somewhat less than 6,000 British and German regulars. *At this council, and on the following day, three options were discussed:*

(1) To leave a small force of about 800 men to guard the line of communications and the supplies, and march the remainder of the army to attack the left flank and rear of the American position.

(2) To send out a reconnaissance in force of some 1,500 regulars and a few Canadians to see to what extent the hill on the left of the American position had been fortified since the last battle, with a view to attacking the next day – or in the event of the hill proving to be strongly held to withdraw the army to Fish Kill.

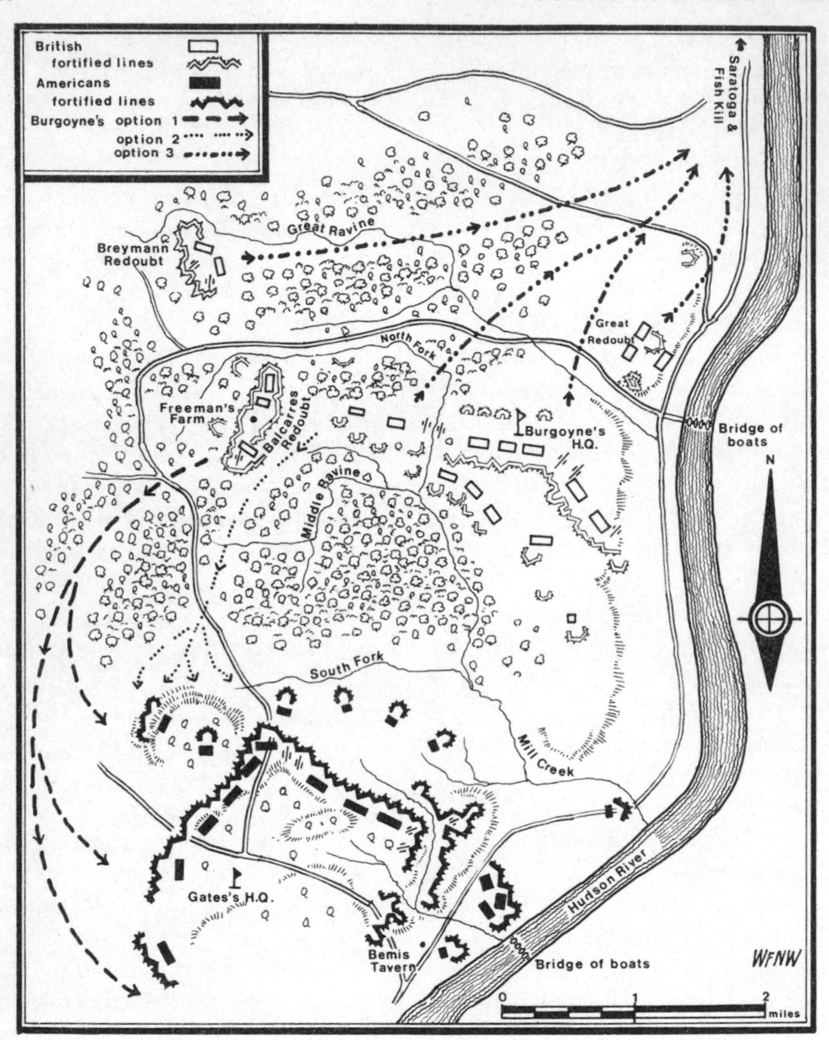

36 The Battle of Bemis Heights: 4 and 5 October, Burgoyne's options

(3) To withdraw at once to the original position at Fish Kill, which could be consolidated and held more strongly than their present position, and there await news of Clinton, and try to re-open communications northward.

The first option, posed by Burgoyne, was altogether too daring for the generals to accept, and it was indeed fraught with extreme danger. It would have been madness to send the entire army blundering around the American left flank over ground that had not been reconnoitred, and against defences whose strength was unknown. Moreover, while this was happening it would have been easy for the enemy to smother the 800 men guarding the supplies, destroy the bridges over the two creeks and thereby imperil the army's line of retreat.

To withdraw to Fish Kill was a sound proposition, for the heights above the creek could be turned into a strong defensive position quite quickly, but Burgoyne would not consider retreat before he had had one further attempt at defeating the enemy. And so a reconnaissance in force (option 2) was decided upon. This plan was a little better than the first option, but a large, unwieldy reconnaissance force often fails to attain its objective, and is liable to draw upon itself retribution in the form of an enemy attack many times its size – and this is just what happened. However, in defending this decision before Parliament Burgoyne said that he did not advance his whole line because those troops remaining in camp operated as an effective check against Gates's right going to the assistance of his left, or from advancing along the river stretch and falling on Burgoyne's rear. It therefore looks as though he always intended this force to attack if the circumstances seemed favourable. In either case it was a gambler's last throw. But Burgoyne had been a gamester all his life; he must have thoroughly approved the great Montrose's adage, 'He either fears his fate too much, or his deserts are small, that puts it not unto the touch to win or lose it all.'

Shortly before midday on 7 October this strong reconnaissance, accompanied by Burgoyne himself, marched from the lines. It had been carefully chosen for mobility and firepower, and comprised Simon Fraser's light infantry and grenadier companies, the 24th Regiment, the rangers under Fraser's nephew, Alexander, and the German jaegers, grenadiers and chasseurs from Fraser's original Advanced Corps. General Hamilton sent a small composite force from those men who had borne the brunt of the fighting on 19 September, and there were Canadians and loyalists as well. There were, therefore, nearer 2,000 than 1,500 men, and the force was supported by two twelve-pounders, six six-pounders and two eight-inch howitzers.

After marching for about three-quarters of an hour the advanced guard were west of the Barber farm, beyond the redoubts and north of the Middle Ravine, where there were two small clearings. Here the force halted, and about 150 irregulars were despatched to reconnoitre Gates's extreme left. The opportunity thus afforded was taken to send foragers to bring in the unharvested wheat, while the senior officers vainly sought to spy the enemy from the top of two log cabins. The front of the line thus formed was fairly open, but both flanks rested on woods; although these offered some protection they also enabled the enemy to approach fairly close unobserved.

First reports coming to Gates indicated a foraging party only, but Major Wilkinson, who was sent to investigate, reported a large force, and Gates (understandably) thought he was about to be attacked by the whole British army. The Americans opened what became known as the battle of Bemis Heights shortly after 2.30 p.m. with a simple but very effective plan. The British line had just started to advance from its temporary halt, when General Poor, approaching from dead ground, fell upon their left; for the next three hours Britons, Germans and Canadians were locked in deadly grapple with an enemy who far outnumbered them at every point of attack.

Gates's plan was for Morgan's riflemen, supported by Dearborn's light infantry, to make a wide detour and attack the extreme right of the British line; General Enoch Poor's brigade was launched against a detachment of British grenadiers and parts of Hamilton's division on the left of the line, while General Learned's brigade was to attack the German grenadiers in the centre once the flank attacks were under way. Poor's men had the shortest route and were therefore the first to go into action. Major Acland's grenadiers fought magnificently, but the numbers were too great for them and they had to give ground, losing as a prisoner their badly wounded commander.

Shortly after Poor attacked Morgan's riflemen opened up a devastating fire on Fraser's right, which necessitated the 24th changing front. This attack was every bit as fierce and powerful as that which Acland's men had had to cope with, and despite the efforts of Balcarres his soldiers were forced to fall back. Realizing the seriousness of the situation and hoping to save the guns, Burgoyne sent his aide, Sir Francis Clarke, to order the artillery to retire, but he fell badly wounded and the order never got through. Fraser, prominent on a grey charger and exposing himself with reckless courage, strove to rally his men and preserve the right flank. But he was unable to stop the turning movement, and inevitably a marksman drew a bead on him, causing a wound that within a few hours was to prove mortal.

The withdrawal on both flanks left the Germans in the centre very exposed to the onslaught of Learned's men, who were soon to be reinforced by General Ten Broeck with 3,000 Albany militia. These brave Germans, standing at bay, showed great discipline and fury, but it was clear to Burgoyne (whose bullet-ridden coat and hat testified to his front-line courage) that defeat, naked and brutal, stared him in the face, and that the whole line must withdraw at once behind the fortifications. It was about 5 p.m.; the British had suffered some 400 casualties and lost eight cannon, but they were now protected by their strong defence works. It seemed as though for the time being they were safe. But one man thought otherwise.

General Benedict Arnold, that impetuous and unpredictable man, had had a flaming row with Gates some days before the battle, and had been dismissed his command; but he lingered on in the camp, and the sight of the battle worked on him like a tonic. When the battle was at its height he was out before anyone could stop him, and now he roared around the field like one possessed, taking charge of first one brigade and then another, hurling them against the British entrenchments with himself always in the lead, totally regardless of the amount of lead that was flying around him. In no time he had raised the temper of the troops to the highest point of expectation. Acting with sense as well as zest he quickly saw that to continue to throw troops against the strongly defended Balcarres Redoubt (into which almost all the British had withdrawn) was wasteful, so he turned his attention to the much less strongly held Breymann Redoubt.

Here he found General Learned's brigade in the process of forming up for an attack, and taking charge – for neither Learned nor Poor were thrusting commanders – he stormed the Germans. They fought with their customary courage, but when their commander was killed they abandoned the position. Benedict Arnold had had his horse shot under him, and his leg crushed, which effectively ended his part in the campaign.

The remnants of Burgoyne's army were now threatened in flank and rear, but darkness came in time. The fight was broken off, and the British withdrew to the area of the Great Redoubt. Their casualties had been grievous: of about 2,000 men actually engaged Burgoyne had lost 176 men killed, more than 200 had been wounded, and 240 had been captured, together with eight cannon. Saddest of all, both personally for Burgoyne and for the army, was the loss of Simon Fraser. The American casualties were no more than 200.

On 8 October Gates's men occupied the line previously held by the British, and opened up a lively cannonade on the Great Redoubt. Burgoyne saw very clearly that his position was untenable; in trying to resolve the fearful dilemma that faced him and the army *he had three options:*

(1) To withdraw to a more easily defended position upstream at Fish Kill.

(2) To endeavour to get back to Ticonderoga.

In considering these two options Burgoyne would have known that any form of withdrawal necessitated abandoning the wounded, and there were also the women to be thought of, for the way back was along almost impossible tracks, the weather was setting in wet and the food supply was running short.

(3) To surrender.

It was probably now that Burgoyne first realized that unless something little short of a miracle occurred he would soon have to surrender – but he would not have dwelt on that unpalatable possibility for long. To begin with Clinton might work the miracle – as yet there had been no news as to how his attack on the Highlands had fared – and there was, he thought, still a very slender chance of getting back to Ticonderoga. Dismissing surrender, Burgoyne decided to keep his options open in respect of options 1 and 2. He decided to fall back to Fish Kill, where the ground above the creek was favourable for defence, and where the army had prepared a position after crossing the Hudson in early September. From there he would send a force up the west bank of the river to reconnoitre the crossing opposite Fort Edward, which led to the Ticonderoga road. The wounded were to be left in the hospital tents under the care of the medical officer. The women marched with the column.

It was 9 p.m. on the 8th when the advanced guard started to leave the camp, and two hours later the rearguard marched. Although scarcely more than a musket shot away, the Americans made no attempt to interfere. The night was dark and wet, the men were dispirited, and the mud was appalling. At 6 a.m. on the 9th Burgoyne ordered a halt, and the column did not get under way again until 4 p.m. Von Riedesel and others criticized Burgoyne for this dangerously long halt, but it was probably necessary to allow the bateaux with the stores to close up, as well as to rest the men, for a pace of one mile an hour told its own tale. Late in the evening the army reached the Fish Kill, forded the stream and sank exhausted into their old positions.

Gates, characteristically, had proceeded with caution. A general more imbued with the offensive spirit would have attacked Burgoyne on the 8th, and finished off the campaign. But Gates, fully aware that he held all the cards, was in no hurry to administer the *coup de grâce*. He had, however, sent Brigadier-General Fellows with 1,300 militia on a roundabout route to block the track at Fish Kill – which he failed to do – and he also had troops operating on the left of the British army and the east bank of the Hudson. But the main army did not leave camp until the late morning of the 10th. However, making better speed than the weary British had done, they were before Burgoyne's position that afternoon.

It was a strong position in depth on rising ground that faced them, measuring about one and a half miles by one and a half; eight guns were sited to fire across the river, where Fellows had taken his men, and other troops were massing. The remaining guns had a good field of fire to their front (Burgoyne having previously ordered the destruction of Schuyler's fine house for this purpose), but on the right of the line the dense country offered a covered approach, and the position could be enfiladed by fire from across the Hudson.

Burgoyne had been using Colonel Sutherland, with six companies of his 47th Regiment, as a sort of advanced reconnaissance force; he was now on his way up the west bank to reconnoitre the crossing near Fort Edward and to drop off pioneers to mend the broken bridges across streams leading into the Hudson. When Burgoyne saw the American army approaching he hastily recalled Sutherland (who as it happened was on a fruitless errand, because Stark – back in the field once more – had already blocked the Fort Edward crossing); but his march north had been reported to Gates, who as a result appears to have formed the wrong impression that there was only a rearguard left for him to deal with. *In these circumstances Gates had four options:*

(1) To mount a frontal attack at once so as to overwhelm the rearguard before it melted away in the night.

(2) To await the morning, and if the rearguard was still there to attack frontally.

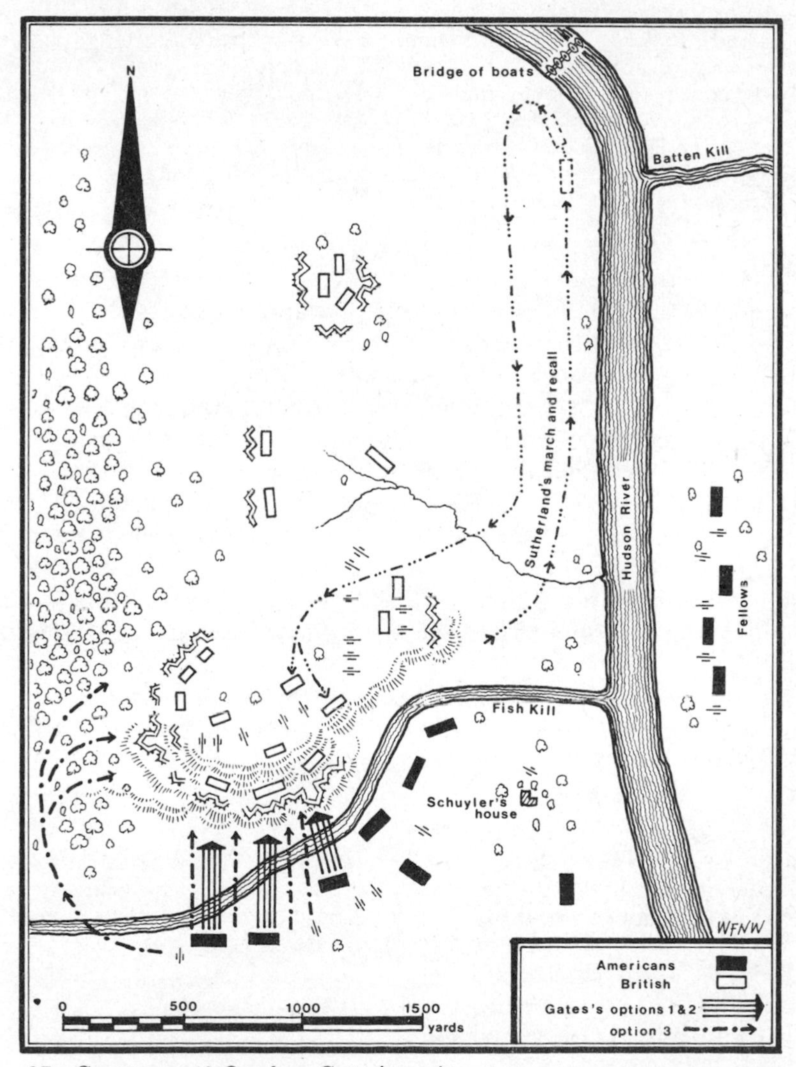

37 Saratoga: 10 October, Gates's options

(3) To await the morning and then mount a more elaborate attack against the enemy's flank and centre.

(4) Before taking any action to reconnoitre the position more thoroughly, and ascertain its exact strength.

Gates decided to await the morning, which turned out to be wet and very misty – but he could see that the enemy were still in position. Although convinced that it was only the rearguard that faced him (for the Americans had been unable to gauge the strength of Sutherland's force and had not seen it returning), he was not prepared to take any chances and so decided to launch a flank and frontal attack (option 3). This was a surprising conversion to the offensive spirit, but the decision he took was a wrong one, for to attack 'blind' through the fog was to court disaster. The information he had received that led him to think he had only a rearguard to deal with was not conclusive – and became distinctly suspect when that 'rearguard' was still in position the next morning. Gates may have found it difficult to believe that Burgoyne would stand and fight, but he should have sent out a reconnaissance to discover exactly what he had to deal with and then made a plan.

Morgan's riflemen and Learned's brigade were sent on a circuit through the woods to turn the enemy's right flank, while Gates (advancing this time with the foremost troops) accompanied the brigades of Generals Nixon and Glover. The leading troops were across the Fish Kill, when a deserter was taken who informed the brigade commander that he was approaching the whole British army strongly entrenched. The attack was called off in the nick of time, and the troops withdrawn under a fire made more effective in a clearing mist. Had the attack gone in the British guns would have taken a fearful toll; it was Gates's lucky day, and he knew it. There would be no further adventures of this kind; he could settle down to bombard the position, and starve the enemy into surrender.

For six days Burgoyne's soldiers, and the ladies who had bravely marched with them, had to endure a heavy pounding from the American cannon. *On 12 October Burgoyne summoned a council of war and put forward five options:*

(1) To remain in their present position – which might be strong enough to repel an attack – and await news of Clinton.

(2) To withdraw up the west bank with the guns, repairing the bridges as they went.

(3) To withdraw stealthily by night, without the guns or baggage, each man to carry six days' hard rations.

(4) To attack the enemy.

(5) To skirt round the enemy's extended left, and march rapidly for Albany.

It is interesting to note that at this council Burgoyne did not give his generals a choice of surrendering – always the optimist trying to compel events beyond any fortune that men might expect of Providence. There was some support for the fifth option, but it was uncertain how far Gates's left extended, and all the others – except abandoning the position by night without transport or guns (option 3) – were rightly considered impracticable. Withdrawal by night was therefore decided upon, but before darkness set in scouts arrived to say that it was too late. The trap was closed; the enemy were on all sides, and to sneak away unobserved even by night was no longer possible.

Foiled in this last hope of escape, the next day Burgoyne called a further, and extended, council. At this he explained the situation – if indeed any explanation was necessary, for the stench of death on the hill, and the knowledge that the enemy had at least 12,000 men to their 5,000, spoke for themselves – and asked if they thought that an army in such a situation could honourably surrender. The answer was a unanimous 'yes'; and under a flag of truce Gates was asked for terms.

Burgoyne rejected Gates's demand for unconditional surrender, and insisted that his troops should march out of camp with the honours of war, and be given free passage back to Britain on the understanding that they would not serve in America again during the war. He also insisted that the terms agreed upon should be called a convention and not a capitulation. To all these Gates agreed – probably because, not knowing the outcome of Clinton's little campaign, he was anxious to get matters settled quickly – and after some not very creditable attempts by Burgoyne at procrastination, the Convention of Saratoga was signed on 16 October.

The next day Burgoyne, wearing full dress uniform, led his men out of camp to lay down their arms and surrender. In victory Gates behaved with moderation and magnanimity. There was to be no gloating by the victors over the discomfiture of the vanquished, for Gates confined most of his soldiers to their quarters. Burgoyne, surrendering his sword, said, 'The fortune of war, General Gates, has made me your prisoner', to which Gates replied, 'I shall always be ready to bear testimony that it has not been through any fault of your Excellency.'* Burgoyne's sword was returned to him; such chivalry and courtesy smooth the rough edges of war.

Gates was right, this disaster was not principally the fault of Burgoyne. He may not have been a brilliant general, but there have been many worse, who have handled armies in easier circumstances with far less distinction. The invasion from Canada failed through

* F. J. Hudleston, *Gentleman Johnny Burgoyne*, p. 212.

lack of unity in command. In the absence of a general staff Lord George Germain had made himself responsible, under keen scrutiny from the King, for the grand strategy. But in his feeble attempts to carry it out he failed to realize that there were insufficient troops in America for two major campaigns – an advance from the north and the capture of Philadelphia – to be successful. And until it was too late he failed to impress upon Howe (and then it was not an order) the need to give Burgoyne effectual support. Up to the very end Howe himself failed to appreciate the seriousness of Burgoyne's situation; instead of sending troops north, he actually ordered them south from New York, leaving Clinton with the vaguest of instructions. In the event, as Jane Clark has pointed out,* Howe's capture of Philadelphia availed Burgoyne nothing, and Clinton's valiant, but inadequate, effort merely discouraged Burgoyne from retreating while he still had the chance, and gained him better terms in a Convention that was never kept.

But none of this should detract from the skill and courage of the American soldiers, who in fair fight had beaten well-trained British and German regular troops. It is true that their numbers were greatly superior, but theirs was a new army without tradition, training or discipline. They were fortunate in their officers, and also in their commanding general, for Gates for all his caution was a competent strategist, and only once did his usually sure-footed judgment fail him.

The surrender at Saratoga was the turning point of the war. Not so much because the French were soon (by a treaty of February 1778) to recognize American independence, and become – later with the Dutch and Spanish – their active allies; nor because of the French muskets that were said to be superior to Brown Bess, for many of these had been landed at Portsmouth before the campaign began, but because a whole British army had been defeated and forced to surrender. It was this fact that gave the American soldier that psychological boost which he badly needed. He had met and defeated his British counterpart, and in doing so he had gained confidence in his officers, in himself and in the ultimate triumph of his cause. Territorially, with the withdrawal of the British from Crown Point and Ticonderoga, and the relinquishment of the Highlands by Clinton, it was only in Rhode Island that they now had any hold north of New York to the Canadian border.

The War of the American Revolution was to be waged on land for a further four years (and at sea longer than that) before another, and far greater, surrender took place at Yorktown. General Clinton had by then become the British Supreme Commander, but it was Lord Cornwallis who on 17 October 1781 (the anniversary of Burgoyne's surrender) sent a drummer to beat the chamade, and two days later surrendered more

* *The American Historical Review*, Volume XXXV, 1929–30.

than 8,000 soldiers and sailors. It was the greatest disaster that British arms had ever suffered, and was to remain so for more than 150 years.

By a treaty signed on 3 September 1783 Great Britain officially recognized the independence of the United States of America; there were few who could measure the consequences, or foretell how her power, wealth and influence would in less than 200 years dominate the world.

8 The Waterloo Campaign
1–17 June 1815

NAPOLEON escaped from Elba at the end of February 1815, and arrived in France, near Cannes, on 1 March, with 1,000 men under command. His sword had slept in its scabbard for scarcely ten months. The Allies (Britain, Austria, Prussia and Russia) were still arguing over the future of Europe at the Congress of Vienna when news was received that the former Emperor of France was on the march again. Their avenging armies were quickly ordered to be ready to enter France by 1 July, and in a carefully co-ordinated plan to converge on Paris for the annihilation of the vast army that Europe's greatest troublemaker was sure to raise. The Duke of Wellington and Field-Marshal Prince Blücher were to invade through Belgium, Prince Schwarzenberg, commanding the Austrian army, would come in from the east, and Count Barclay de Tolly and his Russians would form the back-up army advancing through Saarbrucken. It was a sound plan, but it would require a genius to mastermind it successfully.

Meanwhile Napoleon advanced in triumph across France. He took over a regular army of some 150,000 first-line soldiers, and he set about increasing it with an urgency and brilliance reminiscent of former days. Two hundred battalions were recruited, all the veterans were recalled to arms, the factories and workshops were galvanized to produce equipment and war materials on an unprecedented time-scale, and the Treasury set about funding this gigantic reorganization.

By the beginning of June over 400,000 men had been recruited, conscripted or recalled, giving the Emperor slightly less than 200,000 first-line troops, and an auxiliary army of around 218,000 men. But many of these troops were needed to guard the French frontiers, some had to be left in Paris, and 20,000 were necessary to quell an uprising in La Vendée. The *Armée du Nord*, which was to operate in Belgium, totalled no more than 116,000 men and 350 cannon (see Appendix II), although it would seem that this number could have been safely increased at the expense of some of the garrison troops. Of this army the élite was, of course, the Imperial Guard, many of whom had remained faithful to

Napoleon during his exile; all answered the call to rejoin their regiments.

The majority of the French infantry was equipped with smooth-bored, muzzle-loading muskets, although there were some rifled barrels. The infantryman carried flints, $\frac{3}{4}$-inch iron balls and powder cartridges. The rate of fire from a well-trained man could be two rounds a minute, and the weapons had a range of about 200 yards, although the killing range was very much less. Napoleon had greatly increased the number of light cavalry regiments, and considerably reduced the heavy cavalry. The dragoons became cavalrymen, instead of mounted infantry, and carried swords rather than the usual sabres or lances, with pistols at the saddle.

Most of the French guns were twelve-pounders, firing round shot, canister or grape. The effective range for round shot was somewhat less than 2,000 yards, while canister and grape would range only about 300 yards. The twelve-pounder was a cumbersome weapon to fire and the team did well to get off two rounds in a minute. There were also a few lighter and more accurate pieces – three to four and eight to twelve pounders. The artillery regiments were organized as an integral part of each army corps, as were the engineers.

At the beginning of the campaign, Napoleon could either advance into Belgium, and attempt to beat the nearest Allied armies to him (the Anglo-Dutch and Prussians), or he could stand on the defensive and give battle to each of his enemies as they attacked him, in much the same way as he had done in the 1814 campaign.

The Duke of Wellington gave it as his opinion that Napoleon would have been best advised to do the latter, for it had the additional advantage of gaining time for further recruitment and reorganization. But in this instance it is difficult to agree with Wellington; a strike through Belgium before the Allies were fully mustered had much to commend it. By a stealthy concentration on this the most vulnerable of French frontiers, Napoleon had every chance of defeating Wellington and Blücher in detail. This might bring the Netherlanders to declare for France, the British might temporarily withdraw from the conflict, and such victories, gained off French soil, would enormously increase the country's morale and enable the Emperor with a greatly increased army to pose a threat to Vienna, while engaging and destroying the Russian advance.

Napoleon was well aware of the political dangers in Paris, and must have felt that success alone could keep the wolves at bay. The Allies would always have the advantage of numbers if Napoleon gave them time to implement their plan, but with the use of interior lines it was just possible that the Emperor, with superior mobility and leadership, could destroy the Seventh Coalition before it brought its massive weight to bear.

Having decided to make a pre-emptive strike against the Anglo-Dutch and Prussian armies, Napoleon had four options as to the route he should take to enter Belgium and make his attack. In considering these options he would have been aware that the flat countryside of Belgium was naturally divided by the great rivers that traversed it – the Lys, Schelde, Sambre and Meuse – and that in the gaps formed by these rivers there were a number of fortresses, such as Courtrai, Oudenarde, Nieuport, Ypres, Mons and Tournai, and the less strongly fortified towns of Charleroi and Namur. Napoleon had a choice of two approaches from the west, another from the south and a fourth from the south-east. *His four options were:*

(1) To base his army on Lille and come in between the rivers Lys and Schelde.

(2) To advance from Valenciennes on Mons and Brussels. The object of these two lines of advance would be to defeat the Anglo-Dutch army and drive it back to Antwerp.

(3) To centre on Beaumont or Philippeville, and cross the frontier between the Meuse and the Sambre with a view to forcing a wedge between Wellington and Blücher.

(4) To cross the frontier in the Givet area, make for Namur and fall upon the left of the Prussians, cutting their line of retreat to the east.

Napoleon chose option 3, chiefly because his strategic plan demanded it. However, it also had many topographical advantages: it offered open, undulating ground, with numerous good roads, and there were three places where the Sambre could be crossed in the neighbourhood of Charleroi. This part of Belgium was less heavily fortified than the west, and indeed an attack on either flank if not quickly successful must hasten the junction of the two Allied armies.

Napoleon entered Belgium with less than half as many men as the Allies were capable of bringing against him. His strategy therefore was dictated by the need to defeat each of his enemies separately. To do this he planned to advance with two wings, and a reserve (chiefly composed of the Guard) which would be brought into action on either wing according to circumstances. He also intended that should it be necessary he would withdraw troops from one wing to strengthen the reserve and the wing where the decisive action was to take place. Hold one army, smash the other, and then turn with full force to crush the one that is being held: like all good plans it was simple, yet sound. Moreover, in its initial stages it was carried out with an exactitude which has remained for all time an exemplar.

The concentration of the six corps, which began on 4 June, on the banks of the Meuse and Sambre, was one of Napoleon's finest manoeuvres. Personal attention to the minutely detailed schedule, and a complete

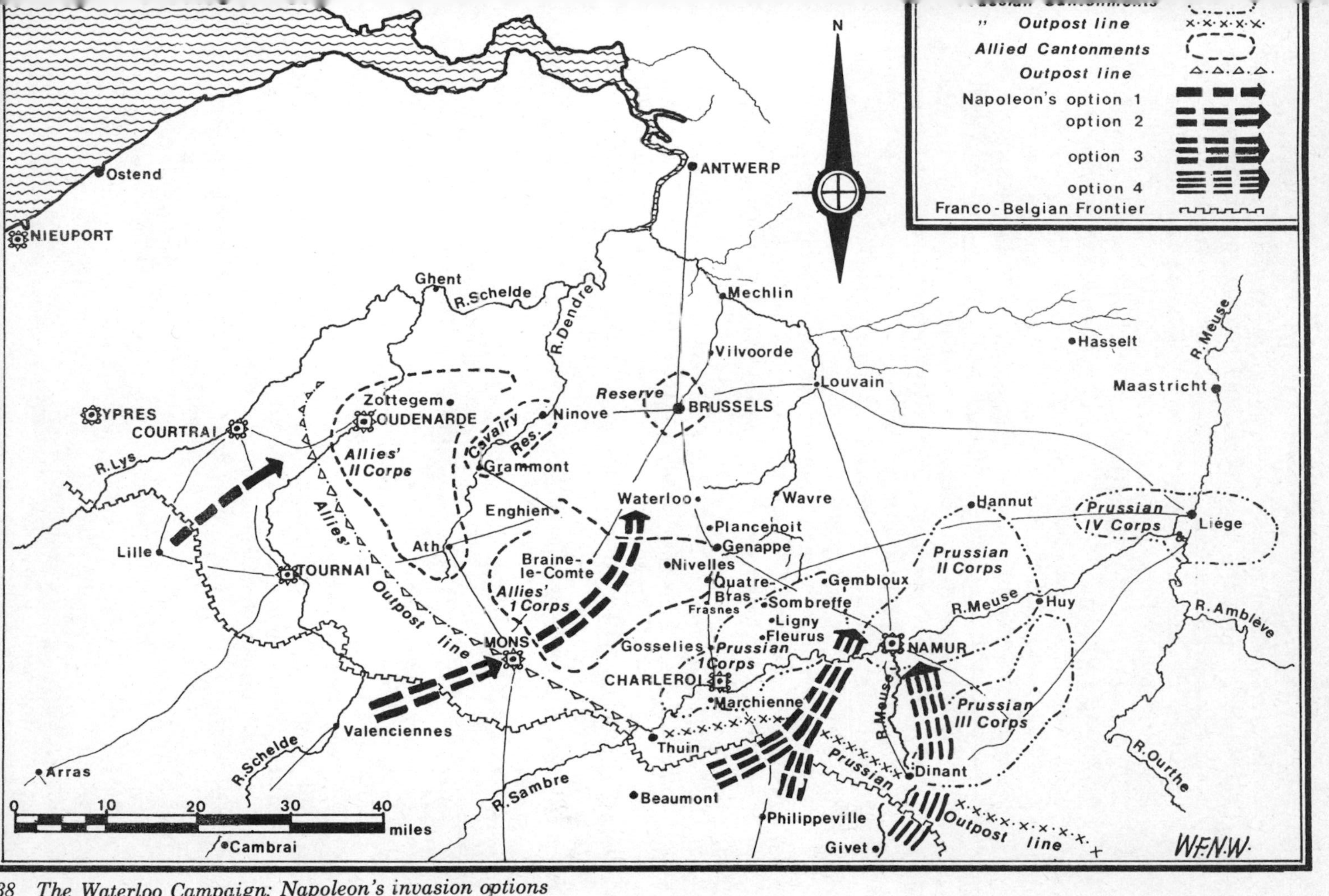

38 *The Waterloo Campaign: Napoleon's invasion options*

security clampdown from Paris to the ports, resulted in more than 100,000 men, who had been garrisoned in some instances vast distances apart, concentrating in ten days on a sixteen-mile stretch of the Belgian frontier, without the enemy having any precise knowledge of where that frontier would be crossed.

If one is to fault Napoleon at this stage of the campaign it can only be for leaving too large a force in France, and perhaps more importantly for leaving some good senior officers there. That Soult was a very inadequate substitute for Berthier (who had killed himself) as chief of staff was hardly Napoleon's fault, but why leave Suchet and Davout in minor posts, and in a childish miff exclude Murat, his greatest cavalry commander? Napoleon needed all the good generals he had, for while the lower echelons were mostly inspired by total loyalty to the Emperor, the senior ranks were seamed with envy, and there were those who diffused an atmosphere of uncertainty and even defeatism.

And what of the men on whom Napoleon was about to pounce? Their combined strength at the beginning of the campaign amounted to 244,400 all ranks and 558 cannon. The Anglo-Dutch army was 102,500 strong with 258 cannon, and the Prussians had 141,900 men and 300 guns (for unit details see Appendix II, p. 292). A comparison of the weapons of the Allied and French armies is scarcely necessary, for the differences were not too great. But the British infantryman handled his musket more efficiently than his French opposite number, and the formation in which the French infantry came to the attack put them at a further disadvantage. Every man in both ranks of a British battalion in lines two deep could load and fire with ease, whereas the French usually advanced in massed close column of battalions; a clumsy formation incapable of developing effective fire power.

The dispositions of the two Allied armies at the beginning of hostilities were calculated to facilitate the Emperor's strategic plan. The easternmost Prussian cantonment at Liège was 100 miles from Wellington's most western cantonment in the Oudenarde area, and although Wellington had his sector of the front well covered with outposts, and his communications with Britain secure, insufficient allowance was made for the rapid reporting of any French move. Lack of timely information nearly proved Wellington's undoing, for he had made no provision for holding any French advance, pending the concentration of his whole army. He reckoned to be able to concentrate 60,000 men at any given point within his cantonment area in the space of twenty-four hours of receiving news of Napoleon's crossing the frontier, but in the event Napoleon did not give him that time because of the delay in transmitting information. From a strategic (but not political) point of view Nivelles would have been a better centre for his headquarters than Brussels, because of quicker communication with the frontier and the Prussians.

The Prussian cantonments were also dangerously dispersed, but their front was less exposed than that which the Anglo-Dutch had to watch, and Blücher had made some preparation for the concentration of his army on Fleurus in the event of surprise, by stationing Zeiten's I Corps on the line of the Sambre in the Thuin–Charleroi area. On 14 June the Allied armies were still widely dispersed in their cantonments, and on that very day Wellington had written to the Czar saying that he expected to take the offensive at the end of the month. Nor was Blücher entirely convinced that Napoleon would attack across the frontier. Nevertheless, the line of his probable advance in the event of this happening had been correctly appreciated, and a plan made for an Allied concentration at Gosselies and Fleurus.

On the night of 14–15 June Blücher got information that the French were preparing to attack his outposts in strength, and he ordered Zeiten to fall back fighting on Fleurus. He further ensured that II and III Corps would be in a position to fight in the area of Sombreffe by 16 June. It was not until 3 p.m. on the 15th that Wellington heard that the Prussians had been attacked at Thuin; this in fact had occurred at dawn, and by the time Wellington first received the news Napoleon was in possession of Charleroi, the three bridges across the Sambre and the roads leading north. An unfortunate breakdown in the Duke's usually excellent secret service was the cause of this delay, and when the information did reach him it was incomplete. In consequence Wellington, without full information, was still uncertain of what action to take. *There were two possible options:*

(1) He could issue preparatory orders for troops to assemble at divisional headquarters and be in readiness to march. Provisional assembly points were given to corps commanders.

(2) He could send a suggestion to Blücher that as it would now seem impossible for the Anglo-Dutch army to make a junction with the Prussians as far forward as the Gosselies–Fleurus area, in time to give battle with a combined force, he (Blücher) should disengage and fall back to the Waterloo–Wavre line, where a junction of the armies could safely be made covering Brussels.

Wellington decided on the first option, and issued orders for provisional assembly points for I Corps to be at Nivelles, Braine-le-Comte and Enghien, and for II Corps at Ath, Grammont, Zottegem and Oudenarde; the cavalry were to concentrate at Ninove, the Brunswickers of the reserve were to collect that night on the Brussels–Vilvoorde road, and the Nassau Brigade was to assemble on the Brussels–Louvain road. He was thus prepared for any eventuality once he knew Napoleon's exact line of advance.

This decision has been fiercely debated. Because later on the Duke is alleged to have said that Napoleon had humbugged him it is sometimes thought that he had been taken by surprise and out-generalled. This is not the case, except in so far as that neither Wellington nor Blücher thought that Napoleon meant to take the offensive so soon; Wellington had determined to make no move until he knew what Napoleon was up to, for he always thought that he had the time necessary for a large-scale concentration.

Nevertheless, when that time was denied him (and even allowing for the breakdown in communications this would not have happened if he had carried out a partial concentration, or had had his headquarters farther forward), the Duke was exposing the army to a grave risk of being defeated in detail by this late concentration so far forward as Quatre-Bras, for owing to distances the troops could arrive only in comparatively small numbers.

On 15 June Napoleon was across the Sambre and ideally placed to attack Blücher before his concentration was completed, and then turn on Wellington. With no troops east of Nivelles the Brussels road was clear; it would seem that the Allies were doing all they could to ensure that Napoleon's plan met with success.

Napoleon's crossing of the Sambre on the 15th was stoutly opposed by Zeiten's Prussians; an action at Marchienne caused a three-hour delay on the left, and on the right more delay was caused when General de Bourmont, in command of the leading division of I Corps, deserted to the enemy with five of his staff – incidentally getting a very frigid reception from Blücher for his pains. This corps (Gérard's) did not manage to cross the Sambre on the 15th, and delayed the commencement of the battle the next day, with serious consequences.

Marshal Ney only arrived from Paris at 3 p.m. on the 15th; he was immediately given command of the left wing, with orders to push on to Quatre-Bras. He had soon captured Gosselies from the Prussians – Wellington's concentration point – but his leading cavalry were held up near Frasnes by a battalion of Prince Bernhard's Nassau Brigade and a Dutch horse artillery battery (the nearest and first Anglo-Dutch troops to appear – and without orders from headquarters). It was 8 p.m. before these troops were driven back on Quatre-Bras, and Ney decided not to pursue. On this occasion he was undoubtedly right, because his wing was well extended, he was new to the command, the ground was unstudied and the strength of the enemy was unknown. On the two following days, however, it is less easy to excuse Ney's tardiness.

On the right, Marshal Grouchy and General Vandamme, later joined by the Emperor, drove back the Prussians. Zeiten, handling his troops with great skill, fell back on Fleurus. At nightfall on the 15th the situation was eminently satisfactory for Napoleon: no more than a handful of Wellington's troops had put in an appearance, and Blücher still had only one corps (and that one had fought hard all day) at his concentration point, although Pirch's and Thielemann's were close by.

The only cause for complaint was that Gérard was the wrong side of the river. It must have seemed to Napoleon that he had the enemy firmly in his grip, and so long as he tightened it – and quickly – on the morrow, what a cataract of misfortune would befall them through their inability to combine.

But on the morning of the 16th things began to go wrong, and they were never really to go right again. Grouchy, commanding the right wing, quickly pushed Zeiten out of Fleurus, and reported to the Emperor a large concentration of Prussians in the area of Sombreffe. But Napoleon refused to believe that they were in sufficient numbers to demand immediate attention, and there was news from Paris that took up time. He did, however, send an order to Ney at 8 a.m. to take and hold Quatre-Bras, and to send a detachment of 8,000 troops to Marbais for possible use on the right wing, in accordance with the original strategy. Napoleon went himself to Fleurus at 11 a.m. where he learned that Ney, for various totally inadequate reasons, had made no advance on Quatre-Bras; another order was then despatched to him.

Meanwhile the Emperor, climbing the mill at Fleurus, took stock of the Prussian position. At this time Blücher had two corps (those of Zeiten and Pirch) in position. Zeiten's was forward, occupying the position circumscribed by the villages of Brye, St Amand and Ligny, with its right resting on Wagnelée. These villages and the small stream at Ligny formed a bastion that gave some protection to the position, but Wellington, when he visited it with Blücher during the morning, was not impressed – and said so. Pirch's corps formed up in reserve of Zeiten's between Sombreffe and Brye. It would seem that Thielemann's corps, which was to occupy the left of the Prussian line, was not in position at the time of Napoleon's reconnaissance, for in a message to Ney at 2 p.m. he spoke of only one Prussian corps being present – the reserve corps may not have been easily discernible.

On completing his reconnaissance Napoleon issued orders for the attack. *He had three possible options:*

(1) To countermand Ney's advance, and order Kellermann's cavalry (who were with Ney) to carry out a holding operation north of Frasnes, while Ney's troops swung on to Marbais, attacking Blücher's right while Napoleon attacked him frontally.

(2) To urge Ney to mount a strong attack on Quatre-Bras, where Wellington had by midday massed a fair number of troops, and having cleared Quatre-Bras to turn towards Marbais and Brye to co-operate with Napoleon.

(3) To take one corps from Ney to roll up Blücher's right, while using the rest of Ney's force, including Kellermann's cavalry, to hold the Anglo-Dutch and cover the Charleroi road.

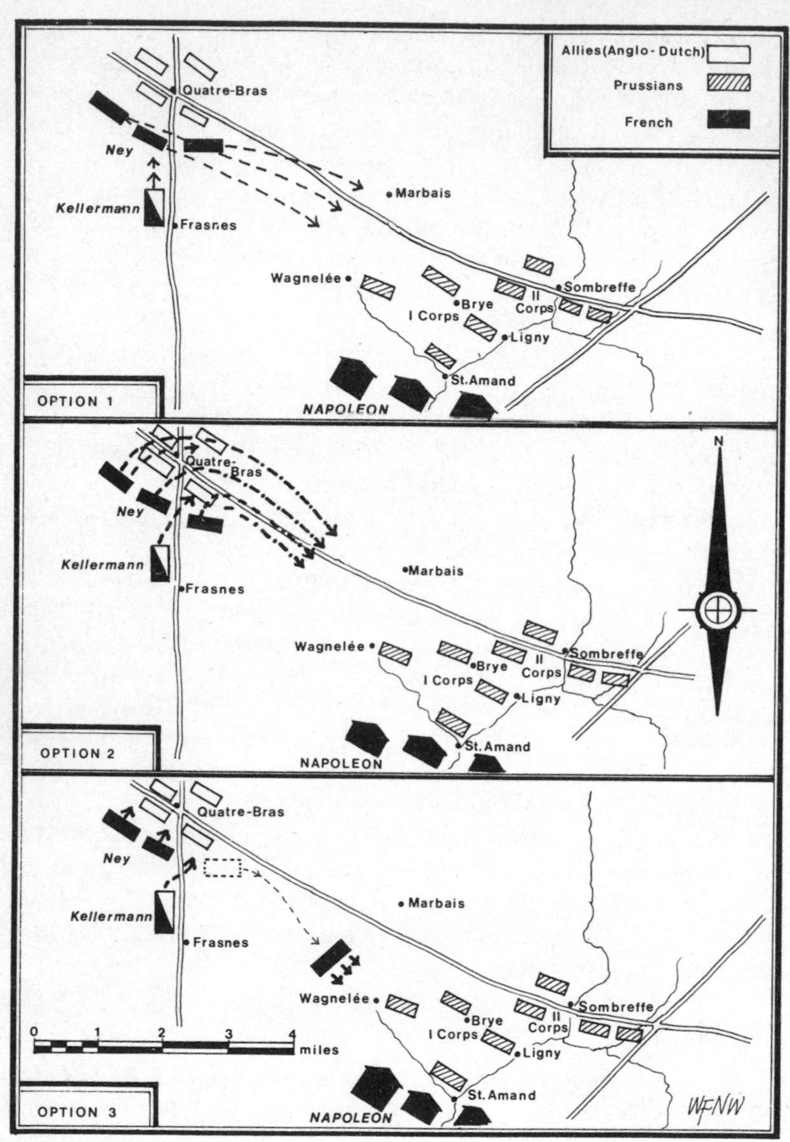

39 *The Waterloo Campaign: 16 June, Ligny and Quatre Bras, Napoleon's options*

Napoleon, like the good general he was, always guarded his line of retreat, and for this reason he dismissed the first option. Furthermore, by the time his reconnaissance was completed he had every reason to believe that Ney was already closing in on Quatre-Bras. The third option would certainly safeguard his line of retreat, but the one he adopted (option 2) would offer the best chance of a double success. For if Ney could throw Wellington's troops out of Quatre-Bras and drive them up the Brussels road, he could then swing in on the main fight against the Prussians, thereby ensuring a complete annihilation of Blücher's men, and allowing the whole force to turn on Wellington afterwards.

The plan was very sound, but its execution was faulty because Ney did not move in to the attack until 2 p.m., and even then he did not realize that his was to be a subsidiary operation, while the Emperor with the right wing and the reserve dealt a decisive blow against the Prussians. A good part of the blame must rest with Napoleon, for the orders Ney received did not make it crystal clear that the main battle was to be fought on the right – but a Marshal of France should have been able to use more intelligence than Ney did on this occasion.

The battle of Ligny began at about 2.30 p.m. on 16 June It was a stern test for the Prussian army, which was a new, untried army, inexperienced and not fully trained. In six hours' bitter fighting, much of it the treacherous and very dangerous house-to-house kind, these young soldiers – especially those of I and II Corps who bore the brunt of it – acquitted themselves most honourably. Napoleon, with his III and IV Corps (Vandamme and Gérard respectively) concentrated on the Prussian right and centre; the Prussian left was hardly bothered. The stubborn defence (houses and even villages changed hands more than once) must have surprised Napoleon if he thought he was operating against only one corps, and around 6 p.m. he ordered the Guard, accompanied by Milhaud's magnificent cuirassiers, to thrust at the very heart of the defence, across the stream and into Ligny.

But before they had time to mount their attack these troops were halted, because word was received of a large force in the left rear of the French, which Napoleon feared might be a part of the Anglo-Dutch army. It was shortly discovered to be Count d'Erlon's Corps. The absurd peregrinations of this Corps, marching between the two wings and in the end never striking a blow, is well known. The fault lay principally with Ney, who for most of the day never grasped his proper role; when in a temper at 6 p.m. he countermanded an order, said to have been from the Emperor, he must have known that it was then too late for d'Erlon to be of use to him. But d'Erlon himself, considering his seniority, was not very clever, for he was on the point of delivering a decisive blow against the Prussian right (his cavalry had already deployed for action) when he decided to about-turn at this late hour and march towards Quatre-Bras.

When d'Erlon's Corps had been identified the Imperial Guard was sent into the fray and stormed its way through the Prussian centre and reserve. When night came the Prussians were badly mauled in the centre but still holding on both flanks, and in control of the Sombreffe–Quatre-Bras road. It was a victory for the French, but an incomplete one. Napoleon had started the fight too late (Gérard's Corps being delayed on the Sambre), and without d'Erlon, or any support from Ney, he had insufficient troops to make a kill. Nevertheless, he thought he had done so, and this was to prove an unfortunate error of judgment.

By the time d'Erlon had rejoined Ney, the Marshal had broken off the engagement at Quatre-Bras; both the French and the Anglo-Dutch held exactly the same positions as they had in the morning. But the French had suffered 4,300 casualties, the Anglo-Dutch 4,700. At the beginning of the battle Ney had had an overwhelming superiority of troops (19,000 infantry, 3,500 cavalry and sixty-four cannon against the Prince of Orange's 8,000 men and sixteen cannon), and between 2 and 4 p.m. the situation for Wellington was critical; however, as more troops arrived they were hastily thrown in to stem the French onslaught. By 6 p.m. the Duke had numerical superiority and was able to go over to the attack, pressing the French back inch by inch. Around 8 p.m. the fighting died down, and the day ended in an indecisive stalemate.

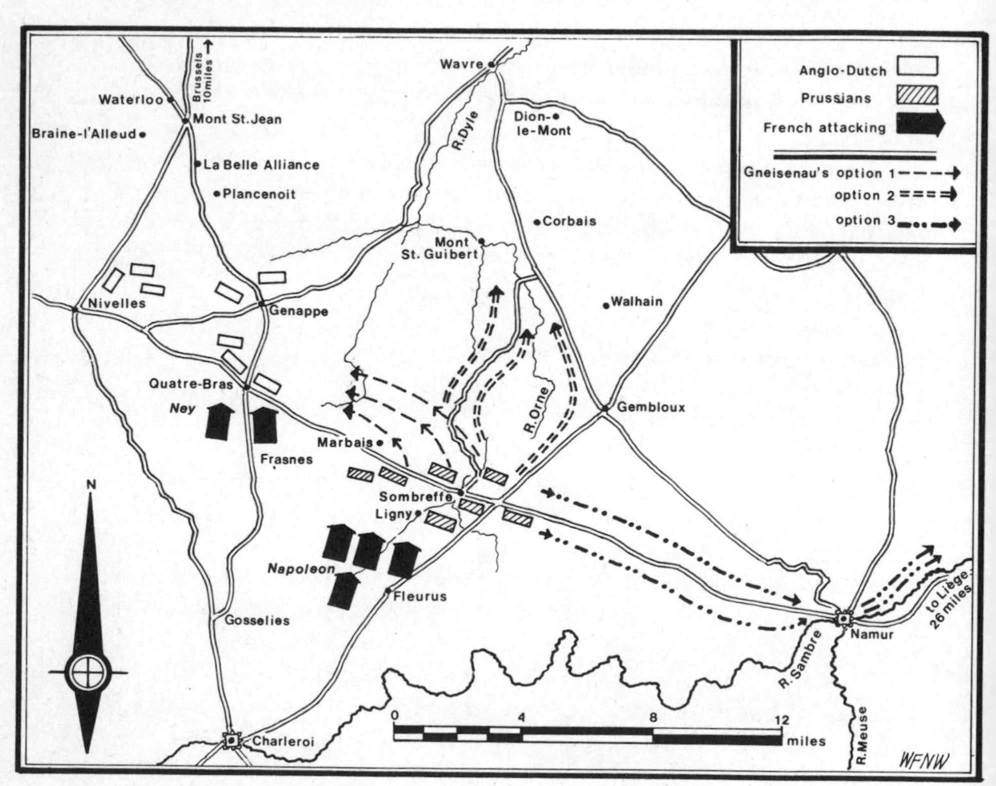

40 The Waterloo Campaign: 16 June, Ligny, Gneisenau's options

During the battle of Ligny that gallant septuagenarian, Field-Marshal Blücher, had been unhorsed while leading a cavalry charge, and his whereabouts was unknown to the Prussian headquarters staff when night put an end to the fighting. This put the onus on his chief-of-staff, General Count von Gneisenau, to decide what action the Prussian army should take. It was abundantly clear to him that the army must disengage in order to regroup before again becoming a cohesive fighting force. The best line of withdrawal, however, was not so clear-cut. Unlike his chief, who had been somewhat sceptical all along of the British determination to co-operate, *Von Gneisenau had three options:*

(1) To swing the army westwards in an attempt to join forces with Wellington, which would not be easy in view of the wide dispersion of the Prussians that night.

(2) To march north along uncharted roads, away from his principal line of retreat, and also from von Bülow's IV Corps, which was still in the Liège area. In so doing his army would be exposed to any follow-up attack of Napoleon's, but this course offered a reasonable chance of keeping in touch with the Anglo-Dutch.

(3) To ensure that the King of Prussia's order to use the army with care was complied with by withdrawing along the natural lifeline through Namur and Liège.

This decision had to be taken by von Gneisenau before any news had been received from Wellington as to the fate of the Anglo-Dutch army, for the Allies were out of touch with each other until early on the morning of the 17th. The first option was almost certainly too hazardous and difficult to accomplish without knowledge of what had happened at Quatre-Bras. The third choice would have been the easiest to take, for von Gneisenau must have realized the importance of safeguarding the army from any further immediate attack before he could regroup. It is therefore much to his credit that with little hesitation he opted for the second. When Blücher was eventually found the old warrior predictably endorsed this decision. It is interesting to note that Napoleon felt almost sure that the Prussians would adopt option 3, and, owing to a part of the Prussian army being in temporary confusion and starting to withdraw down the Namur road in front of Count Pajol's cavalry, Napoleon continued to think this for many hours.

On the morning of 17 June the Allied armies were still some way apart. The Prussians were heading north on the line Gembloux–Wavre, while Wellington, with around 26,000 men and seventy guns, was facing Ney in front of Quatre-Bras. Now that d'Erlon's Corps had rejoined him, Ney outnumbered Wellington, but he did not find out what had happened at Ligny until about 8 a.m. Then he did nothing, for Napoleon was in no hurry to send him orders. At least one distinguished historian of the battle* has exonerated Ney for this inaction on the grounds that if Napoleon had been defeated at Ligny Ney's force would have been inadequate to deal with the Allies, and if Napoleon had been victorious an attack by Ney on the Anglo-Dutch would have encouraged Wellington to withdraw. But Ney's orders on the previous day were to get to grips with Wellington and hang on to him; it is very difficult to carry out a successful withdrawal when the pressure is really on. Ney, with superior numbers, cannot be excused for not advancing to contact, by which time he would have learned the result of Ligny and received orders from Napoleon.

However, on this day Napoleon's grand strategy still held good, and victory was well within his grasp. There were now three options open to him for delivering a knock-out blow. At the time that he had to decide which of the three to adopt he was uncertain of the whereabouts of Blücher's army, but he should have realized – although he did not seem to – that speed was essential. *These were the options:*

(1) He could detach a comparatively small force – say 2,000 cavalry and a division of infantry – to shadow Blücher, while he brought the remainder of the right wing and reserve to join forces with Ney to crush Wellington's troops at Quatre-Bras, and defeat the rest of the Anglo-Dutch army before Blücher (if he was so minded) could influence the battle.

(2) He could follow the Prussians with the right wing, ordering the commander to gain contact with them and neutralize any attempt they might make to form a junction with the Anglo-Dutch. Meanwhile, he himself, with the reserve, could join Ney at Quatre-Bras and defeat Wellington.

(3) He could leave Ney to keep Wellington in check, while he pursued the Prussians with the whole of the right wing and the reserve, to make quite certain that they were unable to play any further part in the campaign.

Napoleon chose option 2. Had it been promptly and properly carried out it could very easily have won him the battle. It had the merit of giving Napoleon time to defeat Wellington either at Quatre-Bras on 17 June, or the next day if he should slip away from Quatre-Bras. But it was necessary for the Prussians to be pursued vigorously and completely neutralized. Thirty-three thousand men – the size of the force detailed for this task – should have been sufficient to keep in check an army that had just suffered a severe defeat.

Both the other two options have their champions, and either might have succeeded. The disadvantage in option 1 (which General Shaw Kennedy, who took part in the battle, thought Napoleon should have adopted) was that if Wellington evaded Ney and the decisive battle against the Anglo-Dutch had to be delayed until the 18th, the

* W. Siborne, *History of the War in France and Belgium*, pp. 160, 161.

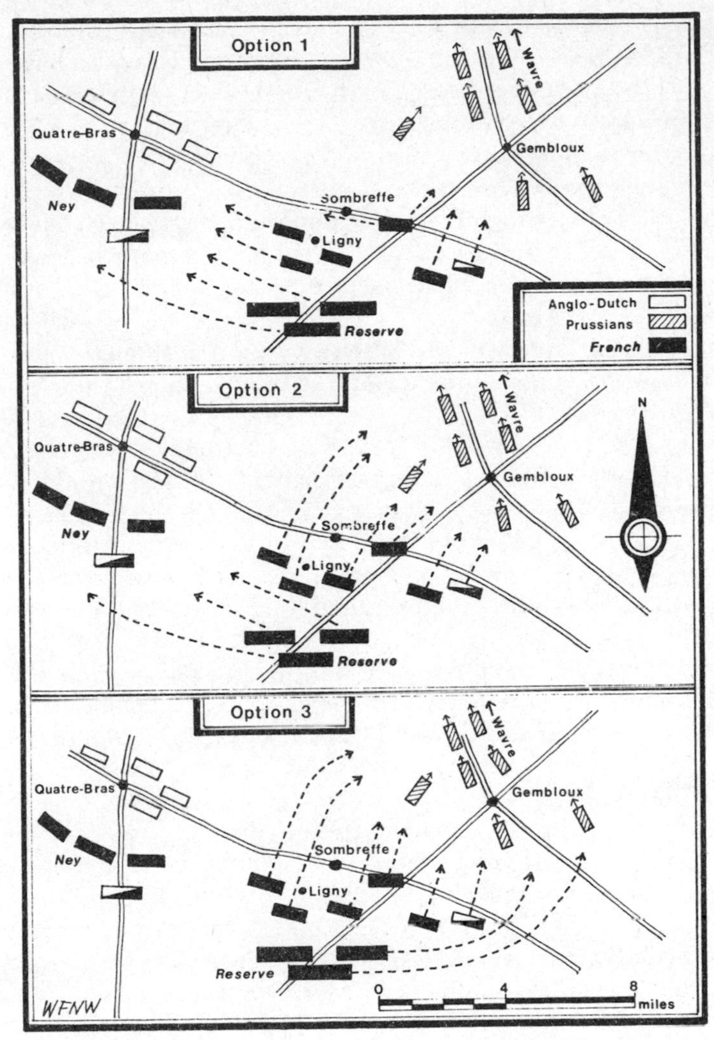

41 The Waterloo Campaign: 17 June, Ligny, Napoleon's options

small force shadowing Blücher would not have been sufficient to keep the Prussians at bay. The third option offered the closest approach to the original strategy, and could have been the best. But in the circumstances it was quite impracticable. The Prussians had to be followed at once and brought to battle speedily, or else Wellington might overwhelm Ney; on the night of the 16th, and even until quite late on the 17th, the French did not know where the Prussians were – and anyway the Emperor was convinced that he had already put them out of play.

Having chosen what was probably the best option open to him in the circumstances, it was unfortunate (from the French point of view) that the speed and resolution necessary to bring the campaign to a successful conclusion were everywhere lacking. To begin with liaison between the two wings of the French army was very bad on the night of 16–17 June; neither Ney nor Napoleon learned the result of the other's battle until quite late on the morning of the 17th, and then Ney – as we have seen – did not feel justified in taking the initiative. Even when he received Napoleon's message at 9 a.m. it was not made completely clear to him what his exact role was to be; nor did a second despatch, dated 'in front of Ligny, at noon', greatly clarify the situation, although it did order him to attack the Anglo-Dutch at Quatre-Bras – but by that time it was too late.

It is easy to blame Ney for irresolution, and for allowing the Anglo-Dutch to slip away, but the fault was principally Napoleon's. The old Napoleon would have been astir by 4 a.m., and clear and concise orders would have been given to both Ney and Grouchy shortly afterwards. As it was, neither of these two commanders received instructions before 9 a.m., and then they were ambiguous and lacked urgency. The Waterloo campaign was lost on the morning of 17 June when the Emperor, through a remarkable and most uncharacteristic dilatoriness, let slip the last golden opportunity to defeat the Allied armies in detail.

Shortly after 10 a.m. Wellington commenced his withdrawal from Quatre-Bras to comply with the parallel movement of the Prussians. It was a brilliantly executed operation, with Lord Uxbridge's cavalry and two horse artillery troops covering the infantry. Not a breath of wind stirred the leaves, the atmosphere was sultry, and deepening storm clouds blotted out the sun, with lurid flashes of lightning presaging the deluge that was soon to drench man and beast. Ney did nothing to hamper the early movements; Napoleon reached Marbais about 1 p.m., and hearing no sound of guns rode on to Quatre-Bras, where he was appalled to find that the Anglo-Dutch had extricated themselves from a most compromising position. At this late hour the Emperor ordered an immediate pursuit, but the combination of spirited rearguard actions and appalling weather enabled the Anglo-Dutch to fall back intact on the preselected position at Mont St Jean.

On the right wing Grouchy was every bit as tardy as Ney. His orders (dictated to, and signed by, Grand-Marshal Bertrand) were to concentrate on Gembloux, seek out Blücher and discover whether he intended to unite with Wellington, safeguard his line of retreat and keep communication with the left wing. But they were far too vague and badly expressed – they even failed to stress Grouchy's most important role, which was to keep Blücher from uniting with Wellington. Nevertheless, Grouchy showed little intelligence and less initiative in attempting to execute them. He advanced his whole force very slowly on Gembloux, not arriving there himself until 7 p.m., and when his leading cavalry commander, Exelman, reported the enemy in the vicinity he took no action to confirm the main Prussian line of withdrawal, but sat in Gembloux that night, and until eight o'clock the next morning, out of touch with both the Prussians and the French.

Dusk was falling when Napoleon reached Le Caillou, a small farmstead on the Charleroi–Brussels road. His army occupied a ridge a little farther on, near La Belle Alliance, and two ridges beyond them the moiled and weary Anglo-Dutch troops, now free from pursuit, sought what shelter a blanket or, for the lucky ones, a bivouac could provide, and waited for the morrow. The rain, persistent, dreary and drenching, turned the ground into a sodden morass.

9 The Waterloo Campaign

18 June 1815

AFTER a cold, drenching night a drying dawn came at last for the three armies. Out of a steely grey sky, rinsed by the night's downpour, the sun shone fitfully and the clouds flew like pennons, heralding nothing worse than occasional showers. Fires were lit along the ridge at Mont St Jean, and the air was filled with the popping of muskets as the troops dried out their weapons.

In the early hours of that morning (18 June) Wellington had received a letter from Blücher promising that one corps would set off at daybreak marching towards him, that that would be followed a little later by a second corps, and that two more would be in readiness to march. Thus assured of this essential co-operation (which was on a larger scale than he had expected), Wellington knew that he could stand and fight.

A master of the defensive battle, he made full tactical use of the topographical features of the Mont St Jean position. On his left the marshy ground of the Lasne – made worse by the recent rain – gave some protection, to his right was the Château of Hougoumont, in the centre of the line was the farmstead of La Haye Sainte, and east of the main Brussels–Charleroi road were the twin farms of Papelotte and Ter-la-Haye. All these buildings could be occupied to form defensive bastions. The dominant feature of the position was the east–west ridge of low heights that crossed the main Brussels *chaussée* at right angles, and almost on the crest of the ridge the *chaussée* was itself crossed by a smaller, unpaved country road leading from Wavre in the east to the Brussels–Nivelles road three-quarters of a mile away to the west. This ridge had minor indentations to the east of the main road, and beyond Hougoumont on Wellington's right the country was open and easily accessible.

Wellington's defence was based on reverse slope tactics, with the greater part of the Anglo-Dutch army drawn up along and immediately to the north of the country road, as it were in a crescent with the right horn curling forward at Hougoumont, and the left on the slightly less protruding farm buildings of Papelotte and Ter-la-Haye. Immediately to the east of the *chaussée* the country road was lined by two banked up holly hedges, and in parts it was deeply sunk. These features, together with the fairly gentle undulation on the north side of the ridge, gave good protection to troops on the reverse slopes and considerably facilitated lateral communication. The defence was so arranged that before the French could get to grips with their enemy they had to survive the accurate marksmanship of the skirmishers, the grape, case and round shot of the artillery and then, when they least expected it, successive volleys from the muskets of those troops on the reverse slope.

Wellington had around 85,000 men and 156 cannon available with which to fight the battle. Although he could see the large mass of troops facing him on the ridge by La Belle Alliance, numbering a little over 70,000 men and 250 cannon, he was uncertain how many troops Grouchy had and where they were. In allotting his troops their defensive positions he had to consider the danger to his comparatively open right flank, and to anticipate whether Napoleon would try to turn it, or launch his entire army in a frontal assault. *He had three options:*

(1) To man the defensive position with his entire force.

(2) To leave a proportion of his army to guard his right flank and line of withdrawal on Ostend.

(3) To leave troops on his extreme right, but call them in if it became clear that there was no outflanking threat.

Wellington chose the second option. It is a decision that has been strongly challenged by historians, and by at least one officer who fought in the battle. The consensus condemns Wellington for keeping a detachment under Prince Frederick of the Netherlands, which included General Colville's 4th British Division (less Colonel Mitchell's Brigade), in the Hal-Tubize area, some ten miles from the battlefield. It is true that out of the total of 17,000 men and thirty cannon there was only one British and one Hanoverian brigade (totalling 5,445 men), but even so when, with hindsight, one knows what 'a near run thing' the battle was, one can only think that Wellington – at about 6 p.m. – must have wished that he had had those men in reserve.

Nevertheless, the decision has found support in some quarters; among modern writers, Jac Weller demolishes criticism.* While admitting that those 17,000 men would have been a real help as matters turned out, he argues that it is easy to be wise after the event, and that at the time Wellington was quite right to guard against Napoleon developing a left hook – and he points out that many French analyists thought afterwards that this would have been Napoleon's best move, † for

*Jac Weller, *Wellington at Waterloo*, 1967, pp. 183, 184.

† Weller quotes General Jomini's *Campaign of Waterloo*, pp. 209–10. But in fact Jomini is here referring to strategic possibilities at the beginning of the campaign; on pp. 156–7 he stresses the tactical advantage of a massive attack on the left and centre of Wellington's line at Mont St Jean – not a wide sweep to the left.

it would have posed a threat to Wellington's communications with Ostend. But once engaged at Mont St Jean Wellington would have found it very difficult to fall back in any other direction than due north, in which case the Hal force would have served little purpose. Weller further states, 'He [Wellington] certainly did not forget this force, which had an extremely important job to do if Napoleon were to manoeuvre to his left. . . .' This is the pith of the matter. But it appears that in the heat of battle Wellington did forget them; by 12 p.m. it must have been very obvious that any threat of a left hook was illusory, and if Prince Frederick's aide (who was with Wellington) had been sent to summon the troops they would have been in the line at the time of the crisis.

In short, it might conceivably have been right in the first instance to leave this force to guard the threat to the right flank, which Wellington believed to be very real, but, as General Shaw Kennedy says, 'Colville should, early on the morning of the 18th, have been ordered to march to Waterloo if he had no information of the advance of the enemy on Hal' – option 3.

The Duke of Wellington's position from flank to flank covered some three and a half miles, but not much more than two miles was held in depth. On the right flank, which Wellington quite correctly considered to be the most vulnerable, he placed General Chassé's 3rd Dutch-Belgian Division at Braine l'Alleud, which is some 1,300 yards to the north-west of Hougoumont. In the event this was outside the battle area, but initially the division was there to secure the extreme right, and to form a bridge between the main army and Prince Frederick's force. At the right end of the actual battle line was General Clinton's 2nd British Division placed *en potence* as a further safeguard against a flank attack, or should that not develop as a tactical reserve.

Lord Hill had charge of the troops to the west of the Nivelles road, the Prince of Orange was entrusted with the centre sector between the Nivelles and Charleroi roads, and General Picton was made responsible for the eastern section of the line. Picton's left (the extreme left of the Allied line) was held by General Vivian's 6th Cavalry Brigade. The rest of the cavalry was massed behind the infantry (see battle plan).

It was immediately apparent to Wellington's perceptive eye that Hougoumont was a key position on the right, and the château and grounds were held by troops of the Brigade of Guards, Prince Bernhard of Saxe-Weimer's Nassauers, and a company of Hanoverian riflemen. An equally important bastion of the defence, the key to any frontal attack, was La Haye Sainte, but Wellington did not seem at first to recognize this fact, for no steps were taken to strengthen the buildings and the garrison consisted only of Major Baring's 2nd Light Battalion of Baron Ompteda's King's German Legion, some 360 men. To quote Shaw Kennedy again, 'The most important mistake which the Duke of

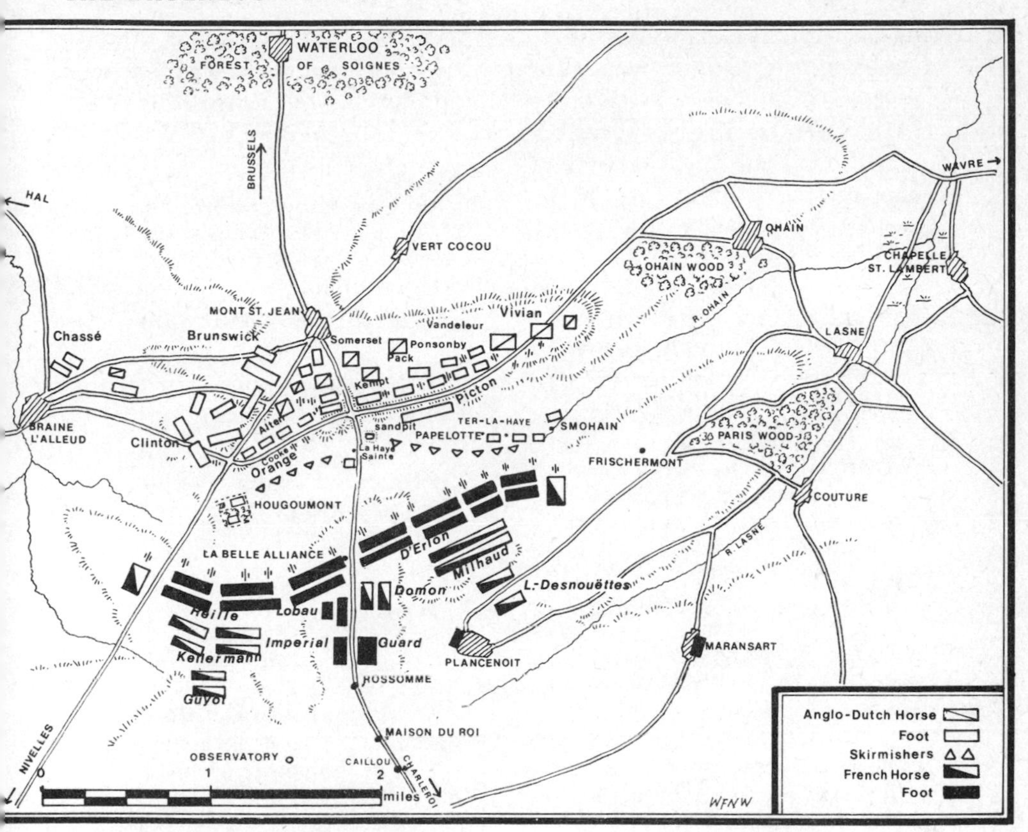

42 The Waterloo Campaign: 18 June, Waterloo, the armies deployed

Wellington committed as to the actual fighting of the battle of Waterloo, was his overlooking the vast importance of retaining possession, at any cost, of the farm and enclosures of La Haye Sainte.'*

Across the undulating plain, less than a mile to the north, Napoleon drew up his army for battle. Trumpets and drums could be clearly heard, while shouts of '*Vive l'Empereur*' added a thrilling accompaniment. Cuirasses glinted in the sun, plumed helmets, bearskin caps, jackets of scarlet faced with greens and purples, and crowned with gold epaulettes, caught the eye. A marvellous mass of colour cloaked a magnificent fighting machine.

The army deployed in three lines. In the front line were Reille's (I) and d'Erlon's (II) Corps stretching from the Nivelles road in the west across the Charleroi road as far as Frischermont. Two hundred yards behind the front line came twenty-four squadrons of Kellermann's cuirassiers, behind them and on their right Lobau's VI Corps, and behind them

*General Sir James Shaw Kennedy, *Notes on the Battle of Waterloo*, 1865, p. 174.

again, in tight formation, a further twenty-four squadrons of Milhaud's cuirassiers. The third (reserve) line comprised the 'Invincibles', headed by twenty squadrons each of Grenadier Guards and Guard Chasseurs, then came the Young, Middle and Old Guard – each of eight infantry battalions, four on either side of the Charleroi road, drawn up six lines deep (for battle plan, see p. 149). *In planning his attack Napoleon had five options:*

(1) He could attempt to smash Wellington's centre with a tremendous punch, at the same time bringing pressure on the flanks.

(2) He could decide on a frontal attack, but in two or more consecutive phases against differing points of Wellington's line.

(3) He could deliver a left hook at Wellington's right flank.

(4) He could launch a major offensive against the left of the Anglo-Dutch line, and roll it away from Blücher.

(5) He could attack the left of the line, and at the same time assail the centre.

When Napoleon was planning his attack he was of the opinion that the Prussian army was too far from the field, and too hard smitten, to play any significant part on the 18th. Nevertheless, he still had to maintain the original strategy of keeping the Allied armies apart, and he had to defeat Wellington decisively and quickly. Some attention has already been paid to option 3; one of the reasons that Napoleon would have been unlikely to adopt it was because it would have tended (if successful) to roll back the Anglo-Dutch on to the Prussians. Conversely option 4 would have had the desired effect of prizing the Allies apart, and giving Napoleon direct contact with Grouchy's force. But it had a double disadvantage: by striking with the main part of the army as far east as Frischermont the line of retreat could have been cut, and the terrain in the Frischermont–Chapelle St Lambert area was unfavourable for a massive attack. Nevertheless, there are those who thought it might have been the best plan, and that by choosing to deliver a purely frontal attack Napoleon sensibly diminished his chances of victory.

Having decided to deliver a central hammer-blow, intended to shatter and drive back Wellington's whole line in headlong retreat through the Forêt de Soignes, Napoleon had the choice of options 1, 2 and 5. To develop the main thrust at the centre alone (option 1) offered fewer chances of quick success than did option 5 – a manoeuvre that had been used successfully at Wagram. Accordingly, orders issued from Rossomme at 11 a.m. made provision for d'Erlon's I Corps to attack on a front from just west of the Charleroi road to the left of Wellington's line, while Reille's II Corps facing Wellington's right centre was to 'advance so as to keep at the level of Count d'Erlon'. As it was not anticipated that the attack would go in before 1 p.m., the twelve-pounders of II and VI Corps were to group with those of I Corps (ninety-four guns) in a preliminary bombardment that would shower shot and shell of a most intensive and destructive kind along a two-mile stretch of Wellington's front.

The attack thus ordered would have had every chance of success had it been executed according to plan. But in the event Napoleon started the

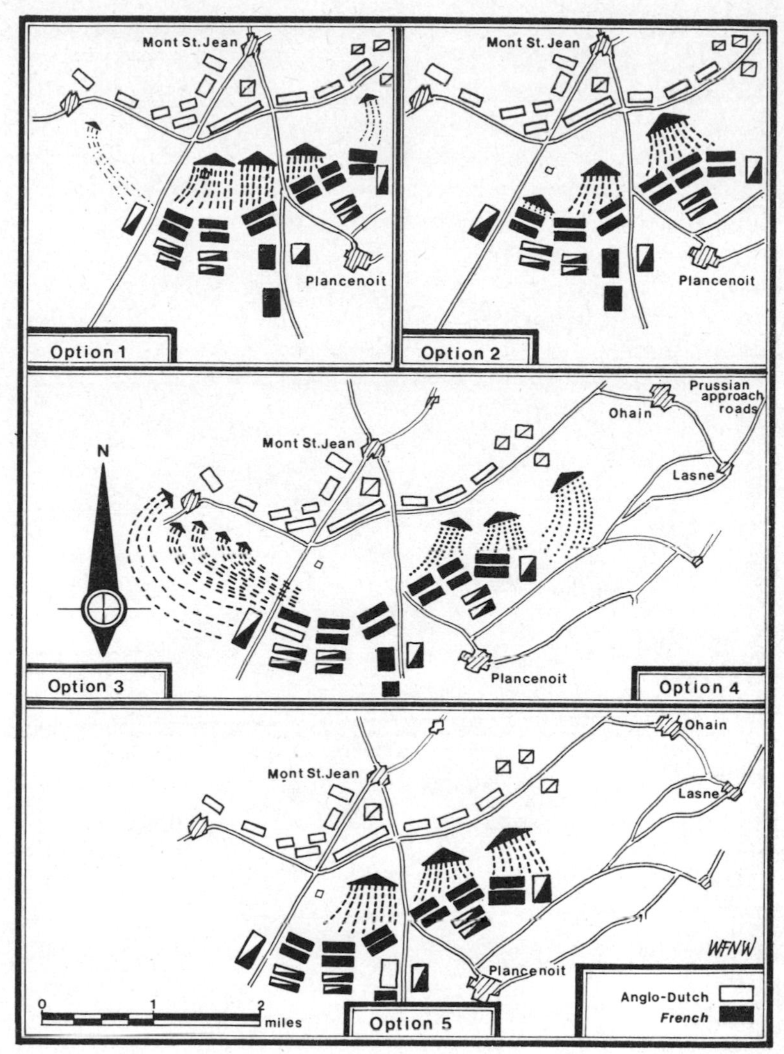

43 The Waterloo Campaign: 18 June, Waterloo, Napoleon's options

battle with an isolated attack against Hougoumont – which Prince Jerome persisted with longer than was intended – and the day thereafter developed into a series of isolated attacks (more in keeping with option 2 than the intended option 5), and isolated attacks do not often succeed. Moreover, when d'Erlon's main thrust did go in at about 1.45 p.m., insufficient attention was paid to the extreme left of the Anglo-Dutch line, where a breakthrough might have been made.

While the two principal contenders, whose blood-letting on the morrow would decide the destiny of Europe, got what sleep they could on the sodden fields around Mont St Jean and La Belle Alliance, the third great army, which was to play such a decisive role, was at last assembling in the area of Wavre some ten miles to the east. All four corps were now concentrated; von Bülow's IV Corps, which had not fought at Ligny, spent the night of the 17th at Dion-le-Mont to the south-east of Wavre, and Pirch's II Corps was also east of the Dyle at Aisémont, but I and III Corps were on the left bank of the river at Bierges and La Bavette respectively. Of these, I Corps had borne the brunt of the fight at Ligny, where III Corps had not been heavily engaged.

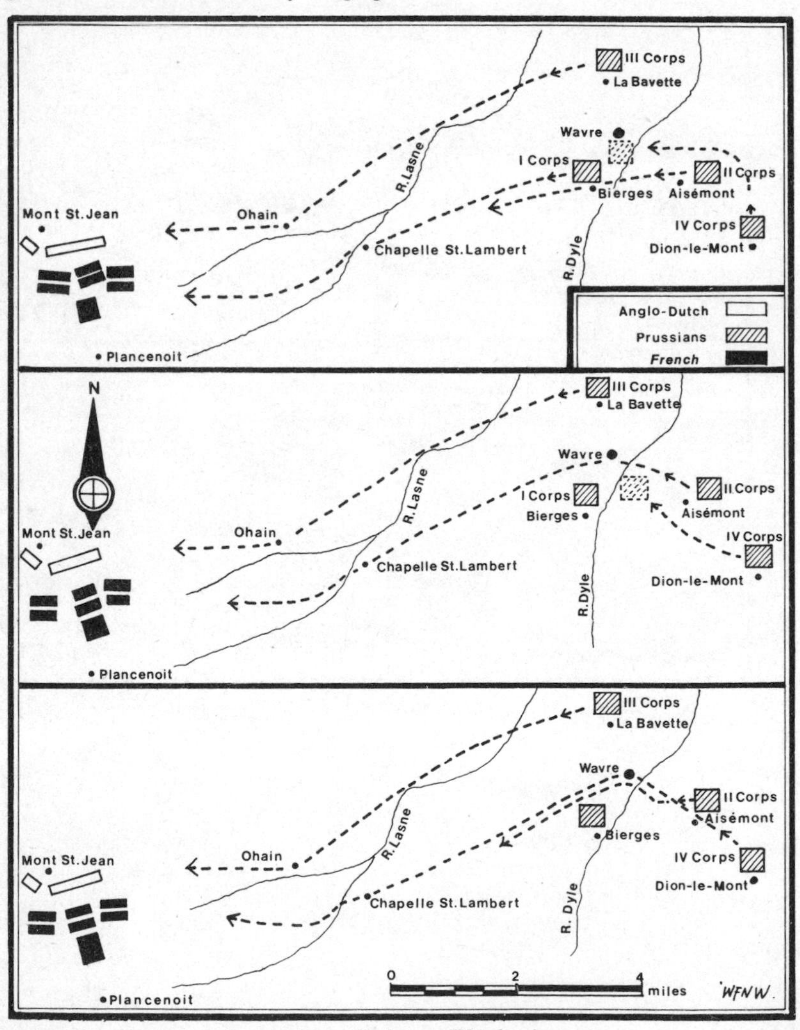

44 The Waterloo Campaign: 18 June, Blücher's options

As we have seen, that night Blücher despatched a letter to Wellington promising support the next day. He and his chief-of-staff, von Gneisenau, now had a difficult problem in deciding their order of march so as to ensure that their troops could be of maximum assistance. *In making their decision there were three options open to them:*

(1) To make the fullest use of the two corps already on the left bank of the Dyle, by sending one at dawn via Ohain to join with Wellington, and the other to march via Chapelle St Lambert and the Lasne valley to attack the French in flank. Pirch (II Corps) could back up the St Lambert corps, and von Bülow (IV Corps) – the farthest from the battlefield – could remain to cover Wavre.

(2) To leave two corps to cover Wavre (probably I, which had had the most fighting, and IV, because of its distance from the battlefield) until it was certain that no French force was advancing up the left bank of the Dyle, and send the remaining two as for option 1, i.e. one via Ohain, the other via St Lambert.

(3) To spearhead the attack with the only completely fresh corps (von Bülow's), ordering it to start marching at daybreak, cross the Dyle, pass through the bottleneck of Wavre and make for St Lambert and the right flank of the French. II Corps could then march as back-up to IV Corps. Of the two corps already across the Dyle one could march via Ohain for Wellington, and the other be kept to cover Wavre.

When considering which of these options the Prussian generals should have favoured, it must be remembered that both Blücher and von Gneisenau were determined to defeat Napoleon, and both knew that this could be done only in conjunction with Wellington's army. But von Gneisenau was cautious; he was not anxious to risk all until he was sure that Wellington had every intention of standing firm, for if the Prussians marched across the very difficult country that stretched from the deep valley of the Dyle through Chapelle St Lambert and the Paris Wood, only to find that Wellington had withdrawn farther north, or been defeated, their army would perish.

This undoubtedly coloured von Gneisenau's thinking (and he it was who did most of the thinking), for either of the first two options would have ensured that Wellington received help long before those anxious evening hours. But it was decided to adopt option 3, and von Bülow was told to proceed with great care, covering his force behind St Lambert until he was sure that battle had been joined at Mont St Jean. Blücher had always intended to spearhead his advance with von Bülow's corps (indeed he mentioned that fact in his letter to Wellington), but in so doing there ensued a further delay. This corps was the farthest from the battlefield, and although it started at about 5 a.m. it had to cross the Dyle and pass through Wavre, crossing the advance of Zeiten (who had been detailed to march via Ohain) and of Pirch, whose corps were delayed until von Bülow cleared them. To make matters worse, when von Bülow's Corps was partially through Wavre a fire broke out and quickly spread, causing the main body of the corps to suspend their march for a while.

By about midday, therefore, the

Prussian IV Corps had reached the St Lambert area, but I and III Corps were only just leaving Bierges and La Bavette respectively. One can perhaps understand the extreme caution displayed in this decision, but from the purely tactical point of view it was of course entirely wrong.

Napoleon was ready to start the battle early, but the fact that his troops were scattered and had to be fed made it difficult to launch any attack until around 9 a.m. However, there then occurred an inexplicable and most damaging delay. Inexplicable, because Napoleon in Book IX from St Helena gave only the condition of the ground as his reason for postponing operations until 11.30. But in that short space of time a partially obscured sun could not possibly have made much difference to the going, and those lost hours were important. It seems more likely that the Emperor was over-confident of his ability to brush the Anglo-Dutch off Mont St Jean in an afternoon, and that he was still convinced there would be no Prussian presence. Therefore the order of battle, so perfectly executed, and the royal review could be proceeded with at leisure.

It was around 11.30 a.m. when Reille's gunners opened up a concentrated barrage against the right of Wellington's line. Reille had been ordered to carry out a diversionary probing action against Hougoumont, and he entrusted the attack to Prince Jerome's division, supported by Piré's lancers. Immediately a fierce engagement ensued when the leading French troops tried to drive the Nassauers and Hanoverians from the wood in front of the Château. Gradually the Germans were driven out of the wood, and a counter-attack from the orchard by the light companies of the Brigade of Guards under Lord Saltoun was only partially successful. By midday the situation on the defenders' right flank had become critical, and the 3rd Guards in the lane were being forced back towards the Great Gate of the Château. There followed a tremendous struggle for possession of the Gate; after tense minutes of hand-to-hand grapple, and great heroism on both sides, the gate was firmly closed, and the few Frenchmen who had fought their way through lay dead.

By 1.30 the pressure was to some extent lifted. Two thousand Guardsmen and Hanoverians had kept 10,000 of some of Napoleon's best troops from the prize they had striven to gain. The French still occupied the orchard, but the commander of the 2nd Guards Brigade was determined to evict them. But now the great thrust against the Anglo-Dutch centre, which Ney had been preparing while Hougoumont was under attack, was about to begin.

However, at about 1 p.m. an incident had occurred that was to delay

Ney's launching of this attack. Napoleon and his staff had been anxiously watching what appeared to be a large body of troops away to the north-east of the heights of St Lambert. Identification through glasses was not possible, so a patrol was ordered to investigate, but before it could complete its mission a captured Prussian officer was brought in who was willing to inform the Emperor that the troops belonged to von Bülow's IV Corps, and that Blücher's entire army had spent the previous night in the Wavre area. The presence of 30,000 fresh troops on his immediate right, with the possibility of more to come, and no exact news of Grouchy, gave a new and most unpleasant slant to the problem confronting the French army. *On receipt of this information Napoleon had three options:*

(1) To detach a cavalry screen of, say, one division to watch and delay the Prussians, while using his entire strength in a decisive blow to breach the Anglo-Dutch line before any Prussian intervention could occur.

(2) To detach a stronger holding force of one corps and two cavalry divisions, with orders to engage when Grouchy attacked the Prussian corps from the east.

(3) To seize the opportunity of occupying the Paris Wood with a division of infantry and two divisions of cavalry in support, in order to obstruct von Bülow's Corps when debouching from the very steep and narrow defile of St Lambert, thus diverting it away from the immediate battle.

Napoleon chose the second option, and it could be said that it was the worst of the three. By so doing he lost the use of more than 10,000 infantry and cavalry, who instead of marking time on the right flank for three hours could have been used, together with the Guard, for an offensive *à l'outrance* on the Anglo-Dutch line, which might well have succeeded. It has to be admitted, however, that at the time Napoleon did not know the exact Prussian position, or that it would be between 4 and 5 p.m. before their threat became serious. To seize the Paris Wood might have been his best move, for if this had been done in sufficient strength and under a resolute commander, such as Lobau, the Prussians must have been driven towards Zeiten, who did not enter the fray until 7 p.m., by which time, without the necessity of having to fight round Plancenoit, Napoleon might have won the battle.

There was a possible fourth option, which everyone knows Napoleon did not adopt, and almost certainly did not for one moment contemplate – but it was a feasible operation. With Grouchy now unlikely to influence the main battle, and with the certain prospect of being faced with an enemy on two fronts, Napoleon could have broken off the engagement. Ney with, say, 40,000 men could have held the Anglo-Dutch, who were not geared for an immediate offensive, while Napoleon withdrew to his right, and ordered Grouchy to join him on the Dyle. Ney, falling back on Genappe, would have been reunited with the main army the next day. The strategic plan of keeping the Allied

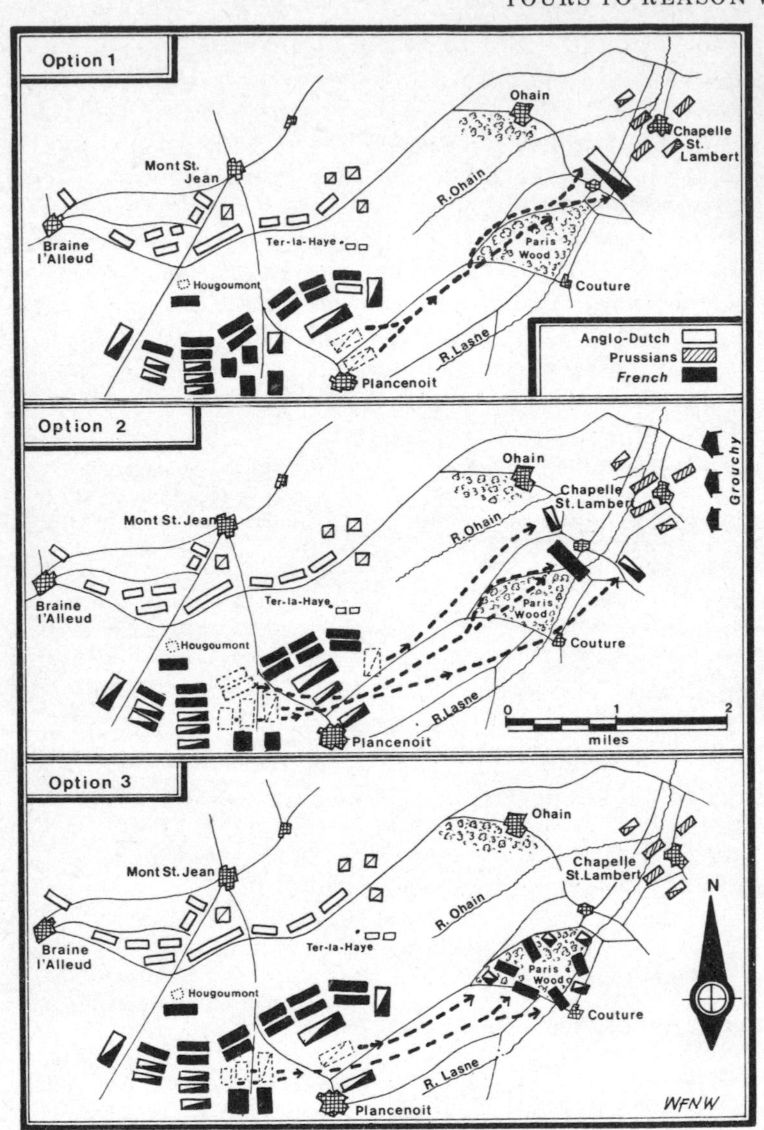

45 The Waterloo Campaign: 18 June, Waterloo 1 p.m., Napoleon's options

armies apart would then have been in shreds (but that was already the case), and thereafter Napoleon would have been numerically inferior by some 35,000 men (see effective strength, Appendix II). However, he would still have been at the head of a powerful army, instead or risking all when by 1.30 p.m. the dice were clearly loaded against him; but Napoleon would never have reached the pinnacle of Europe had he been a general given to drawing back.

With Napoleon's decision taken, Domon's and Subervie's cavalry divisions were immediately sent to observe and guard his right flank; while Lobau's VI Corps took up an intermediary position, from where it could contain the Prussians between Ohain and Lasne. Napoleon no longer delayed Ney's grand advance against the left centre of the Anglo-Dutch line. Soon after 1 p.m. the seventy-eight guns, which were to support the 17,000 or so men of General d'Erlon's I Corps, thundered into action, and some three-quarters of an hour later the corps advanced to the sound of the drums beating the *pas de charge*. This great mass of infantry, partially concealed by the thick smoke that now filled the valley, advanced in four divisions echeloned from just west of the Charleroi road to the left flank of the Anglo-Dutch line. On d'Erlon's left General Bachelu's division of II Corps was in support, while Prince Jerome renewed his attack on Hougoumont.

As the massive onslaught rolled forward it proved to be a most desperate affair. Before the hordes of fiercely determined Frenchmen could close with the defenders, whole battalions had been shorn away under the flail of Wellington's guns; gaps appeared in the closely packed ranks, but these were quickly closed and soon La Haye Sainte and the sandpit on the other side of the road were enveloped. The 95th companies in the sandpit were outflanked and forced to fall back on their battalion, but Baring's men, although they lost the orchard, managed to retain the farm buildings. General Bylandt's Dutch-Belgian brigade, whose men lacked that inner courage which stems from long tradition, beat a hasty retreat to the rear, and the French gained the ridge. But Picton's thin line closed up, and remained unshaken at this very critical time. Pack's and Kempt's brigades of this division shivered like standing corn in the wind – but remained standing, and poured a hail of lead into the oncoming French. The gallant Picton ordered a charge and fell with a bullet through the head, and across the *chaussée* the situation was just as desperate, for such were their numbers that nothing seemed able to check the enemy. But help was at hand.

No one is quite certain who ordered the heavy cavalry to charge at this moment, but what mattered is that Sir William Ponsonby's Union Brigade, together with squadrons from the Household Brigade, leapt into action and – at great cost – saved a most dangerous situation. The going was not easy, for the horses had to scramble in and out of the sunken road and negotiate the holly hedge with piles of bodies all around. The charge soon became something of a shambles, but with the advantage of the slope the heavy British horses soon swept away the enemy cuirassiers, and in the terrible mêlée of hack and thrust, turn and trample, the French infantry, taken completely by surprise, and in a formation totally unsuitable to resisting cavalry, disengaged themselves and retreated down the hill. The charge was carried too far, and ended in disaster. It tore the entrails out of the heavy cavalry, but it saved the line, which Wellington now set about reorganizing and strengthening.

It is time to look again at Marshal Grouchy, who had left Gembloux at about 8 a.m., which was at least three hours later than he should have done – although even so he would not have caught the main Prussian army along the route he took. For some reason, when he did move, Grouchy manoeuvred to his right, marching, with two corps head to tail along the same road, via Sart-à-Walhain. In his despatch to Napoleon the previous night he had correctly surmised that a junction between Wellington and Blücher was likely; he therefore intended marching through Corbais to Wavre. Even that would have been too far to the right in view of his instructions to keep contact. Had he advanced at dawn via Mont St Guibert and Mousty, he would have had every chance of intercepting Bülow's corps.

Grouchy arrived at Sart-à-Walhain at 10 a.m., and an hour later sent a despatch to Napoleon saying that he had information that the main body of the Prussians had moved out of Wavre and were seven miles to the north-east, and requesting orders. Shortly afterwards he sat down to lunch, from which meal he was rudely disturbed by Gérard and his chief-of-staff, who reported the distant, but very definite sound of cannonade.

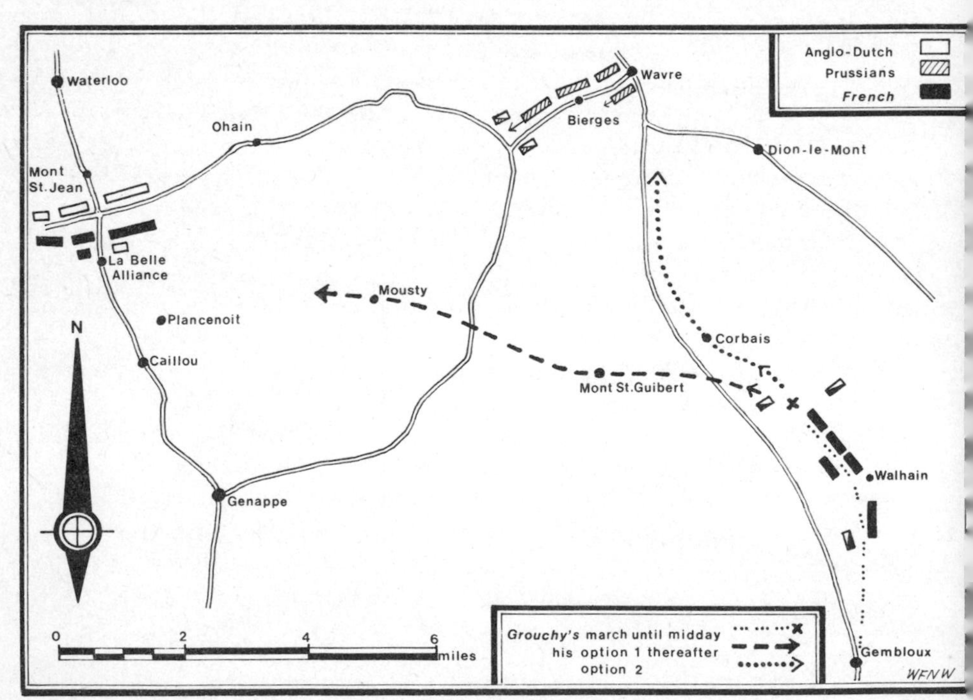

46 The Waterloo Campaign: 18 June, midday, Grouchy's options

Grouchy insisted that it was only a rearguard action, but as the rumbling rolled up to them, re-echoed across the fields and intensified, it clearly denoted that a major engagement was taking place about eight or nine miles away. *Grouchy was faced with two options:*

(1) He could immediately march to the sound of the guns.

(2) He could carry out the last instructions received from Napoleon, which – written from Le Caillou at 10 a.m. on the 18th – read: '. . . you should direct your movements on Wavre so as to come nearer to us, to establish operational and liaison contact with us . . .'

Gérard begged the Marshal to adopt option 1, and his urgent supplications were reinforced when Exelman's aide-de-camp arrived to say that reports indicated that the Prussian army was closing on Wellington. Matters almost came to blows when Gérard asked leave to take his corps towards the battlefield, and Grouchy – quite rightly – refused to divide his force and risk both parts being crushed by superior numbers. And so he continued towards Wavre, arriving there at 4 p.m., where he fought an independent action against Thielemann's corps of 16,000 men and thirty-six guns. A desperate fight lasted until 11 p.m., when the French at last succeeded in opening the road to Mont St Jean – but by then the only living Frenchmen on the ridge were prisoners of war.

A general in Grouchy's position can seldom be wrong if he marches to the aid of his superior commander when he realizes that the latter is hotly engaged and his own precise role is uncertain. But in his argument with Gérard, Grouchy pointed out that had Napoleon wanted him at La Belle Alliance, he would surely have said so in his despatch written at Le Caillou when he knew the extent of the opposition. This is a telling point in favour of Grouchy's decision, and clearly shows that Napoleon never expected the Prussians to reach the battlefield. Furthermore, Napoleon compounded this error when at 1.30 p.m., while the battle was raging, he ordered Soult to despatch another directove to Grouchy approving his march upon Wavre.

It is a matter for conjecture whether Gérard and Vandamme could have turned defeat into victory had they arrived on the battlefield by 6.30 p.m., which would have been possible, but there is no doubt that, whereas Grouchy may have violated a principle in not marching to the sound of the guns, the major mistake was Napoleon's in not giving him positive orders.

By 3.30 p.m. d'Erlon had reassembled some of his battalions after their gruelling attack, which had been halted by the charge of the British heavy cavalry, and now Napoleon ordered Ney to resume the offensive on La Haye Sainte. The French twelve-pounders once more pulsated furiously, and under cover of this destructive fire Ney, at the head of Quiot's brigade, led the new onslaught. But the attack was not pressed

home with sufficient resolution to dislodge the garrison, which had recently been reinforced.

Meanwhile, batteries were being brought up to the west of the Charleroi road, and the terrible storm of iron that hurtled through the air and thudded around Wellington's right centre forewarned the Duke that yet another full-scale attack would shortly begin. So much was obvious, but the form it took astounded every experienced commander in the Anglo-Dutch army. It would not be fair to say that Napoleon had an option to attack with cavalry or with cavalry supported by infantry, for from his observation post near Rossomme he could not see what Ney was doing in a hollow part of the valley. And Ney, thinking that Wellington's order to his infantry to retire a short distance to minimize the effect of shot and shell signalled a partial withdrawal of the whole line, felt he could accomplish a speedy victory through the use of unsupported heavy cavalry.

The magnificent French cavalry charges, carried out by a total of some 15,000 horsemen at times riding boot to boot in a seemingly solid phalanx, led by the intrepid Ney, who in the course of the seven or so separate charges had four horses shot under him, is one of the great epics of this majestic, but very terrible battle. At the command 'prepare to meet cavalry' Wellington's infantry immediately formed square, and before the mass of cavalry could come to grips they had to ride through a hail of whistling ball; but sheer weight of numbers bore them on

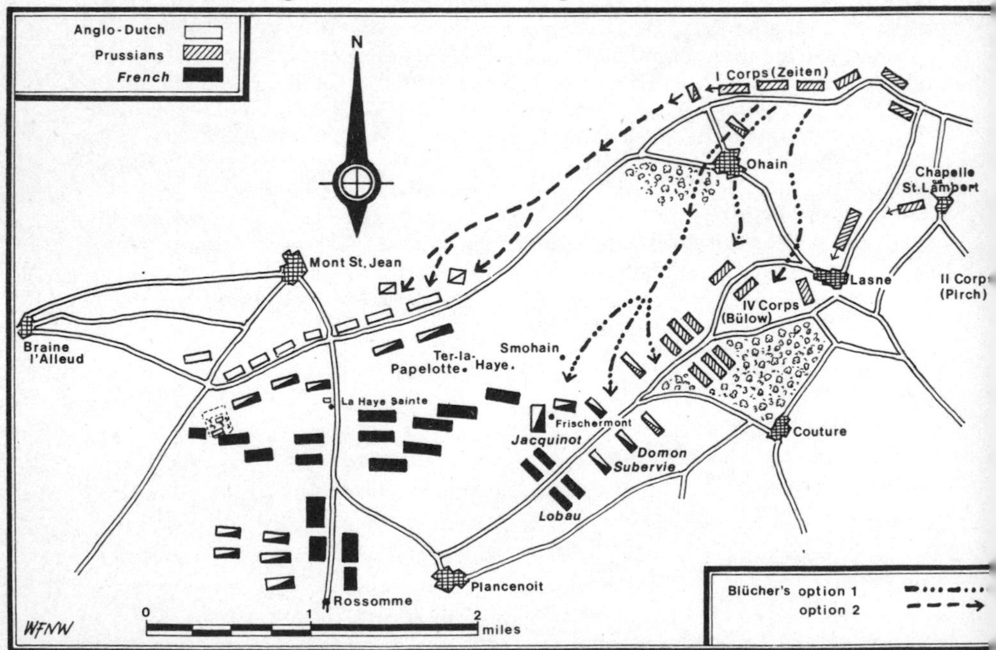

47 *The Waterloo Campaign: 18 June, 4.30 p.m. Blücher's options*

relentlessly towards the waiting squares, which before Ney eventually owned defeat had become alarmingly thin. Nor were Lord Uxbridge's men without a part to play, for a furious cavalry battle raged between the squares. During the two hours that the inexhaustible Ney led his warriors up the slope in charge and counter-charge the scene of destruction, and the noise, were appalling; dying and mutilated soldiers lay everywhere, and above the rattle of musketry and crash of cannon horses reared and screamed.

Towards the end Ney did what he should have done earlier; he attacked with infantry and some light field pieces in support. But this combined offensive had been left too late, for the splendid French horsemen were no longer capable of any further sustained effort. Shattered and ruined in senseless unsupported attacks the survivors rode down the slope for the last time. It was nearly 6 p.m., but there was to be no pause for the weary and hard-tried Anglo-Dutch. The French artillery was sounding the overture for the one positive success that came to Napoleon during the closing hours of the battle: the capture of La Haye Sainte.

But before this took place an interesting situation had developed on Napoleon's right flank, where at 4.30 p.m. the Prussian IV Corps was about to launch its attack on the French cavalry and Lobau's VI Corps, who were in position to defend Plancenoit. Blücher did not wish to attack before his full force was assembled, and, owing to the delay earlier in the day caused by the confused order of march, II Corps had not appeared, and von Bülow was short of two brigades. But the position on Wellington's front seemed so critical to Prussian observers east of Frischermont that Blücher determined on immediate action.

He ordered IV Corps to move to its left in the direction of La Belle Alliance, and a stern fight soon developed with Lobau's VI Corps on the heights south of Frischermont and on the heights of Plancenoit. It was quickly apparent to Blücher that Plancenoit was the key to his contribution to the final outcome of the battle. He had to take and hold the village preparatory to a thrust at Napoleon's right flank, but he felt he needed more troops. *He hoped II Corps would soon be up with him, and he had an option in the use of I Corps:*

(1) He could order Zeiten to change direction and march to his assistance, so that he might bite more deeply into Napoleon's flank.	**(2) He could leave Zeiten to continue his march to join Wellington's left flank.**

Blücher considered that option 1 would best serve the Allied cause, and he therefore despatched an order to Zeiten to turn south and cross the Smohain brook in the direction of Frischermont, in order to join the attack on Plancenoit and Napoleon's right.

The order to march south and join Blücher arrived with I Corps when its advanced units were west of Ohain, and the 2nd Brigade leading the main body was at or near that place. The time was around 6 p.m. Zeiten's chief-of-staff, Colonel von Reiche, had earlier been consulting with General von Müffling (the Prussian liaison officer on Wellington's staff) on the arrangements for committing Zeiten's leading troops when they reached the left of the Anglo-Dutch line, which would be very soon. On receiving Blücher's orders von Reiche directed the leading Prussian troops to retrace their steps to the Frischermont road and there await Zeiten's orders. Meanwhile Müffling had ridden over to Zeiten, and found the latter faced with a very difficult decision. *These were his options:*

(1) Should he obey the army commander's categorical order to march to his aid, when it was obvious that the English left was in grave peril, with Papelotte and Ter-La-Haye just lost.

(2) Should he disregard those orders and continue his march to join the Anglo-Dutch left?

General Müffling undoubtedly performed a very great service to Wellington and the Allied cause when he made the strongest representations to Zeiten that unless he joined the fight on Wellington's left the battle would undoubtedly be lost. To reinforce this plea was the visible plight of the Anglo-Dutch line, and Zeiten was persuaded that his duty lay in adhering to his original instructions and hastening to the assistance of Wellington's left. It must have been a difficult decision to take, but even if events had turned out other than they did it was certainly the correct one.

Despite his fears, Blücher was able to make a decisive contribution to the battle without having the additional weight of Zeiten's corps. The fighting for Placenoit was very bitter, and the village changed hands three times; the Prussians took it from Lobau's men at 6 p.m., but were driven out by the Young Guard, only to retake it from them at about the same time as La Haye Sainte fell. Napoleon then threw in two battalions of his best troops, the Old Guard, to take it once more for the French, but to stem the Prussian tide he had had to commit a large portion of his much needed reserve. And by the time Napoleon had brought up the Guard for his final throw against the Anglo-Dutch, the Prussian I, II and IV Corps had joined forces and were ready to inflict the last overwhelming disaster that was to befall the French army as it endeavoured to disengage and retreat.

While Blücher's Prussians were closing in on Napoleon's right flank, and the fighting round Plancenoit was at its height, Wellington's line was experiencing its most critical moments. After the massive infantry and cavalry attacks had been thrown back amid such fearful carnage, it must have been almost unbearable for the severely mauled Anglo-Dutch troops to be faced yet again by a new flood of strength welling from depths seemingly plumbed to the uttermost. But shortly after 6 p.m. infantry and cavalry, preceded by a whole swarm of tirailleurs, advanced on the key position of La Haye Sainte.

Baring's men had been reinforced, but his constant pleas for ammunition had gone unanswered, and his gallant Germans knew very well that their hour had come. They told their commander, 'No man will desert you – we will fight and die with you.' And so it was, until the last pathetic remnants were ordered to retire. The loss of La Haye Sainte gave the French a considerable advantage, which Ney was not slow to exploit. He brought up a battery to the garden which bombarded at close range the brigades of Kempt and Lambert; his troops occupied the sandpit, and when two battalions of the King's German Legion tried to drive the French infantry back they were cut to pieces by the supporting cuirassiers. Alten's 3rd Division had almost reached the point of disintegration, and farther to the right the situation was just as critical. But the Duke remained quite calm, taking steps to shore up the line at its weakest points, and by his example encouraging all to take courage and stand firm.

Nevertheless, Ney was right when he sent a desperate message to Napoleon asking for reinforcements. He had carefully observed how stretched and battered Wellington's line was; he felt sure that with one big push now he would be through, but his troops had fought themselves to a standstill and more men were needed. *Napoleon, on receiving this urgent appeal for additional troops, had a straightforward choice:*

(1) To adhere to Ney's request.

Napoleon never hesitated. '*Des troups! Où voulez-vous que j'en prenne? Voulez-vous que j'en fasse?*' he exclaimed to Colonel Heymès, Ney's messenger. Certainly there was no way he could have made fresh troops, but he could have taken them from what was left of the reserve. But in spite of committing the Young Guard, most of his reserve cavalry and two battalions of the Old Guard he was still hard pressed on his right, and he was not prepared to uncover completely.

It was almost certainly a wrong decision; Napoleon still had eleven fresh battalions of the Guard available.

(2) To refuse it.

In a little over half an hour seven, or perhaps more, of those battalions (say 4,500 of his finest troops) were to march into battle for the last time, but by then it was too late, for Wellington had had time to reinforce his crumbling line, and was already repulsing the last of Ney's troops on the ridge. Moreover, Zeiten's corps was arriving on the field. With the Prussians closing in, and Grouchy still far away, victory over Wellington could at best be only limited, but had he acceded to Ney's request, Napoleon might have fought his way through to another day.

A little after 7.30 p.m. the sun, which had hidden itself for much of the day, peeped furtively through the clouds, and its dying rays glinted on some 4,500 bayonets of the Imperial Guard as these superb troops, architects of so many victories, formed up for their last parade. High and proud was their bearing, marching into battle in their long blue coats and tall red-plumed bearskin caps. They marched in column on a two-company front, so that each battalion had a frontage of between sixty and seventy men and was at least nine ranks deep.

To meet the Emperor's last great throw Wellington probably had little more than 35,000 men capable of bearing arms, but in that sector of the line where the Guards attacked there were now twice as many brigades as there had been in the morning, and a greater weight of firepower. Just why Ney, to whom Napoleon had handed over these magnificent veterans after they had passed before him in review, swung the columns to their left away from the *chaussée* is not clear. But in so doing he gave them the hardest path to tread over some of the worst churned and most corpse-strewn ground. Moreover, in the smoke and general confusion they lost direction and, instead of delivering their attack at five different points along the Allied front, as they started to ascend the ridge they came together in two columns offering Wellington's gunmen a perfect target.

As the leading column advanced towards Maitland's 1st Guards Brigade, lying down behind the crest in four ranks, the Allied gunners poured canister, grape and shot into the flank of the densely packed French ranks. Men went down like swathes of corn before the reaper, Ney had his fifth horse shot under him, General Michel, commander of the chasseurs, was killed and General Fréant of the grenadiers was wounded. But nothing daunted, the survivors closed ranks and marched on towards the ridge, which appeared to be devoid of their enemy, for all they could see was the occasional group of horsemen.

For those Allied soldiers lying low, with what patience they could muster, these were terrible and tense moments. The *rummadum, dummadum, dum* of the French drummers beating the *pas de charge*, and the vibration of the earth from thousands of tramping feet, clearly indicated that the clash was imminent and would surely be savage. Suddenly the silence behind the ridge was broken, 'Now Maitland, now's your time!' ordered the Commander-in-Chief, and then – carried away by the excitement of the moment – 'Stand up Guards', and as the men rose from the ground, 'Make ready! Fire!'

Some 300 French grenadiers of the Middle Guard fell in the course of a minute, but the fighting all along this sector of the ridge was a desperate affair. A fresh battalion of the Imperial Guard that had escaped the Allied artillery fire, battalions of chasseurs that had been rallied, and troops from d'Erlon's command that were in support, gave the British,

Brunswickers and Dutch-Belgians some of the hardest close-quarter fighting of the whole day. In the end it was Adam's Brigade – stationed on Maitland's right flank – and in particular Colborne's 52nd Regiment of Foot, which gave the *coup de grâce* that broke the splendid spirit and élan of these superb French troops. Advancing his regiment in line four deep, and wheeling it opposite the flank of the still advancing Guard, Colborne's men – soon supported by other battalions of the brigade – emptied their rifles into the flank of the enemy. The French turned to face this new threat, and quickly mowed down 150 soldiers of the 52nd. But it was their last throw, and when Colborne led his men forward in a charge it proved too much for the best of troops, and the terrible and hitherto unheard of cry went up, '*La Garde recule!*'

The end of this battle, one of the toughest, cruellest and most destructive that had yet been fought, came a little after 8 p.m., when the Duke of Wellington raised his hat and waved it three times towards the French; it was the signal for the Allied line to advance. Now, after long hours of patient endurance, every man in the Allied line capable of moving was advancing down the slope in pursuit of a broken enemy.

About 9 p.m. Wellington and Blücher met between La Belle Alliance and Rosomme, and Blücher undertook the pursuit. The remnants of the Anglo-Dutch army lay that night on the field they had so gallantly won. More than 40,000 dead or wounded shared it with them. Soon the moon would rise, throwing long and gruesome shadows as the gouls moved about their grisly business of stripping the dead and killing the wounded.

10 The American Civil War

Background to Chancellorsville

IN the early hours of 12 April 1861 Confederate batteries opened fire on Fort Sumter – a fortress built on a shoal to guard the entrance to Charleston harbour. The next day the Federal commander, Major Robert Anderson, beat the chamade, and on the 14th the small garrison marched out of the fort with the full honours of war. What was to be a long and desperate struggle, with all the tragic consequences of civil war, had now become inevitable.

There is no need to discuss in detail the complex reasons that brought about the secession of the eleven slave-holding states, which led to the war. President Lincoln – who had taken office at the head of a Republican administration in March 1861 – went to war to preserve the Union, and as late as 1863 was saying, 'My paramount object in this struggle is to save the Union, and not either to save or destroy slavery.' But this view was not widely held in the North, where there were many convinced abolitionists who regarded the war primarily as a crusade against slavery. The South may have been justified in believing that their sovereign rights under the constitution had been violated, but *au fond* it was fear of an attack upon slavery that drove them to secede. And as the war dragged on, slavery was to become the dominant moral issue underlying the powerful and capacious arguments that both sides used in support of their beliefs.

The Confederate States of America established their government first in Montgomery, Alabama, and a little later in Richmond, Virginia. The man elected to be President was Jefferson Davis. He had been a regular soldier for seven years and had seen active service in Mexico before becoming a politician and a senator. He and the men who surrounded him believed uncompromisingly in the patrician superiority of the South, and had little regard for Yankee ingenuity and its manufacturing community based (as they held it to be) on commercial greed.

As the country rumbled into war neither side realized how long the struggle would last or how cruel the losses would be. Both sides entered the war totally unprepared. The United States army numbered little more than 13,000 officers and men, and the state militias were outdated and regarded with derision. Both armies therefore had to resort to a *levée en masse*, obtained first from volunteers and later by conscription. There was no difficulty in procuring recruits; indeed the supply of men far outran the arms and equipment available for them. But training was a grave problem, especially for the South. The non-commissioned officers of the old army were mostly either German or Irish, who having no state ties remained loyal to the Government and could form training cadres, and although some 300 officers (about a third of the total) decided that their allegiance to their state came first and resigned their commissions, they were needed to fill technical and staff appointments and were not available for training recruits.

The South was also at a disadvantage when it came to arming and equipping the recruits. The factories and foundries of the North were soon geared to turning out large quantities of arms and equipment, but the South – although by no means totally deficient in industrial power – had to rely for much of its fighting material on what was captured in battle.

When the war had been under way for a year or so weapons, equipment and command structure, while never standardized, began to assume a recognizable pattern. In both armies the organization up to division and corps level was apt to vary according to circumstances, but in general an infantry regiment – commanded by a full colonel – would have ten companies (consisting of about 100 all ranks), of which two were skirmishers; there could be anything from two to six regiments in a brigade, and usually either three or four brigades to a division and three divisions to a corps. Throughout the war the rifles of both sides were muzzle-loaders, with the exception of a small number of breech-loaders used by the cavalry. They were of various types, the most common being the American Springfields and the British Enfields and Whitworths. There were a few Gatling machine guns, but they were never used to much effect.

The Confederate cavalry, led by such men as 'Jeb' Stuart, Bedford Forrest and Turner Ashby, was superb, and for a long time rode rings round its Federal counterpart. Organized, like the infantry, into brigades and divisions with cannon in support and armed (as circumstances permitted) with rifle, carbine, lance or sabre, these magnificent horsemen were the eyes of the army, and on occasions a spearhead of fire and steel. The Federal cavalry was at an early disadvantage in that there were not the same number of natural horsemen in the North, nor could their bloodstock compare with that of the South; but from Gettysburg onwards, under men like General Grierson (a music teacher turned cavalryman) and the brilliant Philip Sheridan, they made their mark in many battles.

The artillery arm in both armies achieved a very high standard, for it seemed to attract a more intelligent and better disciplined type of recruit. Guns came in all shapes and sizes – both rifled and smooth bores – breech and muzzle-loaders. There were six-, twelve-, and thirty-pounders cast in iron and brass, and occasionally much larger siege weapons. The smooth-bore weapons had a range of up to 1,600 yards, and those with rifled bore could be *effective* (provided the weather was clear) up to 3,000 yards. A much used small, high trajectory weapon was the Cohorn mortar, which fired a twenty-four-lb. shell. Not surprisingly with such a very wide range of cannon, the ammunition supply presented a considerable problem, even more to the Confederates than to the Federals, for the former relied to a large extent on captured weapons.

Transport and supply was always a difficulty, chiefly because of the appalling condition of most of the roads, especially in winter. Railways therefore played a very important part in the various campaigns. Except for one brief period the North held command of the sea, and General McClellan was able to make use of sea power in his Peninsular Campaign. Dress is of little importance in this study; it is enough to say that, as in the English civil war, it varied from exaggerated sartorial splendour to rags. But discipline is always important, and with a few notable exceptions there was very little of it on either side, and in its place an easy familiarity which – perhaps surprisingly – quite often worked. Any lack of discipline, however, was more than compensated for by an extremely high degree of courage on the part of all ranks, and a willingness to bear – with scarcely a moan – hardships that were above the average experienced in war.

As in the British army, great reverence was attached to regimental colours; sooner or later every regiment on both sides had its flag to carry as a rallying point in battle. These had not been solemnly consecrated, and were usually made by some ladies' guild and handed to the colonel by a pretty girl – and no doubt none the less sacred for that.

Both sides had problems in the higher command, but those of the North were the greater. When war broke out the overall commander of the United States forces was General Winfield Scott, who was an old and virtually immobile seventy-five. He had been a good general, and he still had a good brain. At the very beginning he saw that although the war could be lost (as it nearly was) in the east it would be won eventually in the west. But the President was right to remove him at an early stage. Throughout the war the North suffered from dissension among its senior generals, and constant obfuscation of command led to frequent changes in the higher echelons. President Lincoln was also Commander-in-Chief, and although he may have interfered unnecessarily at times he was far more resolute and perspicacious than most of his generals.

The outstanding general of the South – and indeed of either army – was Robert E. Lee. The son of a soldier, he himself had been commissioned in the Engineers (later to become a cavalryman) and, like his president, had seen active service in Mexico. When Virginia seceded, Lee with sorrow – for he placed much store by the Union – resigned his commission and offered his services to his beloved state. He was at the time fifty-four years old. Throughout the war he never lost the offensive spirit. Outnumbered in almost every battle he fought, he possessed that true art of generalship which compels the enemy to divide so that he may be struck in detail with a superior force. He was a master of manoeuvre, and displayed powers of leadership which place him high among the great captains of war.

The first major engagement of the war took place on 21 July 1861 when General Scott sent five divisions under General Irvin McDowell to seize the important Manassas Junction just south of Washington. The battle became known as First Bull Run, after the small stream near which it was fought. At the beginning of hostilities Jefferson Davis assumed the role – although not the title – of Commander-in-Chief. Lee, who was eventually (February 1865) appointed to supreme command, started the war as Davis's military adviser, and the field armies had independent commanders. Facing McDowell was General Beauregard, and Scott's plan was for Beauregard to be defeated before General Joseph Johnston could arrive from the Shenandoah Valley and assume overall command. The plan failed, for Johnston was able to join forces with Beauregard and take up a strong defensive position. Both armies were fighting with raw, untrained recruits, and in such circumstances the defenders have the advantage. The Federal troops were, after some initial success, heavily defeated.

The loss of this battle made it obvious to Congress that the war would not be easily won. Half a million volunteers were called for, and command of the army was to pass in November from Scott to General George McClellan. McClellan's strategy was often sound, but his tactics were lamentable; his constant unwarranted caution and hesitation lost him battles that a more resolute general must have won. But he served the North better than most of her generals, for more than any other man he was responsible for training, organizing and inspiring that splendid fighting machine, the Army of the Potomac, whose soldiers loved and admired him.

In the summer before his appointment McClellan had gained a decisive victory over a Confederate army at Rich Mountain in West Virginia. On arriving in Washington he found he had under command slightly less than 200,000 men. General Johnston was still at Manassas, strongly entrenched and with a force – so McClellan understood – not greatly inferior to his own. In fact it was less than half McClellan's, and far less well equipped. McClellan suffered throughout his campaigns

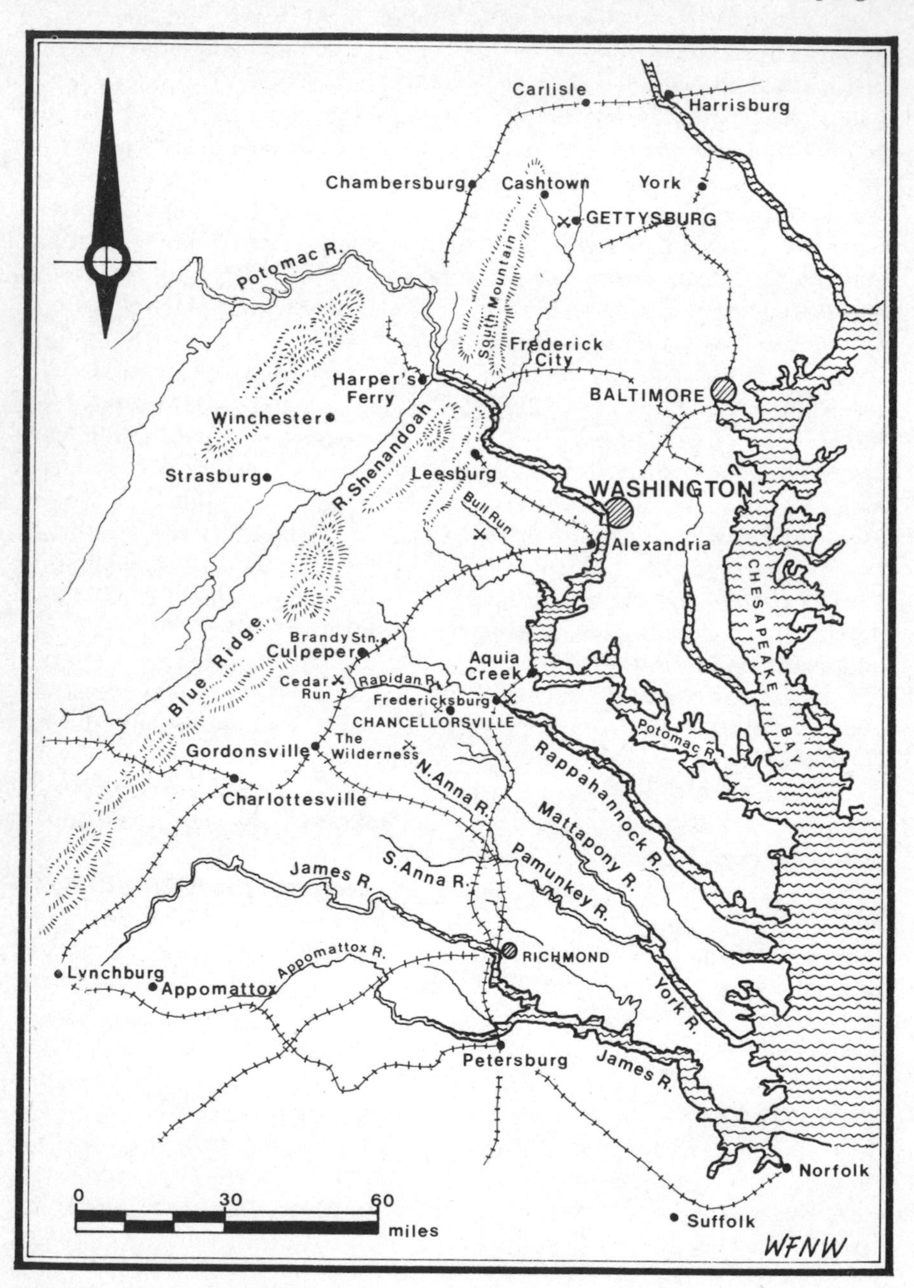

48 The Chancellorsville and Gettysburg Campaigns

from having the great detective Allan Pinkerton as his chief intelligence officer. Pinkerton, world famous for catching criminals, was woefully weak in collecting information. This wrong appreciation, combined with his natural caution, kept McClellan quiet all winter, and when in the spring he did move to attack Johnston did not wait for him, but fell back behind the Rappahannock. McClellan then decided on an amphibious operation. He would transport the army down Chesapeake Bay to Fortress Monroe (still in Federal hands), from where he would march up the peninsula between the James and York rivers and attack Richmond from the east, with the navy guarding his flank. It was a good plan, but success depended on speed, and speed was something McClelland did not understand.

The Peninsular Campaign lasted from the beginning of April until the end of July 1861; for much of that time McClellan was striving to break through the Confederate ring of steel that protected Richmond. At the battle of Seven Pines, General Johnston was defeated and he himself wounded; on 1 June his place as army commander was taken by Lee, and there was born the Army of Northern Virginia, which through the lustre of its victories was for a time to outshine the Army of the Potomac. Gradually Lee pushed McClellan down the peninsula; some of the heaviest fighting (known as the Seven Days Battles) took place during the last week in June, and by the beginning of July McClellan was at Harrison's Landing, exhausted and defeated. A month later Lincoln ordered the army back to Washington.

Apart from the fact that McClellan was outgeneralled by Lee, a contributory cause of his crushing defeat was the threat to Washington posed by General Jackson's army in the Shenandoah Valley, which so alarmed Lincoln that he kept back McDowell's corps from the peninsula. General Thomas ('Stonewall') Jackson was Lee's most able assistant. Tall, bearded, unkempt, always silent and often brooding, he possessed few of the outward graces. Rather did he appear prosaic and austere, although always courteous; a man who kept his own counsel and turned to the Bible as the only firm rock among the shifting sands of sin. But he was a fearless soldier and a great commander, whose men would follow him anywhere. If there is such a thing as reincarnation, those who knew Orde Wingate might also have met Stonewall Jackson.

The battles we shall be examining in detail both took place in the eastern theatre of war; the hard fighting in the west impinges only to a very limited degree on the eastern campaigns of 1862 and 1863. Space will not permit discussion; but it was in the west that the strategy of pushing down the Mississippi and striking east at each state, which was eventually to split and then destroy the Confederacy, was being patiently worked out even as early as 1862. From Kentucky to Louisiana, down the valleys of the great rivers, over mountains, through sprawling forests alternating with sweeps of wild primeval swamps, in searing heat and bitter cold, battles were being won and lost. Thousands of young men perished, generals bickered among themselves, but in the end total success crowned four long years of savage endeavour.

But in 1862 and 1863 the worst of the fighting and the heaviest casualties were to occur in the states of Virginia and Pennsylvania. After McClellan's failure in the Peninsula, Lincoln brought General John Pope across from the western theatre, where he had had some success, to command a new Army of Virginia. He had little enough time to weld his disparate divisions together, nor did the tactless gasconades with which he hoped to animate his troops endear him to them. Jackson defeated a part of his force at Cedar Run on 9 August 1862, and three weeks later, before McClellan's troops – now back from the Peninsula – could materially assist Pope, Lee, with his able assistants Jackson and 'Jeb' Stuart, decisively beat him at the Second Battle of Bull Run. Lee's 2,000 casualties at this battle were 400 more than those suffered by the Federals, but he took 1,500 prisoners, twenty-five guns and thousands of rifles.

In September Lee decided that the best means of relieving the pressure on Richmond would be to invade Federal territory. He had great hopes of attracting recruits in Maryland, but in this he was to be disappointed, partly because on the march the ragged, ill-shod appearance of his troops belied their fighting qualities and scarcely invited volunteers to their ranks. Nevertheless, his plan for the invasion of the North, which he intended should take him into Pennsylvania and beyond, was masterly and very daring, for it involved dividing his army into four parts, so that he could hold the Pennsylvania border and the passes in South Mountain, while the bulk of his force destroyed the Federal garrison at Harper's Ferry, which lay across his lines of communication.

After crossing the Potomac Stuart's cavalry effectively screened Lee's movements, and the over-cautious McClellan – still in command of the Army of the Potomac – could only blunder about, ignorant of his opponent's plan, until the fog was suddenly lifted when, by one of those strange accidents of fate, a private soldier picked up from the roadside an envelope containing a copy of Lee's Secret Order No. 191 outlining all his plans. McClellan now knew how he could defeat Lee's army in detail, and Lee did not know that he knew. Even so, McClellan proceeded with the utmost caution, quite convinced that he was outnumbered. The thirteenth and fourteenth of September saw him force the South Mountain passes, but his approach had been so slow and the defence so stubborn that before he could march south to relieve Harper's Ferry the garrison had surrendered. Lee was forced to fall back to hold a line just west of the Antietam Creek on a low ridge extending north and south of Sharpsburg, and to pray that Jackson and General A. P. Hill would arrive from Harper's Ferry before large numbers of enemy overwhelmed him.

The seventeenth of September was the bloodiest day of the war so far. McClellan's plan of attack was excellent, his execution of it deplorable. He kept attacking piecemeal and invariably failed to reinforce partial success, allowing Lee, who with a greatly inferior force showed consummate generalship, always to be superior at the point of contact. The battle lasted all day, and every attack was resisted with the utmost savagery; by the evening, when A. P. Hill's division had at last arrived to save the day for Lee, the soldiers of both armies had been goaded into a frenzy of death and killing. That night there lay upon the field 21,000 men either dead or wounded, rather more than 16 per cent of the troops engaged.

The next day Lee offered battle again – a superb piece of bluff, which McClellan did not call – and during the night Lee withdrew across the Potomac. It was a Federal victory – if a pyrrhic one – for it put an end to Lee's first invasion of Northern territory, and gave Lincoln the excuse he needed to prepare his Emancipation Proclamation. If this did not, as Lincoln may have hoped, pull the lynchpin from the wheel of revolt, it changed the character of the war. From now on it was a fight against slavery.

On 5 October, almost three weeks after the battle, McClellan went in leisurely pursuit of Lee. But Lincoln had had enough of him, and on 7 November General Ambrose Burnside, who had fought at Antietam, was given the command. It was not a good choice: at the very beginning this big, easy-going, hirsute general was heard to say, 'I am not competent to command such a large army.' That is no mood in which to start a campaign against a general of Lee's calibre, but Lincoln thought Burnside the best man available. Events did nothing to justify his choice.

The Army of Northern Virginia retreated from Maryland in a deplorable condition. Thousands of soldiers were marching barefoot; Jackson reported that 3,000 of his men were without weapons; rations were short and desertions had become extremely serious. The fact that the army was ready to fight again with distinction only three months after Antietam was due to McClellan's dilatoriness and Lee's administrative genius. By December he had organized two corps under Jackson and Longstreet that had been re-equipped and got into good shape. The two armies now faced each other east of the Blue Ridge in the Warrenton–Culpeper area.

Burnside decided to march forty miles to his south-east, cross the Rappahannock at Fredericksburg and lure Lee to battle on ground of his own choosing. Lincoln approved the plan as long as it was executed with speed. It was not. Although Burnside moved fast to start with, a breakdown in administration (fairly common in the Federal army) found him on the banks of the river without pontoons with which to cross it. By the time they did arrive Lee had drawn his army up on the high ground west of the river at Fredericksburg. Burnside, having the advantage of numbers (100,000 to 60,000), forced the river (at considerable cost) and sent his men in wave after wave against the strongly held heights. It was sheer murder; the ground in front of the defences was literally covered with blue uniforms where Federal soldiers lay dead or dying. When the cold December night cast a veil across the battlefield General Burnside had lost 12,647 men.

After the fearful disaster of Fredericksburg, followed in January by an abortive advance which came to a grinding halt in the mud, Burnside – and indeed his lieutenants too – became even more convinced that he was not fit to command an army, and on 25 January Lincoln accepted his resignation. The President remained steadfast in the face of disappointment. Within that slender frame there burned a remorseless fire. He would find another general and extract from him greater skill and daring. His choice this time fell on General Joseph Hooker, who had fought with distinction at Antietam.

During the winter months at the beginning of 1863 both armies regrouped on opposite sides of the Rappahannock. Everyone knew that in the spring the struggle would be renewed. The Northern men, no longer raw recruits but close-knit by active service and animated by a desire for revenge, were confident and cheerful under their new commander. The Southern army, much smaller and far less well equipped, but elated by the quality of their recent performance, prepared with constancy and courage for a further cruel clash.

11 The Battle of Chancellorsville

30 April – 6 May 1863

IN the first two years of the war the Federal army had suffered badly from lack of leadership at the top; a succession of commanding generals had all failed and, no matter how brave the troops, an army is unlikely to win battles if the officer commanding is incompetent. In General Joseph Hooker, Lincoln had good grounds for hoping that at last he had found a man worthy of the troops he was to lead.

Hooker took over an army demoralized and unsure of itself – the desertion rate was about 200 men a day. By the end of March he had licked it into shape and restored its confidence. He was a strict disciplinarian, a good administrator, a hard-drinking, hard-living man, and hitherto he had proved himself to be a brave fighter. He was to lead the Northern army into a battle in which the standard of leadership required was to be very high.

By the beginning of 1863 the Confederate position generally was on the decline. The situation was more serious in the west, where the Federals had control of the Mississippi down to Vicksburg, and large armies were making headway in Tennessee and Arkansas; but even in the east the weight of numbers was slowly pressing the Confederates back. In April it was estimated that the North had some 900,000 men under arms compared with 600,000 Confederates.

That month the Army of Northern Virginia occupied ground stretching from Banks' Ford to Port Royal, on the south bank of the Rappahannock. McLaws' and Anderson's divisions of Longstreet's I Corps occupied the strategic high ground of Lee's Hill opposite Fredericksburg, and the ridge northwards; westward of these troops was the cavalry. Stuart (with 2,400 horsemen) had his main body at Culpeper, and the river crossings were closely watched. Jackson's II Corps was in position along a twelve-mile front from Hamilton's Crossing to Port Royal. He had four divisions totalling about 36,000 men, and 2,000 artillerymen. I Corps had been weakened through the need to send its commander, Longstreet, and two of his divisions to Suffolk to meet a reported threat to Richmond, and to collect food supplies from North Carolina. Two brigades of cavalry were also absent from the army. Lee, who had only recovered from an unpleasant illness towards the middle of April, could therefore muster no more than 62,000 men and perhaps 200 guns, if Longstreet was not able to rejoin his army – and that general, always a difficult subordinate, preferred to spend his time operating against Suffolk with an independent command.

On the other side of the Rappahannock Hooker commanded an army of about 134,000 men* and some 400 guns. Immediately on assuming command he had weeded out unsatisfactory generals, and abolished Burnside's unwieldy organization of Grand Divisions and independent cavalry brigades. There was now a return to army corps, of which there were seven, and a cavalry corps of three divisions. Each infantry corps had three divisions, with the exception of XII Corps (Slocum) which had only two. The cavalry arm numbered more than 11,000 sabres, but their commander, General Stoneman, was ordered to carry out a raid on Lee's lines of communication with a view to forcing him to retreat. In consequence the Federal cavalry (less one brigade of General Pleasonton's division) took virtually no part in the battle, and observation balloons proved a very inadequate substitute for reporting enemy movements. The Army of the Potomac was concentrated immediately north and east of the Rappahannock around Falmouth and along the Stafford Heights.

At the outset of the battle the initiative lay with Hooker, all the cards being stacked heavily in his favour. He had recently had a fairly unhelpful visit from Lincoln, but it had at least impressed upon him the need for a plan that would destroy Lee's army. Richmond, Lincoln told him, could for the moment be forgotten. *He had perhaps four options:*

(1) He could make a frontal attack along the whole line, and pound Lee with his vastly superior numbers. But Lee's defensive position was well laid out, with most approaches covered by cleverly sited artillery units.

(2) He could carry out a long approach march that would enable him to take Lee's army in flank and rear on Lee's right flank.

(3) He could do the same on Lee's left flank. (The right flank of Lee's army could not be easily turned; the left flank might be, but it would involve crossing two rivers and traversing thickly wooded country).

(4) He could pack a hard punch at one end of Lee's extended line, with a view to driving that in and then swinging his force to roll up the centre and

*Colonel Rogers (*The Confederates and Federals at War*) estimates Hooker's total strength at 133,711 men and 404 guns. Colonel Henderson gives a round figure of 130,000 men, one American source puts the total at 138,378 and another at 134,000. Colonel Rogers also gives Lee's army as totalling only 56,444 men, with 228 guns, but he may not have included some 5,000 reinforcements, sent to the army in March.

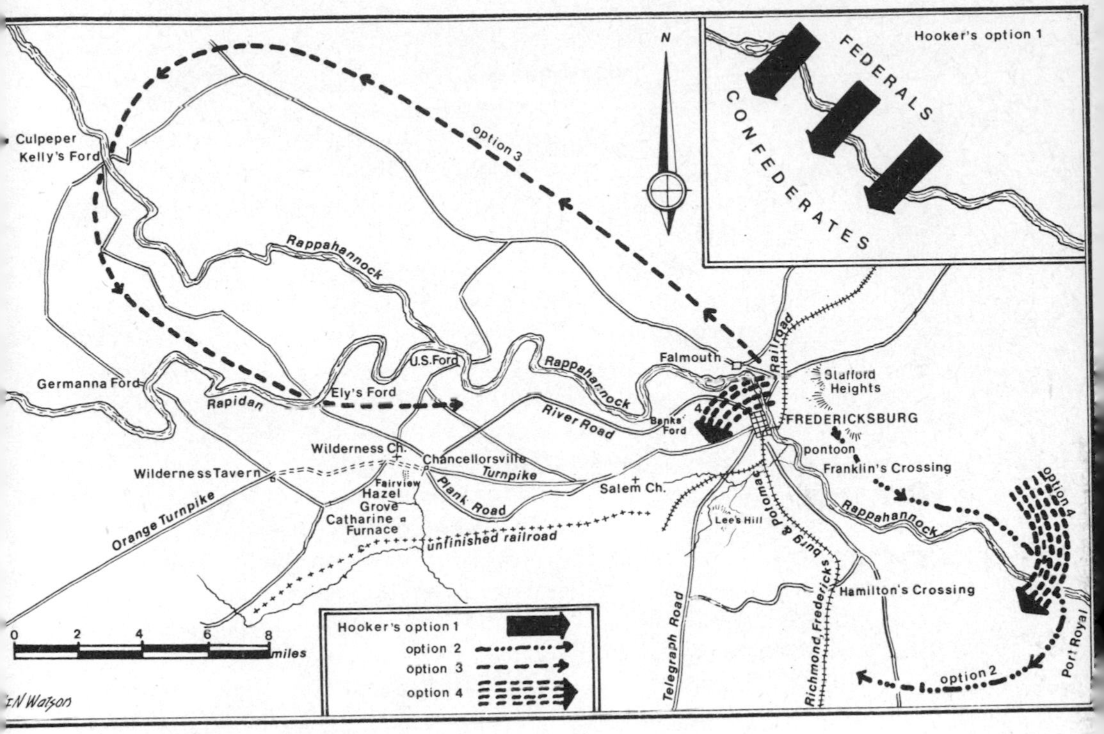

49 The Battle of Chancellorsville: Hooker's first set of options

other flank. The disadvantage here was that Lee's left and centre were very strong in natural defences, and on his right Lee had placed his best and strongest corps. Moreover, although much inferior in numbers, Lee's defensive position was sufficiently flexible to enable him to reinforce threatened points.

Hooker had learned from Fredericksburg that to attack frontally against the strong defensive line, with its many natural barriers, would be disastrous, and the idea of packing a punch at one flank in the hope of rolling it back and swinging inwards could have proved as disastrous as a full frontal attack. He therefore decided to attack Lee in flank (option 3), and rightly concluded that the enemy's left flank offered the best chance, although it would be a long, circuitous and difficult approach.

Hooker had seven corps with which to operate his plan. Initially he himself, with three corps (V, XI and XII) and a cavalry screen, would carry out the flank movement. They would cross the Rappahannock at

Kelly's Ford and the Rapidan at Ely's and Germanna's Fords. Three corps (I, III and VI), under command of General Sedgwick, were to be in readiness to cross the Rappahannock below Fredericksburg to create the impression that a frontal attack was developing, and to divert Lee's attention from the right hook. The II Corps left one division at Falmouth, and had two (less a brigade, which was ordered to United States Ford) on the north bank of the river opposite Banks' Ford. The cavalry – less one brigade – was sent via Culpeper to get right behind Lee and cut his communications.

It was a good plan and Hooker knew it was. He was a general somewhat given to braggadocio, and not averse to staking his reputation by declaring that with an army that was 'the finest on the planet, Lee's boys may as well pack up their haversacks and make for Richmond, and I shall be after them'. Nevertheless, it was a plan not without its dangers – especially with an opponent like Lee. Hooker had split his army into three parts, and in the early stages of the march more than twenty miles divided the main force from Sedgwick. This gave Lee the advantage of interior lines, and a chance perhaps to defeat each section of the Federal army in detail. Hooker had clearly considered this, for Sedgwick had been given orders to press his attack if Lee moved in strength against the main force which was concentrating at the Chancellorsville house. But it is unlikely that the confident Hooker was too worried, for each of his two wings were nearly as strong as the entire Confederate army, and the cavalry threat to Lee's communications would, he felt, force him to retreat.

The plan, with a few minor modifications, worked with smooth precision. The main body of cavalry had been ordered to start their march on 13 April, but an unexpected spate made the Rappahannock impassable for a few days. The main manoeuvre was put into motion on 27 April. The Rappahannock was crossed at Kelly's Ford on the 28th and the night of the 29th, and then V Corps marched to cross the Rapidan at Ely's Ford, and XI and XII Corps crossed that river at Germanna's Ford. By the morning of the 30th all three – after what had proved a tricky and very wet operation – were across.

Sedgwick's feint was successful in that a Confederate force was moved into position to resist an attack, and at this stage II and III Corps were ordered to join the main body in the Chancellorsville area. II Corps moved up to United States Ford and crossed there when the main flanking body had made good the other bank, and III Corps, crossing at the same place, joined Hooker at Chancellorsville on 1 May. Well might that general feel confident that he had Lee's men between the upper and nether millstones.

From the time Hooker started to cross the Rappahannock Lee was aware that something was about to happen on his left flank, but first reports from his cavalry commander, General Stuart, did not indicate the extent of Hooker's movement, for it was not until early on the morning of 30 April, when Stuart's message identifying the three corps and reporting them across the Rapidan was received, that Hooker's intention became clear. The previous day Jackson had sent back to say that Sedgwick was crossing the Rappahannock, so the Confederate commander now knew the full extent of his predicament. Unless he did something fairly quickly he would be crushed between two powerful forces converging on him from the east and west. *Lee had three options.*

(1) He could retreat to the south while there was still time, and take up a strong position behind the North Anna. This is what Longstreet had urged before he departed with his two divisions for Suffolk.

(2) He could act swiftly, making use of his interior lines, and attack one of Hooker's wings with his whole army. But which wing should he attack? Sedgwick's seemed the obvious choice, for his was the smallest, and Lee would

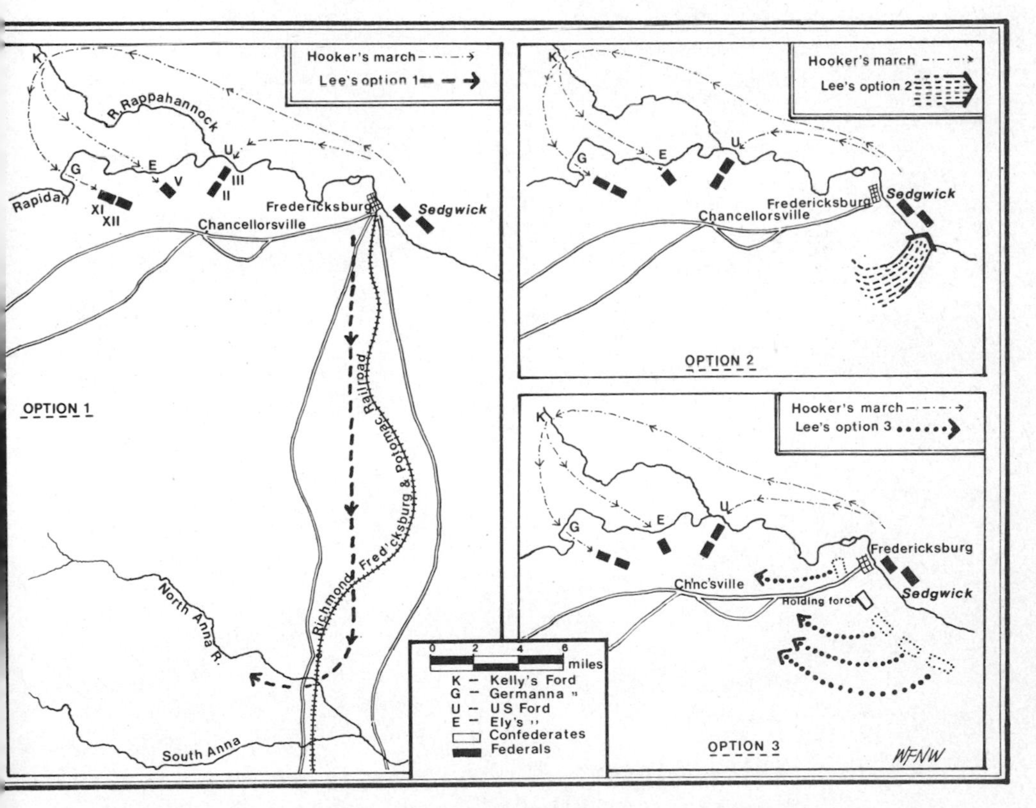

50 The Battle of Chancellorsville: 30 April, Lee's first set of options

have nearly double his numbers. But having defeated Sedgwick could he extricate himself before having to face Hooker?

(3) He could leave a holding force of about 10,000 men facing Sedgwick, and march with his remaining 50,000 to confront Hooker's 75,000 highly confident troops in difficult country around Chancellorsville. This violation of one of the principles of war – that you should not divide your force – would be particularly risky in view of his inferior numbers.

A lesser man than Lee, faced with so dangerous a situation, would have opted to pull back if he could before the pincers closed round him. But Lee once again showed that neither lack of numbers nor imminent peril could disturb that calm confidence that was his aura. He knew well that to retreat would be playing Hooker's game; it is true that Stoneman's cavalry in his rear had so far achieved virtually nothing, but this he did not know at the time, and his retreat could well have been cut off. Experience gained in the flat country bordering the river at Fredericksburg had taught him that the Federal artillery could cause great havoc to an attacking army. And so, without much hesitation, he decided to leave a holding force to act offensively in front of Sedgwick, and swing the rest of his army to meet the oncoming Federals in the Chancellorsville area (option 3). He was taking a huge risk, but it offered the best chance of survival – and to some extent it was a calculated risk, for Lee knew Hooker and would have had an idea as to how the man would react when hit.

Accordingly Jackson was ordered to detach General Early's division from his corps to remain in position before Fredericksburg, and this division was supplemented by one brigade (Barksdale's) taken from General McLaws' division of I Corps. Thus a force of around 11,000 men and fifty guns was positioned along a six-mile front. The rest of the army – two divisions, less one brigade, of I Corps, and Jackson's II Corps less Early's division – was on the march towards Chancellorsville by the early morning of 1 May. The thick early morning river mists rendered the Federal observation balloons valueless, and gained Lee some initial advantage.

The country around the Chancellorsville house, where the major engagement was to be fought, was totally unsuitable for military manoeuvre. The house stood at the eastern edge of a dense forest, which had been cut over and left to grow scrub. It was now a matted jungle of oak coppice, second-growth pine and gnarled, wind-blasted thorn scrub extending approximately six miles in a north–south direction and twelve from east to west. This area was well named the Wilderness. There were creeks and swamps, and a few open spaces amid the dense undergrowth, but these were mostly quite small and offered poor fields of fire. The area was traversed by a number of roads, the principal one being the Turnpike, which ran through the Wilderness, and was joined east of the house (close to the Tabernacle Church) by the Plank Road and continued to Fredericksburg. The River Road, as its name suggests, followed the Rappahannock and came into Fredericksburg from the north. There were other recognized tracks, and an unfinished railway ran from Fredericksburg westwards, some two miles south of the Chancellorsville house. The Turnpike had a fairly rough gravel surface, and the Plank Road was literally made of planks bound and nailed to sleepers.

By 1 May Hooker was busy establishing a secure base along a line that ran roughly north-east and south-west through Chancellorsville, extending in the north to United States Ford on the Rappahannock, and in the south to the area of Hazel Grove. His headquarters were in the house itself. From here he sent forward two divisions on each of three roads – River Road, Plank Road and the Turnpike. Those marching on the River Road encountered no enemy. But as the cavalry screen in front of the leading troops crossed a ridge on the Turnpike, a Confederate advance guard was met and the cavalry were driven back. But the divisional commander (General Sykes) quickly deployed his leading troops and drove off the Confederates. Shortly afterwards the six divisions took up a defensive position on a ridge crossing the three roads, which was on the edge of open country.

Lee had advanced towards Chancellorsville with General Anderson's division of I Corps in the lead; he had two brigades on the Turnpike and three on the Plank Road, with artillery on both roads. McLaws' division (less, of course, the brigade at Fredericksburg) followed Anderson's troops on the Turnpike, and Jackson's Corps followed the leading brigades on the Plank Road. It was the leading brigades of Anderson's division that had the fight with the Federal troops, and shortly afterwards one of the Confederate brigades on the Plank Road drove in skirmishers of the Federal XII Corps. Jackson now came forward in person to take charge of this short but spirited fight; soon McLaws had deployed four brigades against the Federal infantry, which attacked with artillery support.

This contact battle took place just short of two miles from Hooker's headquarters about midday on 1 May, and the Federal general very soon became aware of the situation. He now had to make his first tactical decision, *and he had two options:*

(1) He could carry out the main part of his original plan by advancing in strength against Lee, securing the open ground around the Tabernacle Church and urging Sedgwick to press his attack on Fredericksburg and Marye's Heights, thus crushing the Confederates between his two wings.

(2) He could fall back on to the strong defensive line he had almost completed, stretching from United States Ford to Hazel Grove and the Plank Road, and there await an attack.

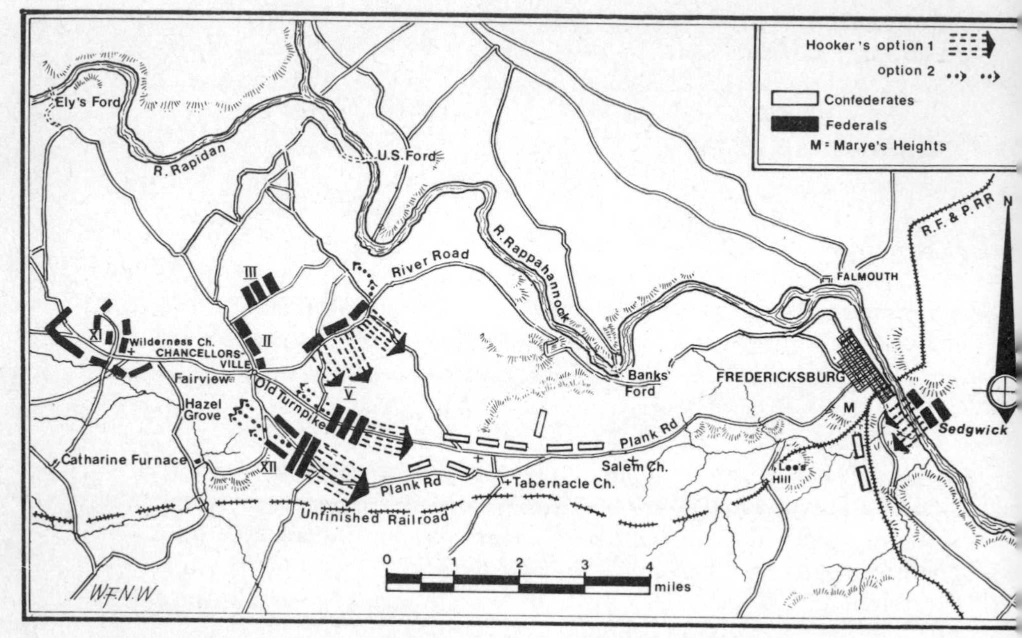

51 The Battle of Chancellorsville: 1 May, Hooker's second set of options

To the consternation of his corps commanders Hooker, on learning that Lee had marched against him and engaged his leading troops, ordered those troops to withdraw to the strong defensive line around Chancellorsville – option 2. Behind his words of high resolve, and inordinate pretensions – even now he was saying that Lee's army had become 'the legitimate property of the Army of the Potomac' – there lurked in Hooker a grave doubt that he could defeat Lee in open battle. Recent experience at Fredericksburg had impressed upon him – with unfortunate results – the value of strong entrenchments. His plan to destroy Lee's army had been masterly, but when his opponent failed to conform his courage failed him; although the position to which he withdrew was exceedingly strong – as Lee himself was to admit – he had surrendered the initiative, and was never to regain it. He had also lost confidence, and when a leader loses confidence loss of morale in his army usually follows.

Hooker manned his defensive line with General Meade's V Corps on the left of the line, resting his flank on the Rappahannock near United States Ford; next to him, in the Chancellorsville house area, came General Slocum's XII Corps, with Birney's division of III Corps on Slocum's right. The rest of III Corps and two divisions of II Corps were in reserve, and Hooker had his right flank refused, with General Howard's XI Corps stretching back to the Orange Plank Road and Talley's clearing. Outposts had been positioned in the woods, which gave some protection to the men still busy clearing fields of fire for rifle and cannon, strengthening parapets and entrenchments, and constructing a formidable abatis with fallen logs.

Lee was right to state in his report that 'the enemy had assumed a position of great natural strength . . .'. Nevertheless, there were weaknesses. The principal one was that without cavalry (Pleasonton's brigade was insufficient) Hooker was completely blind. He might well remain ignorant of Lee's intentions right up to the point of contact. Artillery, although sited to give destructive fire down the main approaches, could not be properly deployed in such jungle conditions. Moreover, Hooker had no plan to go over to the offensive; *reculer pour mieux sauter* could alone justify his recent decision. Instead, believing his six-mile line to be too thinly held, he ordered Sedgwick to send I Corps to Chancellorsville.

On the evening of 1 May Lee, accompanied by Jackson, towards whom

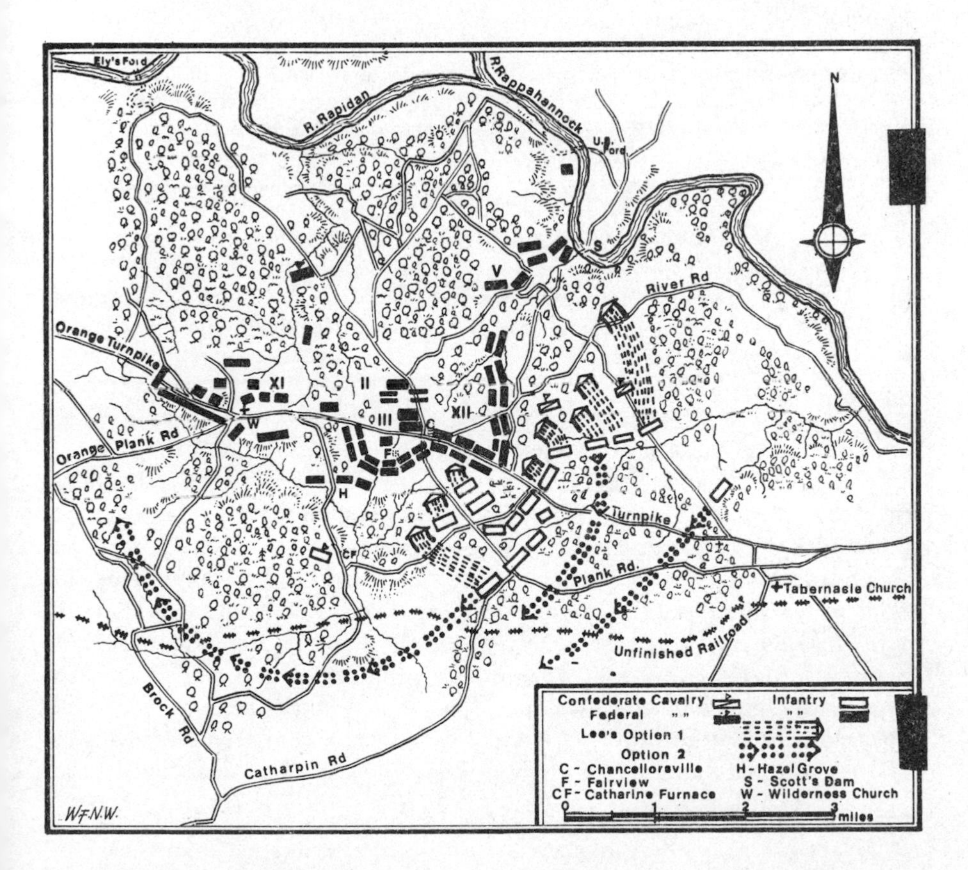

52 *The Battle of Chancellorsville: 1–2 May, Lee's second set of options*

he had developed a great friendship based upon absolute reliance, sat in a clearing by the Plank Road and pored over his maps. Jackson felt that Hooker's nerve had completely gone and that he would withdraw during the night; Lee could not believe this. The first decision, therefore, was quickly taken: if Hooker was waiting to be attacked, he should not be disappointed. But how to succeed against this strong position with inferior numbers? There were only two possible ways – and neither seemed to offer much hope of success. *His options were:*

(1) To attack the position frontally.

(2) To turn a flank.

Lee had personally reconnoitred the left of Hooker's line and found it strong; Jackson reported the centre as being virtually impregnable. The left flank resting on the Rappahannock could not be turned, but the right flank – as reported by Stuart – was in the air. Here was a possibility, but the risks would be appalling. Admittedly Hooker had made it slightly easier for the Confederates by withdrawing into dense country without the use of cavalry, but if the flank march was detected the Confederate army – which would have to be divided into two further parts – could be annihilated in detail. To give the turning force any hope of success the Federal army would have to be distracted from the front during the long march (twelve miles), and actively engaged when the flank attack went in; to co-ordinate these two attacks at so great a distance would be extremely difficult. Finally, the Wilderness had not been properly surveyed and the flanking force, upon whom would depend not only the success of the operation but the continued existence of the Army of Northern Virginia, would be marching along narrow, uncharted tracks.

Lee was not prepared to batter his army against strong defences as Burnside had done at Fredericksburg, so with his usual promptitude of decision he accepted the risks and opted for the flank attack. Undoubtedly the fact that Jackson, so steadfast in battle, would command the flanking force helped him to make this decision. He knew he could rely on his clear judgment and quickness of perception to make the attack a success; nor did he demur when Jackson, on being asked what men he proposed to take, answered 'my whole corps'. This left Lee with only Anderson's and McLaws' divisions, less Wilcox's brigade that had been sent to Banks' Ford* – a total of no more than 16,000 men – to watch, and later engage, the whole Federal front. The cavalry was to cover the flank march.

* This brigade played a very useful role in delaying the Federal I Corps' arrival on the field of battle.

While Anderson's and McLaws' divisions were busily engaged in active defence operations in an attempt to create an impression of strength, Jackson's corps started their long flank march at 7 a.m. His chaplain, the Reverend B. T. Lacy, had once been the incumbent of a parish in the neighbourhood, and his local knowledge made him an invaluable guide. The route selected (now well marked for visitors) was along Furnace Road, south from the crossroads at Catherine Furnace, to where it met a private road running just to the west of Brock Road; here it headed north, and crossed the Orange Plank Road to Wilderness Corner on the Orange Turnpike.

The column had to pass very close to the Federal lines in the Catherine Furnace area, and Jackson left a regiment there to protect the flank and act as rearguard. Despite every precaution, it was almost inevitable that so long a column (the infantry alone covered six miles), with gun carriages, ammunition and ambulance wagons, would be discovered, and this occurred at about 8 a.m. in the Catherine Furnace area. General Birney, commanding the division in Sickles' corps which held the line between XI and XII Corps, immediately reported the fact to his superior and engaged the column with his artillery. This caused considerable confusion and dispersal of the wagons.

Sickles, in his report to Hooker, intimated that the enemy could be extending his front, contemplating a flank attack, or more probably withdrawing in the direction of Gordonsville. *On receipt of this information the army commander operating from a strong base with five corps well concentrated had three options:*

(1) He could take no offensive action, but in case a flank attack was intended he could guard against surprise by sending Pleasonton's cavalry brigade to watch his right flank.

(2) He could launch an immediate attack on the flanking column with divisions from III and XI Corps, leaving II and XII Corps to hold McLaws' and Anderson's divisions.

(3) He could engage the enemy along the whole front, and send orders to Sedgwick to press his attack against Early's division.

Either the second or third option could have brought disaster to Lee, for we have seen the risk that he was taking. The Confederate army was divided into three parts, and to attack it in detail while it was off balance must surely have resulted in success. As it was Hooker took no immediate action beyond sending a warning to Howard to be ready in case of a flank attack. Having dithered for almost three hours he finally decided that Jackson's column was heading south and was part of a general withdrawal on Gordonsville. But even then Hooker gave no orders for an all-out attack by one or two corps; he told Sickles to advance against the column with caution and develop the situation. It was not a decision at all; it was merely a weak commander's compromise.

Sickles carried out his orders, and at about 2.30 p.m. a running fight developed at Catherine Furnace. The 23rd Georgia Regiment, which had been left as rearguard, fought stoutly but was soon overwhelmed by Birney's division. The mêlée became extended when Anderson attacked Birney's flank, forcing him to call for reinforcements. General Barlow's brigade was sent from XI Corps, thus weakening Hooker's right flank. There now developed an interesting situation, which concerned the commanders of the rear troops of Jackson's column. General Thomas commanded the rear brigade and his senior officer, General Archer, the penultimate one. Jackson himself was marching some miles to the front, and preparing to deploy his leading troops. *An immediate, but difficult, decision had to be taken at brigade level.*

(1) Should the column continue the march, being prepared to fight a running rearguard action if necessary?

(2) Should a brigade be sent back to join the fight, and ensure that the wagon train was safely extricated?

(3) Should a larger force be deployed for action?

One must assume that although time was short between the making of the plan and putting the flank march into execution, division and brigade commanders were fully in the picture. The rear brigade commanders would therefore have known that the corps was to launch its attack as soon as possible, and this must have made a decision all the more difficult to take, for time was important. But had Hooker succeeded in driving a wedge between the two flanks of Lee's army the result – if Hooker had acted with strength and resolution – could have been disastrous, and Archer clearly felt that at all costs this must be avoided. Furthermore, he decided that the quickest way to settle the matter and rejoin the marching column was to engage with as much strength as possible, for he took command of the rear brigade and adopted option 3.

Archer ordered General Thomas's brigade and Colonel Brown's artillery battalion to march back with his own brigade and engage the enemy. In his report he says, 'I immediately ordered back my own and Thomas' brigade, but when I arrived at the furnace found that the enemy had already been repulsed . . .' It could be said, therefore, that his efforts were unnecessary, but this added display of force may have influenced Birney not to pursue the attack; in any event it was a bold, and correct, decision on the part of a subordinate commander.

Meanwhile, the head of the column reached the Plank Road at about 2 p.m., and here the cavalry commander, Fitzhugh Lee (a nephew of the army commander), whose men had been acting as flank guard, took Jackson forward to reconnoitre. To the amazement of both men they saw that the Federal lines were totally unprepared, with rifles stacked and the troops lying about playing cards, cooking or sleeping. One can imagine Jackson's feelings at seeing his enemy so totally unprepared. *He was now faced with two options:*

(1) To wait – and it would be a long wait of two or three hours until his whole column had closed up and could be deployed from column of route into line – and then continue to the Turnpike so as to get round the Union right flank. This could well lose surprise, but would ensure a powerful punch.

(2) To order his leading division to deploy at once and go in to the attack.

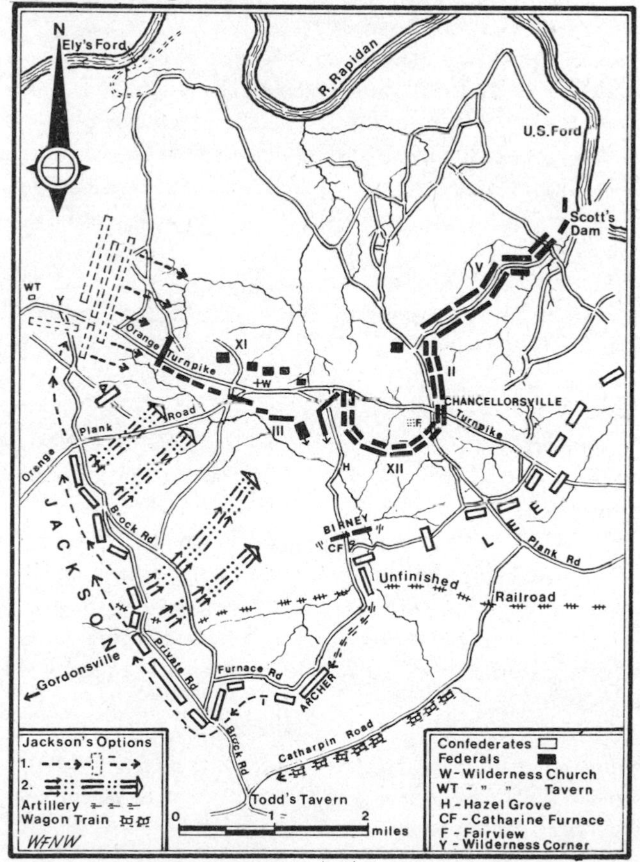

53 The Battle of Chancellorsville: 2 May, Jackson's options

It could not have been an easy decision, and a less percipient general than Jackson might have succumbed to the temptation of falling on the foe, and ensuring surprise for a restricted action; but Jackson knew that this attack had to be a knockout blow and not a piecemeal affair with limited success. He therefore adopted option 1, and took what precautions he could to avoid losing surprise. The only serious objection to this decision – and initially it was not easy for Jackson to be sure of his timing – was that by having to wait more than three hours for his deployment to be completed he was forced to start the battle too late in the day.

It is difficult to understand why Howard's corps was so completely unprepared for battle. Admittedly at 4.10 p.m. Hooker was sending a signal to Sedgwick informing him that 'the enemy is fleeing', but there is no record of Howard having received any message from headquarters since the warning of a possible flank attack, which Hooker sent at 9.30 a.m. Moreover, at about midday the forward picquets had captured two prisoners, who said they were part of a force moving round Howard's flank, and at 2 p.m. there were further warnings from the picquets. One would think that at least a reconnaissance in force would have been ordered. In consequence of this unpreparedness, when the first line of four brigades under General Rodes went in to the attack at about 5.30 p.m. complete surprise was achieved and complete chaos resulted.

Howard was short of one brigade (Barlow's) which had been sent to aid Sickles, but he had twenty regiments of infantry and six batteries; most of this force faced south, but the flank was refused and a part of von Gilsa's brigade of Devens' division faced west.* The ground occupied by the corps was by no means all jungle. There were clearings with good fields of fire round Talley's Farm, Wilderness Church and Dowdall's Tavern (where Howard had his headquarters), but beyond the latter place the forest was dense and dark. With such dash and élan did the initial attack go in that the two flank divisions of XI Corps (those of Devens and Schurz) were swept away after only token resistance, and the advance rolled on like the bore of a tidal river. The open areas round Chancellorsville soon presented a terrible spectacle of overturned wagons and guns abandoned, the sodden earth reddened by the blood of the dead and dying.

As night came on Jackson's attack was beginning to lose its impetus, and the presence of Berry's reserve division restored the situation at least temporarily. In the dark all cats are grey and in the confusion of close-quarter fighting in jungle country mistakes are easily made. At

*The exact number of men facing west is disputed, but it was probably only two regiments and some picquets, with two further regiments in reserve.

about 8.30 p.m. Jackson rode forward to his front line to encourage A. P. Hill to press forward with his brigades before the enemy could erect effective barricades. As he was riding back through the 18th North Carolina Regiment an officer mistook the party for an enemy patrol. A volley was fired and Jackson received three bullets. With great difficulty, and under heavy fire, they got him off the field and into hospital; but his wounds were mortal, and a week later he was dead.

Command of the corps then devolved upon A. P. Hill, but shortly afterwards he too was wounded, and Stuart – who was busy scourging the Federal wagon park around Ely's Ford – was then sent for to take command. Lee's manoeuvre had been brilliantly successful, but it had not achieved very much. Howard's corps had been routed and thoroughly demoralized, but it was frightened more than hurt, as the casualty figures showed at the end of the battle. Before midnight General Reynold's I Corps arrived from Fredericksburg, and this corps brought Hooker's numbers at Chancellorsville to around 85,000.

The Army of the Potomac could now muster sixteen divisions in the Chancellorsville area, and at Fredericksburg General Sedgwick's VI Corps, supplemented by Gibbon's division (less one brigade at Banks' Ford) from II Corps, numbered about 30,000 men. The bulk of the Confederate army faced Hooker at Chancellorsville, but Stuart's II Corps with 26,000 men was separated from Lee's 16,000 by almost two miles; to oppose Sedgwick was General Early with some 11,000 soldiers. The moon was riding high, casting long shadows of trees across the clearings, but bright enough for troops to find their way along the forest tracks and renew the fight. The advantage was still very much with Hooker. *He had at this moment perhaps three options:*

(1) He could order Sedgwick to push Early aside, leaving a division to contain him, and then attack Lee in rear, while he himself engaged both wings of Lee's Chancellorsville troops.

(2) He could shorten his present line, and construct a strong defensive position behind which most of the army could withdraw, but maintain front line troops in the Hazel Grove area. Sedgwick to proceed as for option 1.

(3) He could act defensively from his present position in the Chancellorsville area until Sedgwick had defeated Early's troops and was ready to attack Lee in rear. He would then engage Lee's two wings.

The first option would have given Hooker every chance of completely destroying the Confederate army, which was his directive from Washington. However, for it to succeed it was essential for Hooker himself to act offensively, because Sedgwick would have a long march, and Lee must

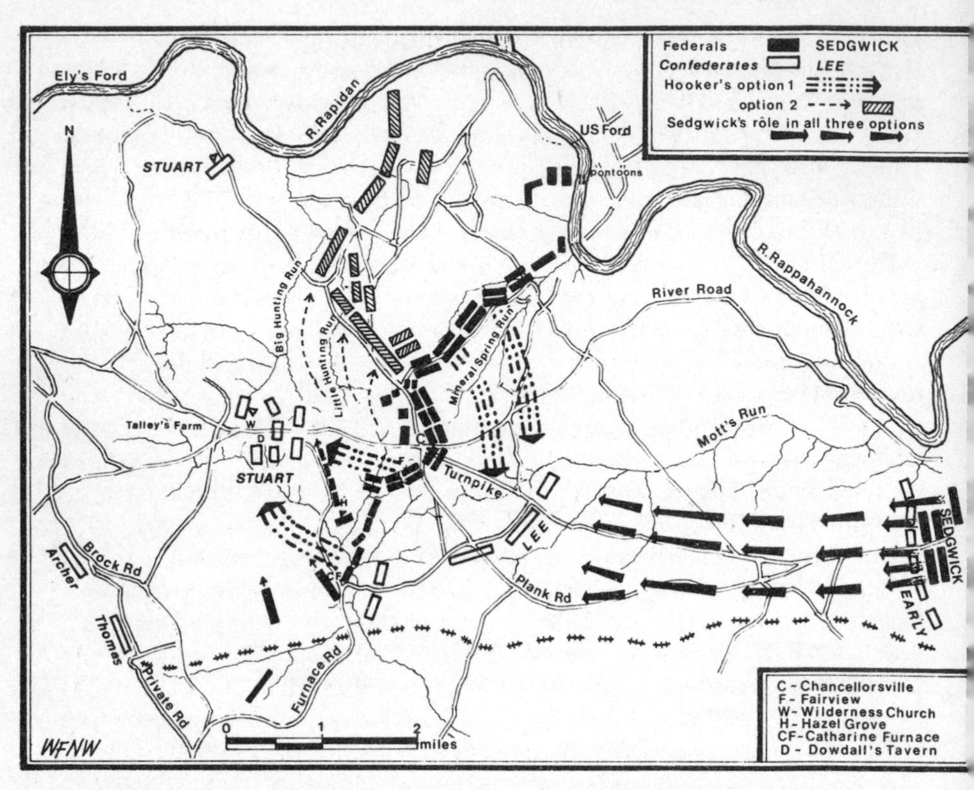

54 The Battle of Chancellorsville: 2 May, Hooker's third set of options

not be allowed to detach a force to take him in flank. Thus options 2 and 3 would not have been advisable. But Hooker had become defence-minded, in spite of the fact that he had an overwhelming superiority of numbers. He decided on the second option, and at 9 p.m. signalled Sedgwick accordingly.

During the night his men were ordered to prepare a shorter and naturally strong defensive line behind Hunting Run and Mineral Spring Run with both flanks resting on the river. I, V and a part of II Corps were withdrawn behind this line, while the remainder of the army was left forward – thus he had two lines of defence. It was certainly not the best decision, but it could conceivably work, provided the forward troops were sufficiently aggressive to keep Stuart from joining forces with Lee.

While Hooker's men were feverishly constructing what was to become in a remarkably short space of time an almost impregnable defensive line, he ordered Sickles to launch an attack by the light of the moon against the right flank of II Corps' advanced position. Although carried out with great courage by the men, the attack was not properly planned. It might well have been successful, for with the loss of their leader and the strain imposed by an intensive night battle, the Confederate II Corps was temporarily subdued. As it was, little ground was regained, and a part of the force lost direction in the forest and was hotly engaged by the Federal troops that had not been informed of the movement.

The rest of the night of 2–3 May was comparatively quiet, but as the stars faded from the dawn sky, and the men stood to, Stuart – no mean successor to Jackson – was ready to attack in column of divisions, deployed in line of brigades. Nor was he slow to take advantage of an inexplicable move by the pusillanimous Hooker, who just before dawn had ordered Sickles to withdraw from the vital Hazel Grove feature. The importance of this salient was immediately recognized by Stuart, who established thirty pieces of artillery on it, and the cross-fire of these guns with those of Anderson and McLaws, which came up from the south and south-east, was eventually able to silence the Federal batteries on Fairview. Furthermore, by withdrawing these forward troops Hooker made it much easier for the two wings of the Confederate army to unite.

But unification was not achieved without some of the hardest close-quarter fighting of the whole war, and for five hours the battle swayed first one way and then the other. Valiant men storming through a hell of whistling metal took the Federal breastworks, only to lose them again in the face of an equally courageous counter-charge. But Hooker for some reason – perhaps a head blow he received when the Chancellorsville house was hit had still further unnerved him – would not reinforce his hard-pressed front line with the two corps of Meade and Reynolds, which formed a reserve line but could so easily have been brought into the battle. As it was, with ammunition failing, and the brushwood (and many of the wounded in it) burning, the Federal infantry slowly and sullenly gave ground. At 10 a.m. on 3 May, the two wings of Lee's army were again united, and therefore their chances of bringing the battle to a victorious conclusion were greatly enhanced.

Hooker was now back in his strongly defended tight semi-circle, arched between the Rapidan and the Rappahannock. Sedgwick had not made an appearance, and neither Hooker nor Lee could be sure of exactly where he was, but Hooker had been moved from the position which made a junction with Sedgwick possible. *Lee now held the advantage, and had two options:*

(1) To launch an all-out attack on the crippled and demoralized Federal army, leaving Early to deal with Sedgwick as best he could.

(2) To hold Hooker in check, and turn the greater part of his army on Sedgwick.

Lee decided to adopt option 1. To drive the Federal army out of their strong defensive position and into the Rappahannock would be a fine and conclusive feat, but it was fraught with difficulty and danger. Certainly the Federals were in a very shaky state, but their defences were strong, and Sedgwick could take Lee in rear at a time when he was fully committed frontally. In fact this nearly happened, for at about midday – when preparations for an assault on Hooker's lines were nearing completion – Lee received a message saying that Sedgwick had broken through at Fredericksburg and taken Marye's Hill. This decided Lee to change to the second option, which could well have been the better of the two.

Jackson's flank attack was one of the great epics of the battle of Chancellorsville, but the fate of the campaign was finally decided by the fighting that occurred to the east on 3 and 4 May, where Sedgwick's VI Corps did battle with Early's division, which was later reinforced by Lee. The sequence of events that led to Lee's decision to swing the bulk of his force against Sedgwick started at 11 p.m. on the night of 2 May, when Sedgwick received the order from Hooker (see p. 190) to cross the river at Fredericksburg, to advance on Chancellorsville so as to be there by daylight, and to leave one division to guard his rear. On this sector of the front at this time the Confederates held a line some seven miles long with five brigades, from Taylor's Hill on the left to Hamilton's Crossing on the right. Sedgwick's Corps was already across the river and carrying out operations in pursuance of one of Hooker's earlier orders – to 'pursue the enemy on the Bowling Green Road'. He now ordered Gibbon to bring his division (less the brigade still at Banks' Ford) across the river at Fredericksburg.

The commander of VI Corps must have been dazed by the weight of contradictory orders he received from the army commander; we are told that on receipt of this last message he even thought of recrossing the river, in order to cross it again at Fredericksburg as ordered! Fortunately – if that report is correct – there were insufficient pontoons available, and instead he marched north so as to destroy the enemy before heading for Chancellorsville.

Sedgwick has been criticized for not fighting his way through Early's less strongly held right flank, and thus getting to Chancellorsville more quickly. Had he done this he might have taken Lee in rear, but would more likely have been taken in rear himself by the bulk of Early's division then holding the strong position opposite Fredericksburg.

The key to the Confederate position was the Marye's Hill ridge; as soon as Early learned that Sedgwick was moving against it he sent Hays with a brigade to reinforce Barksdale. Even so the three-mile line from Taylor's Hill to Howison's House could only be very lightly held.

Barksdale had four regiments occupying ground from Marye's Hill to the vicinity of Howison's, with six guns placed either side of Brompton House, and more artillery on Lee's Hill. When Hays arrived one of his regiments was placed on Barksdale's right and the remainder of the brigade occupied the ridge which ends with Taylor's Hill. A canal ran in front of this position, which gave added protection.

Hooker had omitted to inform Sedgwick of the strength of the enemy he might have to overcome on his way to Chancellorsville, so although he should have been aware of the natural strength of the left and of the position he was uncertain as to how strongly it was held.

In deciding how best to break through the Confederate line he had four options:

(1) A frontal assault on the key position (i.e. the line from Taylor's Hill to Marye's Hill) for which he had two divisions (less a brigade) available, leaving the remainder of his force to hold in check the centre and right.

(2) To turn the northern end of the Confederate position with one division, and attack frontally with another division.

(3) To try to break the key position by a pincer movement, with a division turning the north end of the line and another the southern part of this sector in the area of Hazel Run.

(4) To launch a simultaneous frontal attack on the whole length of the Confederate line.

The military historian is accustomed to sorting out differing opinions, contradictory contemporary accounts and inaccurate information, and the battle of Chancellorsville presents him with a great many such hazards. It cannot be said for certain which option Sedgwick had determined on, but from a careful study of the ground, and an examination of papers and documents, it would seem to have been option 3. However, it is quite certain that at least some regiments of the leading division (Newton's), if not the whole division, put in a frontal attack first which was thrown back with heavy casualties by Confederate troops lining the famous stone wall, from which six months earlier they had destroyed Burnside's army.

It is probable that this attack was only to determine the enemy's strength, and from it Sedgwick would have got the wrong impression, for in fact although the defences were strong the numbers behind them were small. In any event he then decided upon a pincer movement (option 3). He ordered General Gibbon, who had just crossed the river, to turn the left of the position above Taylor's Hill, and General Howe to manoeuvre his division between the two Confederate lines in the area of Hazel Run, and turn the right of the upper sector. But it seems that he had failed to appreciate the difficulties of the ground, for Gibbon was stopped by the canal, and Howe by the depth of the stream, and the fact that he could not manoeuvre between the two parts of the enemy's line. Both divisions

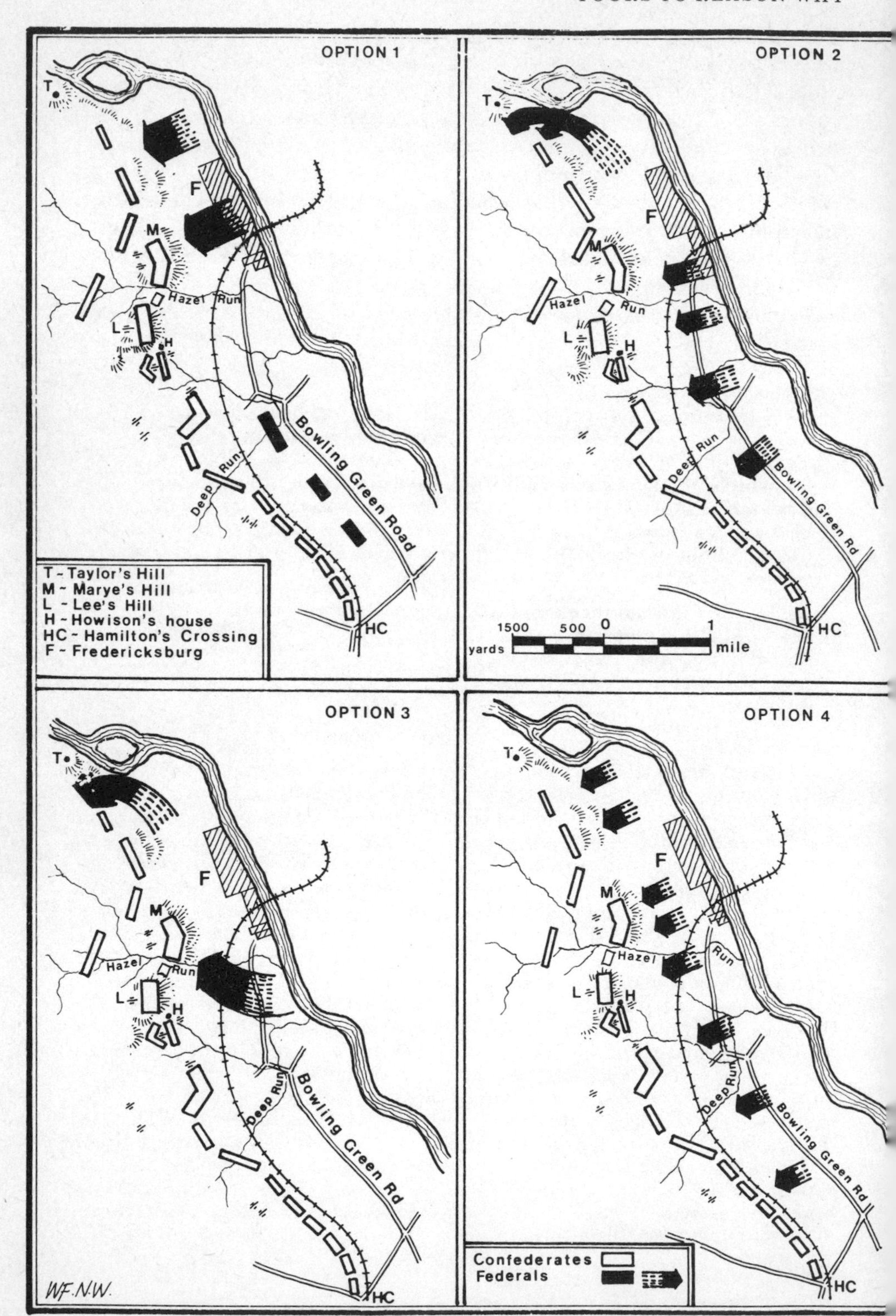

55 *The Battle of Chancellorsville: 3 May, Sedgwick's options*

suffered casualties, and this second failure, together with an incident that may well have occurred after this attack,* convinced Sedgwick that he had no alternative but to adopt option 4.

An interesting situation developed at this critical stage of the battle, and it is necessary to go back in time a little to get it in perspective. It will be remembered that General Wilcox's brigade of Anderson's division had been sent to Banks' Ford on 1 May (see p. 184). This ford was of considerable importance to the Federal army; on the other side of the river to Wilcox was one brigade of Gibbon's division and also Engineer General Benham in charge of the pontoons. The latter had been ordered by Hooker to send all his pontoon boats upstream to United States Ford – Hooker was clearly looking after himself first and Sedgwick not at all – but Benham, who knew his job, reckoned that Hooker could get by with only half his (Benham's) fleet and acted accordingly. This decision was to save VI Corps from complete destruction.

At the same time as Benham was making this wise decision, Wilcox could see how hard-pressed was Early's left flank on Taylor's Hill. *He had two options:*

(1) To march to the sound of the guns.

(2) To remain where he was, carrying out his orders to keep the important ford covered.

Wilcox decided to leave Banks' Ford uncovered and march to the assistance of Hays on the extreme left of the Confederate line. His arrival on Taylor's Hill did not prevent the Confederates being driven off the position; nevertheless it was a bold and correct decision, and as it so happened Wilcox then found himself in a favourable position to harass Sedgwick when he eventually started his march for Chancellorsville.

At 10 a.m. on Sunday 3 May Generals Newton and Howe advanced their regiments in line of battle against the two ridges from Taylor's Hill to Howison's House, to be met with withering fire from the hopelessly outnumbered, but strongly embattled, Confederates. After some desperate fighting the defenders were driven from their position, and with his right now in the air General Early was forced to retire to a new line

*Permission was sought and honourably, but unwisely, granted for the Federals to remove their wounded under a flag of truce, thereby showing them how slender were the Confederate resources.

of defence two miles in rear. From here he sent the message that decided Lee to march to his aid.

Early withdrew in two bodies (the larger using the Military and Telegraph Roads and the smaller the Orange Turnpike) and Sedgwick did not pursue him, for his orders were to make for Chancellorsville once he had cleared the enemy from his path. However, he wasted a considerable amount of time relieving the leading division, and his progress was delayed still further by shell and skirmish from Wilcox. It was therefore about 4 p.m. when he met General McLaws' division near Salem Church. The Confederate line was just inside a thick wood with open fields stretching for about 1,000 yards beyond. The Federal attack had to be made over this open ground in the face of artillery fire and, when the leading troops reached the wood, very concentrated small-arms fire. The fighting was extremely fierce and the casualties heavy. First one side and then the other gained the advantage, until darkness and fog put an end to that day's operations, with the Confederates occupying their former position unmolested, and Sedgwick falling back to a defensive position covering Banks' Ford.

The next morning (4 May) Lee, with three brigades of Anderson's division, arrived and took command. The Confederates now had the advantage of numbers, and Early was preparing to swing right-handed to retake Hazel Run, Marye's Hill and Fredericksburg. Though Hooker, sitting in his fortified horseshoe, had almost lost the battle, *he still had three options:*

(1) To oppose vigorously the small force in front of his lines, and order Sedgwick to continue attacking.

(2) To sit tight and order Sedgwick to do the same, hoping to win a defensive battle.

(3) To reinforce Sedgwick so as to defeat Lee's right wing, which would leave the small Confederate left wing at his mercy.

Hooker chose option 2, which was the only decision that ensured the defeat of his army. Had he adopted option 3, Lee must have been in grave difficulty. From the moment of crisis, about 4 p.m. on Sunday 3 May, until late on the night of 4 May, Hooker had done nothing constructive, but contented himself by sending a series of contradictory messages to Sedgwick.

Although the Confederate reinforcements had assembled before midday on 4 May, it was not until about 6 p.m. that preparations and reconnaissance were completed and the attack ready to be made. There is an interesting eyewitness account of General Lee's unusual lack of calm at the incompetence which caused this delay, thus enabling Sedgwick to escape. The latter, assailed in front and flank, fought stoutly until nightfall. He then sought, and obtained, permission to withdraw across the river, which thanks to General Benham, he accomplished in the early hours of 5 May. When it was too late Hooker countermanded the order permitting him to cross.

The next day Lee left Early's division and Barksdale's brigade to watch the river from Banks' Ford to Fredericksburg, and marched the remainder of his force back to the Chancellorsville area. Heavy and continuous rain made progress slow, and by the time his troops were in position it was too late to mount an attack on Hooker that night. But Lee, despite the fact that he had rather less than 40,000 men to storm a strong defensive position held by about 80,000, was confident that he could win the day. And so was Hooker, for when dawn broke on Wednesday 6 May the Federal lines were deserted. Now that Sedgwick was north of the river, Hooker was not prepared to risk being driven from his defences a third time, for on this occasion with a river behind him, defeat would indeed be catastrophic.

Lee's men, cheated of their prize, were perfectly capable of pursuit, and Lee was not a general to let a beaten foe depart in peace. But the supply situation had been adversely affected by Stoneman's cavalry raids (which apart from this had achieved very little), so an immediate pursuit was out of the question. A short time later – especially after Longstreet had joined him – Lee might have been able to cross the river and winkle the Army of the Potomac out of its camp opposite Fredericksburg, and manoeuvre it into an unfavourable position. But it is doubtful if what appears now to have been a favourable opportunity for a counterstroke was at the time a practicable proposition for Lee. No doubt both armies welcomed a respite from further savage fighting, and the politicians in Richmond were happy to wait upon events in the North, where certain Democrats were now said to be favouring secession.

Chancellorsville was a miniature campaign comprising four battles, and Lee's Confederate army had triumphed. But it was a costly victory. It is true that the Confederate losses in killed and wounded (13,000) were less than the Federals (17,000), but Lee could ill afford to lose so many, while the North could replenish its ranks without too much trouble. Moreover, the South had lost a man who was literally irreplaceable. Stonewall Jackson, with his cool, rational mind, clarity of vision and adventurous spirit, had fought his last battle, and with his going Lee had lost the one lieutenant he knew he could rely upon in any situation. Had he lived, Gettysburg might well have been a Confederate victory.

12 The Battle of Gettysburg

1–5 July 1863

THE battle of Chancellorsville decided nothing; it did not for long even lower the morale of the Army of the Potomac. There was nothing wrong with the fighting spirit and ability of those soldiers, it was just the army commanders who left a lot to be desired. So far there had been four, and perhaps the greatest disappointment had been the present incumbent, for Hooker had come with a high reputation as a fighting general, but when it came to the test he had proved a man of straw.

Meanwhile, the vice was tightening on the Confederates in the west – that theatre of war which was to prove so decisive. So long as General Grant could be kept out of Vicksburg, and the 200-mile stretch of river between that town and Port Hudson could remain in Confederate hands, so long could the Confederacy remain a united nation. But by May Grant had got a fair stranglehold on Vicksburg and had launched a joint land and river operation to wrest the town from the Southerners, and it was partly to relieve the pressure on General Pemberton – the Confederate commander besieged in Vicksburg – that Lee planned his second invasion into Northern territory. There were other considerations that seemed to make this an opportune moment to invade, chief among them being a wrong appreciation of Union manpower, and a mistaken belief that an invasion of the North might have beneficial effects abroad.

Both armies had some reorganization to do. General Hancock succeeded Couch in command of II Corps of the Army of the Potomac, and there were command changes in the Cavalry Corps both in May (when General Stoneman left the Army) and again only two days before Gettysburg. General Hunt had sixty-five batteries under his command, and some 370 guns; in all the Army of the Potomac mustered between 85,000 and 90,000 men. Lee decided to divide his nine divisions into three army corps – instead of the two at Chancellorsville; General Ewell succeeded Jackson in command of II Corps, Longstreet retained the command of I Corps, and A. P. Hill had command of III Corps. General the Reverend Doctor Pendleton – who for the time being found his métier in the manipulation of cannon, rather than in the service of God – commanded fifteen battalions, with four batteries to a battalion and four guns to a battery, which were divided among the three corps. The single cavalry division under Stuart also had thirty guns in five mounted batteries, giving Lee a total of 270 guns, and about 75,000 men.

Towards the end of May Hooker's intelligence informed him that it seemed probable that Lee would soon commence a march towards Culpeper and Warranton with part or all of his army. *This gave Hooker three options:*

(1) He could march for Richmond, which would be virtually unprotected, and draw Lee's army after him.

(2) He could cross the Rappahannock and first deal with any force left in front of Fredericksburg and then pursue the main army.

(3) He could attack Lee in flank with his whole army north of the Rappahannock.

To have adopted the first option might have had the desired effect, but it was a gamble, and Hooker was under constant pressure to keep Washington covered. He decided that in the event of Lee leaving troops at Fredericksburg, he could not pursue the Confederates with an unbeaten force in his rear; Hooker therefore decided on option 2 and informed Lincoln accordingly. But neither the President nor General Halleck (commanding general of the Union forces) would support this decision. They advised, and they were almost certainly right, that even with greatly superior numbers the strong defences round Fredericksburg would occupy Hooker for too long, and allow Lee's main army the freedom of the North. Hooker, therefore, had to content himself with option 3, whether or not Lee left a containing force before Fredericksburg.

Lee began his move north on 3 June, and he did leave troops at Fredericksburg. A. P. Hill's corps was given the task of delaying the Federal army, and in this it was partially successful. Hooker sent troops across the river at Franklin's Crossing in order to gain information, and being dissatisfied with what he learned ordered Pleasonton, his cavalry commander, to carry out a reconnaissance in force towards Culpeper. It was here that the remainder of Lee's army was ordered to concentrate, with the cavalry at nearby Brandy Station. Pleasonton's reconnaissance force met the Confederate cavalry between Brandy Station and Beverley Ford. Stuart, with the stronger force, managed to beat off the attacks of Generals Buford and Gregg in a battle that lasted several hours, but he had been badly mauled, and it was a turning point in the cavalry war, for it was the first occasion on which the Federal horsemen showed themselves to be the equals of the Confederates.

In the battle of Brandy Station Stuart was forced to call for infantry support, which enabled Pleasonton to confirm the forward movement of Lee's army. Hooker accordingly left Fredericksburg on 13 June, and when the last of his troops was clear of the Stafford Heights A. P. Hill marched his corps to join Lee. On the 14th, General Ewell's II Corps engaged and defeated the Federal General Milroy at Winchester, and with Hooker coming up on his right flank Lee now made for the Potomac.

Longstreet advanced along the eastern edge of the Blue Ridge, while Ewell's and Hill's corps moved down the Shenandoah Valley. But at this stage Lee made a serious error in allowing Stuart – with three of his five brigades – to ride round the right flank of Hooker's army to carry out a raid in the Harrisburg area. In consequence the Confederate commander was groping in the dark without precise information of Hooker's movements. Ewell's corps crossed the Potomac at Shepherdstown between 15 and 22 June, Hill's crossed at the same place during the 24th and 26th, and Longstreet crossed at Williamsport also during the 24th and 26th. The three corps were together briefly at Hagerstown, but then pressed on independently. Longstreet and Hill made for Chambersburg, while divisions of Ewell's corps went as far north as Carlisle, and east through Gettysburg to York, where they collected much needed supplies for the army and in so doing roused the wrath of the Pennsylvanians.

Hooker marched to the Potomac on a parallel line to the Confederates, but nearer to Washington, and crossed the Potomac at Edward's Ferry on 25 and 26 June. Farther up the river at Harper's Ferry, General French had a force of 11,000 men. Hooker, like his predecessors in command of the army, never obtained accurate information as to enemy numbers, and as he was under the impression (quite wrongly) that his force was inferior to Lee's, he wished to hive off a corps to join up with French to form a powerful threat to Lee's communications and rear, while the remaining six corps engaged the Confederates frontally.

However, General Halleck (the Commander-in-Chief)* when he learned of this plan, forbade it because he refused to consider the abandoning of Harper's Ferry – even as a temporary measure. Hooker, very understandably, resented this further thwarting of his plans, and sensing that Washington had lost faith in him tendered his resignation, which was promptly accepted.

General Halleck was certainly wrong to veto Hooker's quite sensible plan, and probably wrong to have accepted his resignation on the eve of

* President Lincoln had appointed Major-General Henry W. Halleck to be Commander-in-Chief of the Federal armies at the time of McClellan's failures in 1862. He was succeeded by General Grant in 1864.

a large battle. Meade, who was appointed Hooker's successor, had not been in his commander's close confidence, and did not even know the exact whereabouts of all the corps. However, he had had considerable experience in command, having been present as a divisional commander at Antietam, and a corps commander at Fredericksburg and Chancellorsville. He was said to be quick-tempered and irritable, but his tactics were sound, although, like nearly all his predecessors, he erred on the side of caution.

It was now that Lee suffered from having allowed Stuart to ride off in pursuit of glory (which eluded him) round the right flank of the enemy, for although Lee learned that Hooker was across the Potomac and marching north, he had no idea of his route or order of march. However, by 28 June he realized that a battle was imminent and he ordered his far-flung divisions to concentrate at Cashtown, eight miles west of Gettysburg. By the night of 29 June units from both armies ringed Gettysburg at various distances, but no thought had been given to fighting the battle there; indeed, Meade, who had established his headquarters at Taneytown (fourteen miles from Gettysburg), had reconnoitred a defensive position at nearby Pipe Creek.

But on 30 June General Heth, commanding a division of Hill's Corps, ordered a brigade into Gettysburg to procure shoes for his men, and these troops, on meeting a column of Buford's cavalry (part of a force that Meade had detached under General Reynolds to cut across Lee's expected line of march) just west of the town, retired to report. The following day Hill (Lee had not yet reached Cashtown) ordered two brigades of Heth's division to advance on Gettysburg to test the strength of the enemy in that area, and a mile west of the town they met Buford's cavalry division drawn up on McPherson Ridge. Events had decided the place of battle.

The country around Gettysburg is dominated by a series of low, parallel ridges running north and south. Just west of McPherson Ridge, where the first day's battle was mainly fought, is Willoughby Run, a fairly narrow, fast-flowing but shallow stream, which was no obstacle; half a mile to the east is the long Seminary Ridge, at whose northern end – beyond the town – is Oak Ridge. East of Seminary Ridge – and in places less than a mile distant – is Cemetery Ridge. At its southern end are two rocky outcrops called Round Top and Little Round Top, which dominate the ridge and provided good observation posts; at the northern extremity of the ridge is Cemetery Hill, and here the ridge curves round in the shape of a fish-hook to end at Culp's Hill. Running north and south just to the east of the town is Rock Creek, which at the time of the battle (but not now) was a stream in places ten feet wide and three feet deep with steepish banks, and much of its course through woodland. Within this stern amphitheatre the greatest battle of the war was soon to be fought.

Buford's cavalry drawn up along McPherson Ridge fought dismounted, desperately hoping to keep Heth's men at bay until the infantry could come to their support; for two hours they held their ground with great courage. By 10 a.m. General Reynolds had come on the field in advance of the leading infantry division, which was commanded by General Wadsworth and contained Meredith's famous 'Iron Brigade', and these troops were soon in action. They drove the Confederates from McPherson Ridge, capturing a brigade commander (Archer) and a fair proportion of his brigade, but the Federal army suffered an irreparable loss when Reynolds, riding forward to examine the ground, was killed by a musket ball. He was a fine general, who had recently refused command of the Army of the Potomac on the grounds that Washington interfered too much with generals in the field. He was replaced in command of I Corps by General Abner Doubleday, later to become famous in connection with baseball.*

While this preliminary action was being fought, divisions of both armies were being hastened to the field – those of the Confederate army coming from the west, north and north-east, and the Federals arriving from the south. Early on the scene was General Howard, in advance of his XI Corps, who finding himself, on the death of Reynolds, the senior officer on the field assumed command of the battle. Lee had ordered Ewell to redirect his divisions from Cashtown to Gettysburg, and by midday Rodes' division had occupied Oak Ridge, and was soon heavily engaged by Union troops of I and XI Corps which had taken up a position north of the town. Meanwhile, Heth's division – later reinforced by Pender's – of Hill's Corps kept up the pressure on the Union left, where General Doubleday's I Corps were mostly deployed.

It now became a matter of time as to which army would receive reinforcement first. Howard did his best to hurry forward the corps of Slocum and Sickles, but it was the appearance of General Early of Ewell's Corps along the Harrisburg Road that made the Union right flank untenable. Howard, by about 4 p.m., saw that he had no choice but to order I and XI Corps, whose men were by now beaten and about to break, to take up a position on Cemetery Hill and Cemetery Ridge on the south side of the town. At about this time General Hancock reached Gettysburg with orders from Meade to assume command and to report on the suitability of the ground for defence. But Howard had probably already selected the strong defensive position that the Union forces were to occupy as they eventually reached the field, and Hancock was happy to concur, only noting a weakness on the left flank.

*During the battle Meade relieved Doubleday of his corps and gave the command to Major-General John Newton.

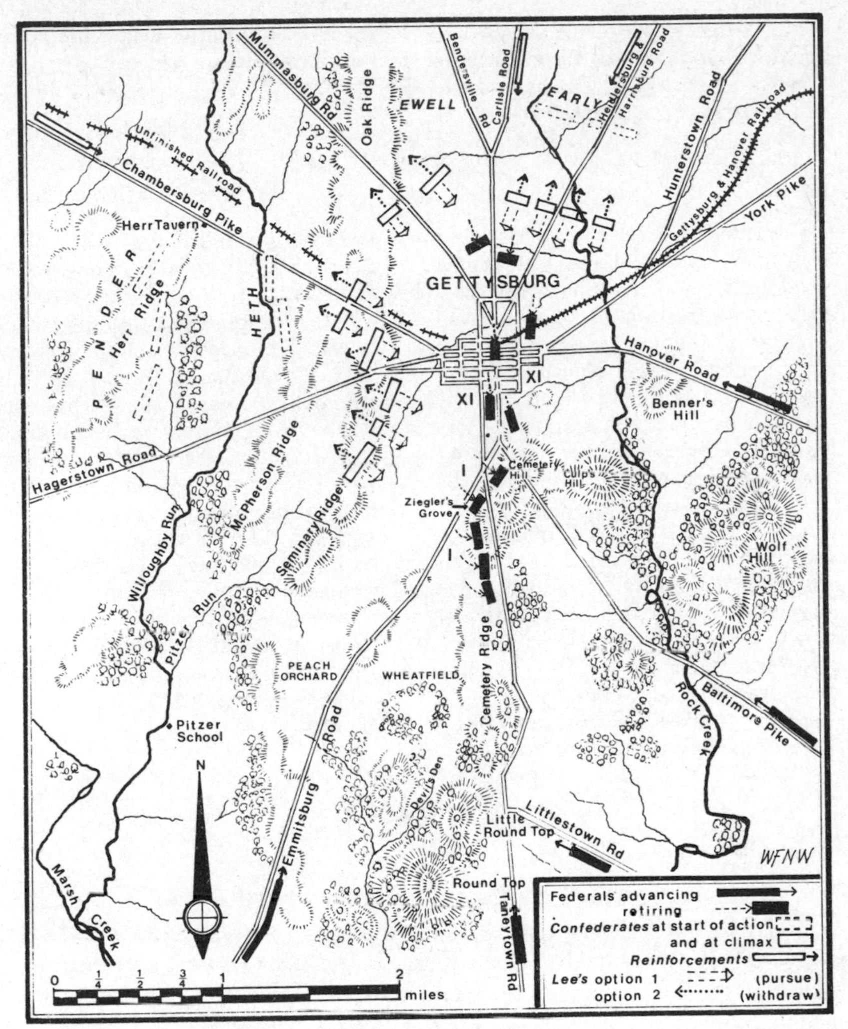

56 The Battle of Gettysburg: 1 July, Lee's options

Lee arrived on the battlefield in time to witness the discomfiture of the Union troops. XI Corps were streaming through Gettysburg to the area of Cemetery Hill, and Doubleday's I Corps (which had been terribly mauled) hardly paused on Seminary Ridge before crossing the intervening fields to Cemetery Ridge, the position Howard had earlier selected as a rallying point. There were still a good few hours of daylight, and Ewell's leading two divisions, although battle weary, were pressing the enemy; Johnson, commanding Ewell's other division, had not yet reached the field. *Lee had to make an immediate decision with two options:*

(1) To order Ewell to pursue the enemy, whom he could see were thoroughly disorganized, although he was uncertain what troops they had in support.

(2) To disengage, and wait to see what happened on the morrow.

Lee decided that the enemy should be pressed to the utmost, and he sent a message to Ewell to that effect. But the message seems to have left the final decision to the discretion of the corps commander – it was a fault of Lee's that he was inclined to trust his lieutenants too much – and in giving Ewell this discretion, Lee let slip a golden opportunity at the end of the first day's fighting.

Ewell was without full information, and his troops had had a hard afternoon's fighting, so he decided against pushing the attack. It is easy to see, with hindsight, that Lee was right and Ewell wrong, for the Union army was still scattered and those troops in the Gettysburg area could have been defeated before any organized defence was ready. Moreover, the vitally important Culp's Hill was unoccupied. Ewell (who was commanding a corps for the first time) is often blamed for his lack of action on the evenings of 1 and 2 July, but in this case at least some of that blame might properly rest with Jubal Early. He commanded the leading division, which towards the end of the fighting on that day was close to Culp's Hill, and although on account of the woods he might not have seen that it was unoccupied, its importance should have been appreciated and a reconnaissance made. This was an occasion when a subordinate commander, making a snap decision, could have greatly influenced events.

The Federal army spent much of the night of 1–2 July consolidating its position; as the divisions reached the field they were guided to positions previously determined by Howard and Hancock. General Meade and his headquarters arrived at 1 a.m., and when dawn broke he could see the strength of the position by riding round it. XI Corps occupied Cemetery Hill in a half-circle reaching back to Culp's Hill, where Wadsworth's division of I Corps faced Johnson's division, which had not been in time to fight with Ewell on the first day. Later in the morning Slocum's XII Corps took up position on Wadsworth's right. The remainder of I Corps and II and III Corps were allotted the line of Cemetery Ridge from Ziegler's Grove to Little Round Top, and V Corps went into reserve until the arrival of VI Corps (which was not until 2 p.m.), when they were moved to the extreme left of the Federal line.

It was a very strong position, but there were those at the time and later who thought it had been jeopardized by the action of General Sickles, commanding III Corps. He found that the ground allotted him was dominated by higher ground 1,000 yards to his front, and he sought

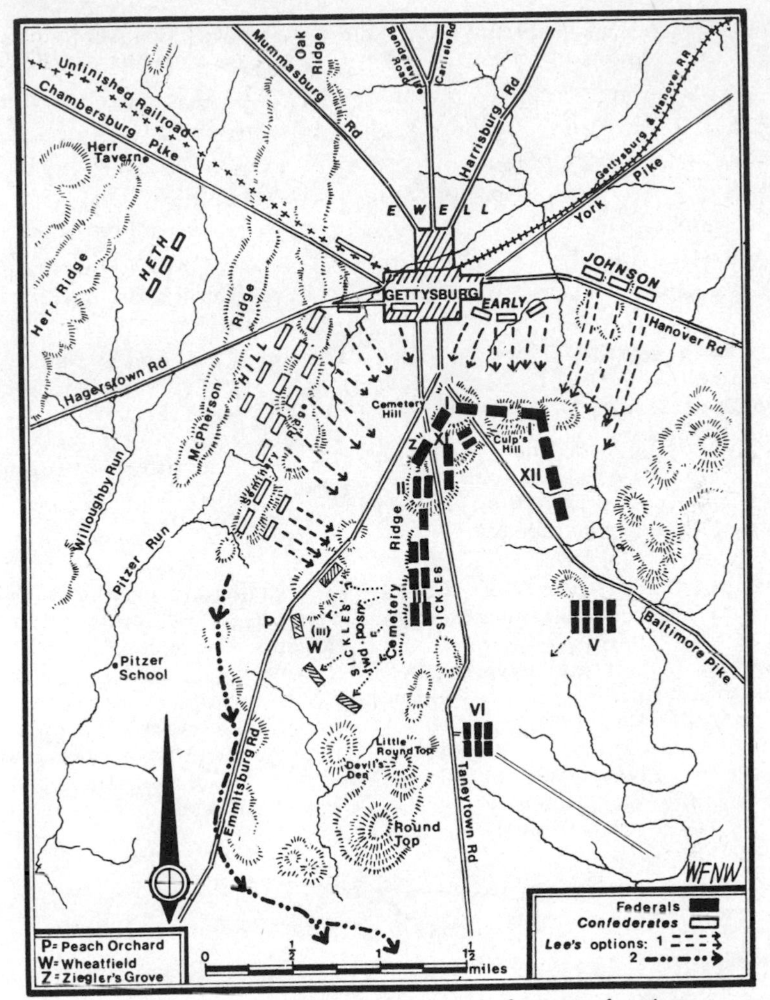

57 The Battle of Gettysburg: 2 July, Lee's first set of options

permission to occupy it. Meade sent General Hunt, his artillery commander, to examine and report. Hunt's report was ambivalent, and Sickles on his own authority advanced his corps to occupy ground just east of the Emmitsburg Road, in the Wheatfield–Peach Orchard area, with his left refused and running back to the Devil's Den. His corps was therefore in a very exposed position and his left flank presented a salient angle for the enemy to exploit. General Meade arrived on the site too late to change the dispositions, but made clear his disapproval. Sickles can be condemned for disobeying orders and making a wrong decision, or praised for showing an initiative that saved the day. Most critics have condemned his action; but strangely General Longstreet, writing some years later to Sickles – who lost a leg when Longstreet's Corps routed his salient – said, 'I believe it is now conceded that the advanced position at the Peach Orchard, taken by your corps and under your orders, saved that battlefield to the Union cause.' Either way it is an interesting study in decision-taking by a subordinate commander.

As soon as the Federal position became clear to Lee and it was obvious that Meade was expecting to be attacked, he had to decide whether or not to oblige him, and if so how. *He had two options:*

(1) To attack Meade before his army became fully concentrated, and he could make good use of interior lines.

(2) To refuse to attack a position that was already strongly held, and becoming stronger every hour, but to march round the left flank of Meade's army and position himself in favourable ground to the south, and there fight a defensive battle.

If we are to believe Longstreet, Lee had always intended on this campaign only to fight a defensive battle; Longstreet had strongly advised option 2, and was surprised that Lee decided on option 1. Longstreet attributed this decision to Lee having been made over-confident by the events of the previous afternoon. This may have been so, but it seems unlikely, because an invading army living on the land has to be aggressive, and usually cannot afford to stand on the defensive. Nevertheless, Lee was taking a great risk in deciding to attack this position, for although morale was high and numbers about equal, he was still without cavalry and therefore without proper information.

Seminary Ridge is a longer feature than Cemetery Ridge, and lies about a mile to the west across a shallow valley. It would have to be occupied and form the start line for any attack. Ewell's and Hill's Corps, which had borne the brunt of the fighting on the previous day, were holding a position at the north end of the ridge (Hill's Corps), and in the Gettysburg area (Ewell's Corps) facing south to cover the Federal right flank, where it bent round from Cemetery Hill to Culp's Hill. Longstreet's troops were fresh, although short of Pickett's division, and they would take the right of the Confederate line. *Having decided to attack, Lee had three options:*

(1) He could make his main thrust on Meade's right, seizing Cemetery Hill and Culp's Hill with Hill's and Ewell's Corps, using Longstreet's divisions to keep the enemy left occupied.

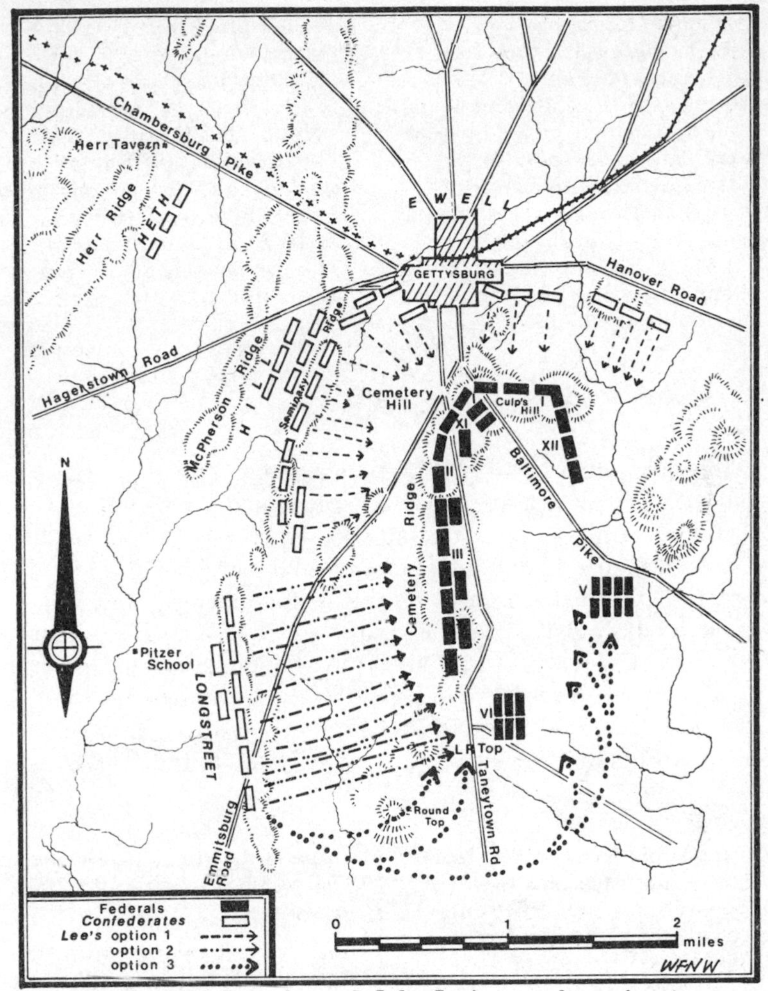

58 The Battle of Gettysburg: 2 July, Lee's second set of options

(2) He could order Longstreet to seize the left of the Union line (and at that time the Round Tops were not occupied), while Hill and Ewell co-operated as soon as they heard Longstreet's guns.

(3) He could circle round Meade's left flank and attack him in rear.

When Lee decided to adopt option 2 it is probable that he thought the Federal left was on the lower end of Cemetery Ridge, for it was unlikely that Sickles had moved forward by then. The ground, although nowhere particularly favourable for attack, was somewhat easier on the left of the enemy line and it was important to secure a foothold on the ridge before the Round Tops were occupied. Furthermore, Longstreet's men were much fresher than the troops of II and III Corps.

A move round the enemy's left flank and an attack in rear might have been the most sensible decision, for in committing his army to a frontal attack with inferior numbers, against an enemy strongly positioned on his own land, Lee was entering upon the wrong sort of fight. But without information as to where the rest of Meade's army was, a flank operation was too dangerous.

The decision to attack the left of the Union line must, therefore, have been the right one, but it will always be debatable whether Sickles' forward move made it easier. Longstreet, who should know, obviously considered that the III Corps salient was at least partially responsible for his narrow failure. But it is not easy to agree with this opinion.

For some reason – which has never been properly explained – Longstreet's attack was very late in going in. This was greatly to the Federals' advantage, for it meant that Sykes' V Corps reached the battlefield in time to fight, and VI Corps (which arrived at 2 p.m.) to form a reserve. Lee intended him to attack in the morning, but Longstreet says that he did not receive the orders until 11 a.m. He left the Herr Tavern on the Chambersburg Pike at about midday, and marched south-west to the Black Horse Tavern on the Hagerstown Road. Crossing this he ascended some high ground and suddenly realized that his column had been spotted by Federal signallers on Little Round Top. *He had two options:*

(1) To double back to the Herr Ridge (a distance of a little more than a mile), and renew his march in dead ground along the west bank of Willoughby Run to the Pitzer School, and here begin his deployment. This might confuse the enemy, and give the Confederates a covered approach, but it would mean losing about two and a half hours.

(2) To go forward along the track to the Pitzer School and be, for part of the time, in view of the Little Round Top observation post.

Longstreet decided to countermarch and remain in dead ground until almost – but not quite – the place of deployment. It was a decision that has seldom been criticized, but there are those (and the present writer is among them) who think that it was a wrong decision. He had already been spotted, and what could the enemy do? Sickles would not have had the time, or the inclination, to fall back. Meade was as ready as he could be, although he might conceivably have put some troops on Little Round Top, but in the event this was done anyway. Meanwhile the time that Longstreet wasted was vital to Meade, whose troops were still arriving on the battlefield.

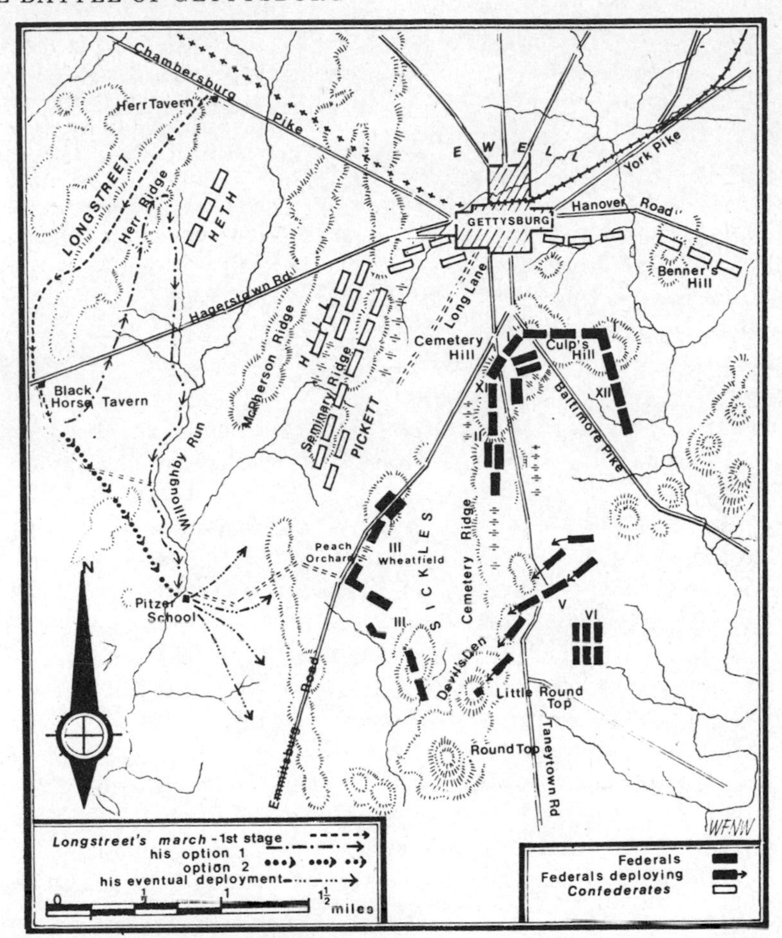

59 The Battle of Gettysburg: 2 July, Longstreet's options

It was about 4 p.m. when Longstreet hurled his 12,000 men against Sickles' III Corps. The fight for the salient was a desperate affair lasting upwards of two hours. Longstreet's artillery pounded the angle of Sickles' line while General Hood, commanding his first division, drove in from the west and the south, and McLaws launched his division also from a westerly direction. The brunt of the attack was borne by Birney's division of III Corps who, after Sickles had had his leg shot away, took command of the corps. Gradually the Federals were forced out of the Peach Orchard and wheatfield, and fell back fighting furiously to a rocky outcrop – aptly named by the soldiers the Devil's Den – and here the hand-to-hand grapple was most savage. Meade, seeing that III Corps was likely to break, ordered Sykes, who commanded some of the best disciplined men in the army, to send reinforcements. At the same time the army commander looked anxiously to his left flank, for Hood's men were past the Devil's Den. He sent one of his staff, General Warren, to examine the ground and report.

Warren ascended Little Round Top, occupied by a signal station. Here he could see Confederate troops forming up in the woods on the west side of the Emmitsburg Road preparatory to attacking Meade's left flank. He sent an urgent message to Meade for a division to be sent to this point; but before Meade could receive the message the situation developed dangerously, and Warren galloped down the hill and, running into his old brigade, advancing to the attack as part of Sykes' Corps, diverted it to Little Round Top. Through this prompt decision, taken entirely on his own initiative, General Warren saved the army's left flank from being turned, which might well have lost Meade the battle.

In the nick of time Meade was able to establish a secure anchorage for his left flank. But the crisis was not over, for in the centre – where Hill had thrown in troops in support of Longstreet – gaps were occurring and Meade, with his reserves fully committed, had to denude the right of his line with one division of II Corps and most of XII Corps in order to stem the advance on his left. This in turn left his right vulnerable, and there here developed one of the most interesting and crucial actions of the whole battle – and an action to which few historians have paid full justice. It also brings out a most intriguing study in leadership and command.

Ewell had been ordered to demonstrate in force against the right of the line (Cemetery Hill held by a greatly weakened XI Corps and Culp's Hill originally held by XII Corps) when and if conditions looked promising for a successful attack. Of Ewell's three divisions, Johnson's had not been committed during the first day's fighting, but the other two – particularly Rodes' – had been hard hit. General Hill had also been ordered to demonstrate against the centre of the enemy's line. Ewell received these orders in the morning, and presumably passed them on to his divisional commanders. Plenty of time, therefore, elapsed for the commanders to recce the ground over which they would have to attack. But it seems that this was not done.

Longstreet opened the attack about 4 p.m.; at this time Rodes' division was in the western part of Gettysburg town. Early in the east of it, and Johnson further east, and north of the Hanover Road. The opening of Longstreet's attack was the signal for a cannonade by batteries firing from Culp's, Cemetery and Benner's Hills. The Confederate battery on the latter hill was silenced, and withdrawn at about 5.30 p.m., and it was not until then that Ewell decided that it was time for his corps to attack. Johnson at this time was at least a mile from his objective (Culp's Hill), and as Ewell had ordered all three divisions to go in simultaneously, it would be a good hour before the attack developed.

In this engagement both Ewell and Rodes seem to have remained somewhat remote from the battle line, and although Rodes planned that his attack should be mounted from a start line along a track known as Long Lane, which paralleled the Emmitsburg Road, it was a twenty-six-year-old brigadier-general called Stephen Ramseur who was put in charge of the actual operation. The attack was not properly co-ordinated within the corps (Rodes' division had the longest approach march, and was not ready to engage at 7.30 p.m. with Johnson's and Early's troops), and although liaison with Hill's left brigade to co-operate was eventually achieved, this information never reached Ewell, who fought the battle believing his right flank to be in the air.

For an hour and a half, in rapidly failing light, Early's and Johnson's divisions fought furiously for a foothold on Cemetery Hill and Culp's Hill respectively. Early's men, in spite of the gallant attacks by the Louisiana Tigers who were virtually annihilated in this action, were thrown back off their hill, but Johnson's division managed to retain a foothold on Culp's Hill. Meanwhile, Ramseur – peering through the dusk at about 9 p.m. – could see the Union batteries positioned on West Cemetery Hill, with two lines of infantry behind formidable breast-works. He had a difficult decision to take. Should he carry out his orders to attack, or should he send back a message to Rodes recommending that the attack be cancelled?

Ramseur decided, after consultation with a neighbouring brigade commander, that the position was too strongly held, and the hour too late for him to launch an attack. He consequently informed Rodes of the position and recommended that the attack should be cancelled. Rodes did so, and Ewell concurred. These decisions were probably correct in circumstances that should not have arisen.

In fact only 2,900 infantrymen, out of the 6,400 that had occupied the position before Longstreet's attack, were still there, and there were no reserves. The position on Culp's Hill was very similar. Had Ewell's divisional commanders carried out preliminary reconnaissance, and co-ordinated their attack in good time, and had Ewell opened his offensive two hours earlier, it seems certain that Cemetery Hill and Culp's Hill would have fallen. Thus on the second day's fighting Lee lost the chance of victory, first by the prompt decision of a Federal general, and secondly by the indecision of one of his corps commanders and a general obfuscation of command within that corps.

The second day's fighting (2 July) ended with the Confederates having gained some ground on their right, but nothing decisive had been accomplished. Meade strengthened his left during the night, and returned to Culp's Hill (where they found their old position occupied by the enemy) the regiments of XII Corps that had been taken to strengthen the centre of the line. He was therefore stronger on each flank than in the centre.

During the night a most interesting event took place in the sparsely furnished, plain room of the farmhouse situated on the reverse slope of Cemetery Ridge, where Meade had his headquarters. Here were summoned all his corps commanders for a council of war. There are many, and conflicting, accounts of what happened, but it appears to have developed into a debate (not dissimilar to those held eighty years later in an American general's headquarters at Anzio), with the army commander giving no lead, but eventually – and reluctantly – agreeing to the majority decision. *Basically Meade had three options:*

(1) To follow the precedent set in all previous battles fought by the Army of the Potomac – with the exception of Antietam – and withdraw following two days' hard and not particularly successful fighting.

(2) To remain on the defensive, and let Lee spend himself attacking a strong position.

(3) To go over to the offensive.

It is difficult to pass judgment on Meade for his behaviour at this council, for – as already mentioned – the accounts of it vary so much. But certainly so far his had not been the guiding hand in the battle. Reynolds had decided to fight at Gettysburg, Howard or Hancock had selected the defensive position, and Warren had saved the left from being rolled up. Now, according to General Doubleday, it was Meade who wanted to withdraw – probably to Pipe Creek – and when his generals disagreed he broke up the council with the words, 'Have it your own way gentlemen, but Gettysburg is no place to fight a battle in.' On the other hand General Gibbon, who was present, says (and he was more likely to be right, for Doubleday had an axe to grind and was not at the Order group) that it was Newton who thought Gettysburg was no place for a battle. In any case, the outcome of this council was that the Union troops would stay on the defensive in their present position (option 2), and it was undoubtedly the correct decision, even if it was not arrived at in a very military manner.

Lee, unlike some of the Union generals, had no thoughts of withdrawal. His options were entirely centred on the best way to reinforce the partial success of the two previous days, for now he had three brigades of Pickett's division in the line, Stuart's cavalry were at last on hand,

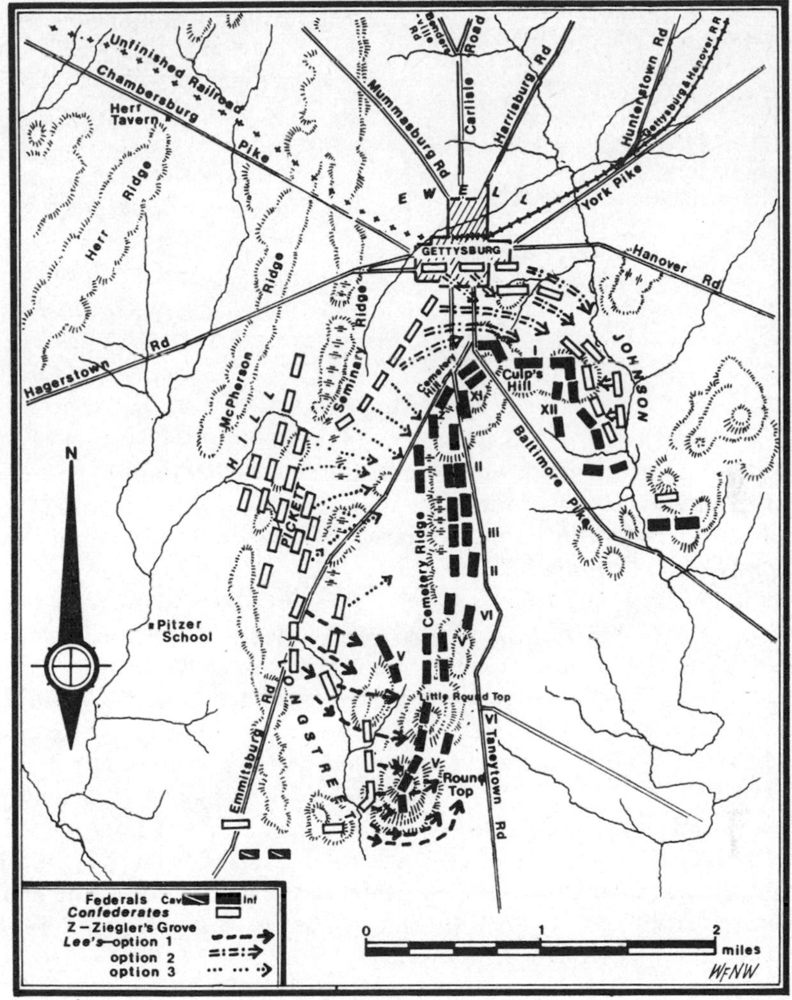

60 The Battle of Gettysburg: 3 July, Lee's options

and Longstreet and Ewell had gained a foothold on each of Meade's flanks. *He had three options:*

(1) To turn Meade's left flank with Longstreet's corps as his main effort, with subsidiary attacks in the centre and right by Hill's and Ewell's corps.

(2) To reinforce Ewell and try to turn the position from the direction of Cemetery Hill and Culp's Hill.

(3) To continue with a frontal attack against the centre of the line, with simultaneous support from Ewell and Hill.

Longstreet was adamant that 15,000 men could never pierce the Federal line, and that to turn their left flank offered the only chance of success. Lee disagreed, and felt that as the attacks the previous day on both flanks had come up against strong opposition the centre must be Meade's weak point. To attack the right flank would have been possible with careful co-ordination, but Cemetery Hill was strongly held. Whichever option Lee decided upon (and in fact it was option 3) the cavalry would need to operate to the east in rear of the Federal line to harass the expected retreat. It is difficult to be certain whether Lee or Longstreet had the right answer. In the event Longstreet was proved right, but this was partly because Lee's timing misfired, and there is no telling whether a flank attack would have been more successful – probably not with the Round Tops securely held.

The battle on the third day was a much more straightforward affair than the manoeuvres on the first two days, and the centrepiece was the magnificent, but unavailing, charge of Pickett's division, with Pettigrew's and Trimble's brigades from Hill's III Corps in support. But Lee's plan went wrong from the start. Ewell has been blamed for being tardy on the first two days, and hasty on the third. It certainly seemed that he jumped the gun on the morning of the third day instead of waiting to synchronize his attack with that of Longstreet's men, and Lee felt that the attack failed in consequence. But on this occasion he had less choice in the matter of timing than on the previous evening, for the men of the Federal XII Corps were determined that Johnson's forward troops on Culp's Hill would not remain there for long after dawn, and a battle was forced on Ewell which lasted from about 4 a.m. until 11 a.m. By then his troops had been decisively defeated, and could be of little further use. There followed two hours of uncanny quietness, which at 1 p.m. was shattered by the thunderous roar of 138 Confederate guns.

Lee had decided to pack his punch at the right centre of Meade's line, and his artillery was to blast that part of the ridge, and all that moved on it, for almost two hours. Nothing like it had been seen, or heard, on the land before. Shells screamed overhead, case shot exploded in the air above the prone troops, and as on this hot July day the sweating Confederate gunners sponged out the reeking barrels and rammed down more shot, a yellow fog of powder smoke and dust blotted out the valley. Meade replied with eighty cannon, but General Hunt, realizing that an attack was imminent, wisely conserved his ammunition. The Confederate artillery blazed on until at last the pyramids of cannonballs and stocks of grape began to run out, and General Alexander got Lee's permission to stop firing. The noise had been the worst part, for many of the guns were firing high, and those soldiers chiefly affected were the headquarters staff who had thought themselves secure on the reverse slope of the ridge.

The bombardment over, the sulphur-laden air began to clear, and at about 3.30 p.m. General Longstreet gave Pickett the order to advance. Twelve thousand men, their bearing high and proud, advanced across the valley towards the waiting enemy on Cemetery Ridge. A clump of trees (which still stands) marked the general direction,* and here the ground was held by Hancock's II Corps. As the lines came into view the Federal artillery opened up first with shot, then with shell and canister. Great holes were ripped in the ranks; the survivors knew that once through this hell a storm of musket fire would lash them like driven rain, but still they struggled on, no longer in neat formation, no longer keeping direction, but a confused mass of very courageous and determined men. As the leading files were mown down, those in the rear came forward; but these men – like all soldiers in close contact with danger, yet not in the firing line – were emotionally, as well as physically, exhausted, and although some reached the ridge they could not possibly remain on it. Moreover, a Federal brigade had now got round their right flank, and Pickett had no choice but to give the order to retire. He himself and one lieutenant-colonel, out of twenty-two officers of field rank, regained the Confederate lines; 67 per cent of the total force that charged with him were casualties.

At much the same time as Lee's infantry was suffering so severely along Cemetery Ridge, his cavalry under Stuart was being roughly handled by General David Gregg's horsemen. A spirited action which lasted for nearly three hours was fought some three miles to the east behind the Federal line; this was no skirmish, but a cavalry battle in the old style with several thousand men hellbent with sabre, supplemented with carbine and shell. In the end Stuart was driven from the field, and quite unable to give any aid to the infantrymen. Stuart can be numbered among the great cavalry leaders of all time, but Gettysburg was not his battle.

Lee could do no more than admit defeat. The remnants of his scattered brigades were pulled back to a line on Seminary Ridge, while the wagons, loaded with wounded, started the long march home. *Meade, the conquering general, had three options:*

(1) To allow the enemy to withdraw unmolested.

(2) To remain on the defensive that night, and counter-attack at dawn.

(3) To counter-attack immediately, while the enemy was still off balance. A fair proportion of Federal troops had not been fully committed.

*There is a school of thought that considers the trees at Ziegler's Grove were the intended focal point, for here the line was at its weakest, but that the leading troops lost direction in the heat of battle. The present writer does not subscribe to this view.

There is a very interesting account extant, written by a senior officer who was in conversation with Meade when he rode on to the ridge immediately after the Confederate troops had been beaten back. It is clear from what he said to this officer that Meade had no idea of the extent of his victory, and thought it quite likely that Lee would re-form and attack again. He therefore gave no orders for an immediate counter-attack, but instructions only for strengthening the line at certain points. Nor did he take any offensive action the next morning, but adopted option 1. He was a cautious general, who had been hit hard,* and Lee was something of a legend. Meade may have been right to wait upon events, certainly until the next day, for if Lee withdrew he would be ready to pounce. He has, however, been severely criticized for missing a great opportunity, and at the time Lincoln was sadly disappointed.

The fourth of July was a black day for the Confederate cause, for that morning Vicksburg was surrendered to General Grant. The Confederation was split in two and it no longer had any troops or fortifications along the Mississippi, for now – in President Lincoln's poignant phrase – 'The father of waters rolls unvexed to the sea'. On the plains of Gettysburg, where the President was soon to utter further immortal words, Lee knew that there was nothing he could do but to recross the Potomac, and in Virginia to mend the broken pieces of his sadly diminished, but still proud and defiant, army. Orders were issued at noon for the withdrawal, but night had fallen before the leading troops set off in torrential rain, which made the roads almost impassable, to secure crossings of the river at Williamsport and Falling Waters.

On the morning of 5 July, when General Meade saw that the Confederate position had been abandoned, *he had three options:*

(1) He could march in immediate pursuit of Lee (at least one corps, VI, was comparatively fresh, having done very little fighting at Gettysburg), and bring him to battle.

(2) He could shadow the retreating Confederates with his cavalry, and make a plan to slip round their flank, by way of the South Mountain pass, and head them off from the Potomac.

(3) He could proceed very cautiously and offer battle when a favourable opportunity occurred.

* The Union losses for the whole battle were a little over 23,000 men killed, wounded and missing, of whom 3,155 had been killed; the Confederates reported 2,592 killed and a total casualty list of 20,451 – but their returns were not complete and their total losses were probably nearer 25,000.

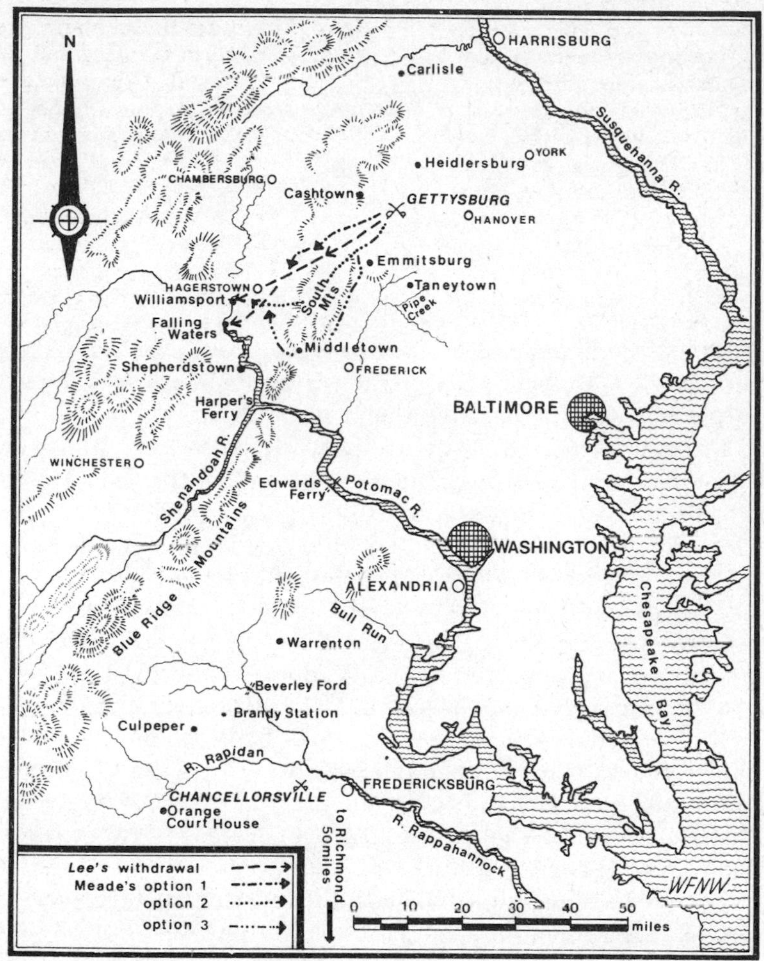

61 The Battle of Gettysburg: 5 July, Meade's options

Meade's thinking – as revealed in his frequent despatches to Washington – during this final phase of the Gettysburg battle takes a good deal of understanding. From his first despatch it is clear that he knew the approximate route that the Confederates were taking to the river, and he quite rightly sent out his cavalry to get confirmation, but he also gave orders for the army to move to Middletown, which by the Emmitsburg Road is some thirty miles from Gettysburg and well to the south-east of the crossing place that Lee was most likely to make for. His decision was a compromise between options 2 and 3, but he carried it out in a very leisurely way and it was 12 July before he came up with Lee. Despite the condition of the roads the Confederates had arrived at Williamsport on 6 and 7 July, but they were unable to cross the river because the recent rains had made it impassable.

On 12 July Meade held one of his beloved councils of war, and found his corps commanders almost unanimous in their desire not to attack. Meade would not overrule them until he had made a thorough reconnaissance the next day. That night, the waters having subsided, Lee made good his escape across the river. President Lincoln was aghast; he and Halleck found Meade's handling of the pursuit inexplicable, and it does seem that he failed to gather the fruits of the bitter struggle in which so many had given their lives.

Once across the river and back in Virginia, the Confederate army camped along the Rapidan near Orange Court House, and the Army of the Potomac near Culpeper. Throughout the summer they growled at each other, but nothing decisive was accomplished.

Although the war was to last for almost two more years. Gettysburg and Vicksburg marked the moment of truth. The Confederates continued to show a bold front and to display great courage under sound leadership, but it was obvious that they could no longer hope to triumph. The Mississippi valley and much of Tennessee were lost, the Army of Northern Virginia had suffered casualties it could not replace, and although it continued to give blow for blow as best it could (and occasionally, as at the battles of the Wilderness and Cold Harbor, to good effect) the end could not be long delayed. Moreover, in the western theatre a spectacular breakthrough had been achieved; General Sherman's armies, utterly irresistible, swept in fury from Chattanooga to Atlanta, and from Atlanta to Savannah.

In March 1864 Lincoln appointed General Ulysses S. Grant to replace Halleck in overall command of the Federal armies. He was the fourth general to hold that position, and even allowing for the fact that in the early days of all wars of long duration reputations are blunted – if not destroyed – by commanding generals being expected to do too much with too little, he was undoubtedly the best. Lincoln had at last found a general who would seek out the enemy, hit him hard and go on hitting him until he disintegrated and dissolved.

Under Grant's orders the Army of the Potomac gradually beat back the Army of Northern Virginia to entrenchments round Petersburg, while Sherman coming up from Savannah defeated General Johnston's troops in North Carolina. Fighting stubbornly all the time, against odds that grew greater every day, the Southerners could no longer stem the dyke; the bitter waters of defeat lapped all around them, and on 9 April 1865 Lee surrendered his army to Grant at Appomattox Court House. On 26 April Johnston surrendered his to Sherman at Bennett's House, near Goldsboro'.

When it was all over Jefferson Davis fled south, through the Carolinas into Georgia, where he was captured. He was taken to Fort Monroe and kept prisoner there for two years. Lee, a sad and tired man, went home to his family; he had behaved with fortitude and dignity in his darkest hour. The peace terms were not harsh, for President Lincoln – the architect of victory and so soon to be the victim of an assassin's bullet – understood the importance of moderation, and mercy for the vanquished. The doctrine of state sovereignty would be destroyed, and the system of slavery abolished – but at a cost of almost 600,000 lives and more than 6,000 million dollars.

In the immediate post-war years, the rapid development of industrial and financial capitalism in the North soon shifted the political power away from the hitherto slave-holding oligarchy of the South. Most of the fighting had taken place in the Southern states and the destruction of buildings and the devastation of the land dealt a crippling blow to Southern economy. There were years of bitterness during the difficult period of reconstruction (traces of which can still be found), but gradually the wounds healed, and henceforth the fortunes of a united nation would broaden through the years.

13 Palestine 1917

The Battles for Gaza

WHEN Turkey entered the war on the German side at the beginning of November 1914, Major-General Sir John Maxwell had recently been appointed commander of the troops in Egypt. The primary task of this force was to keep open the Suez Canal, a vital line of communication. But the strategy later developed was an offensive one, for three principal reasons. First, to protect the whole hundred miles of the Canal would require a very large number of troops, and so the Egyptian Expeditionary Force was in due course ordered to hold the enemy forward in Sinai. Later, as stalemate developed on the main front, a forward attacking policy was desirable in any theatre where it could be achieved, and by 1917 certain political considerations in the Middle East had a definite influence on those directing the overall war strategy in London.

The Turks mobilized over two and a half million men in the course of the war. The true Turk was a bonny fighter, particularly stubborn in defence and most proficient in gunnery, but many of the subject races became less reliable as the war progressed. Their overall commander in the Palestine theatre for the first three years of the war was Djemal Pasha, whose military prowess was no match for his ambition, and like so many of his senior colleagues he resented German interference. The man whose VIII Corps led the first attack on the Canal in February 1915 was also called Djemal, but the mastermind in that, and in the subsequent battles, was a Bavarian colonel called Kress von Kressenstein, who was at the time chief of staff to VIII Corps, but soon to rise to army command.

The Suez Canal is approximately 100 miles long from Port Said to Suez, although by flooding a part of the east bank the Indian troops allotted for its defence had a shorter front to guard. The principal defences were on the west bank, with a few outposts on the east bank. The first Turkish attack by some 20,000 men was an irresolute affair and easily beaten off. After Lord Kitchener's visit in November 1915 the defences were shifted from the west bank to a line about 11,000 yards

62 Palestine, 1916

east of the Canal to ensure that the enemy could not bring it under gunfire. For the remainder of the year, subsequent to the first attack in February, hostile activities were confined to raids by small parties of troops led by Kress; these raids were made largely because the enemy fully realized that any threat to the Canal must keep large numbers of British troops in Egypt.

At the beginning of 1916 the command structure in Egypt was reorganized, and General Sir Archibald Murray was given the command of all the troops, including the newly constituted Egyptian Expeditionary Force. By July he had completed the heavy task of sorting out the surplus units that had built up since the evacuation of Gallipoli, and was able to despatch about 240,000 men to France and other theatres of war, leaving four Territorial divisions, two independent brigades (which with garrison battalions amounted to some 60,000 rifles), one mounted division and one brigade of yeomanry for the defence of Egypt.

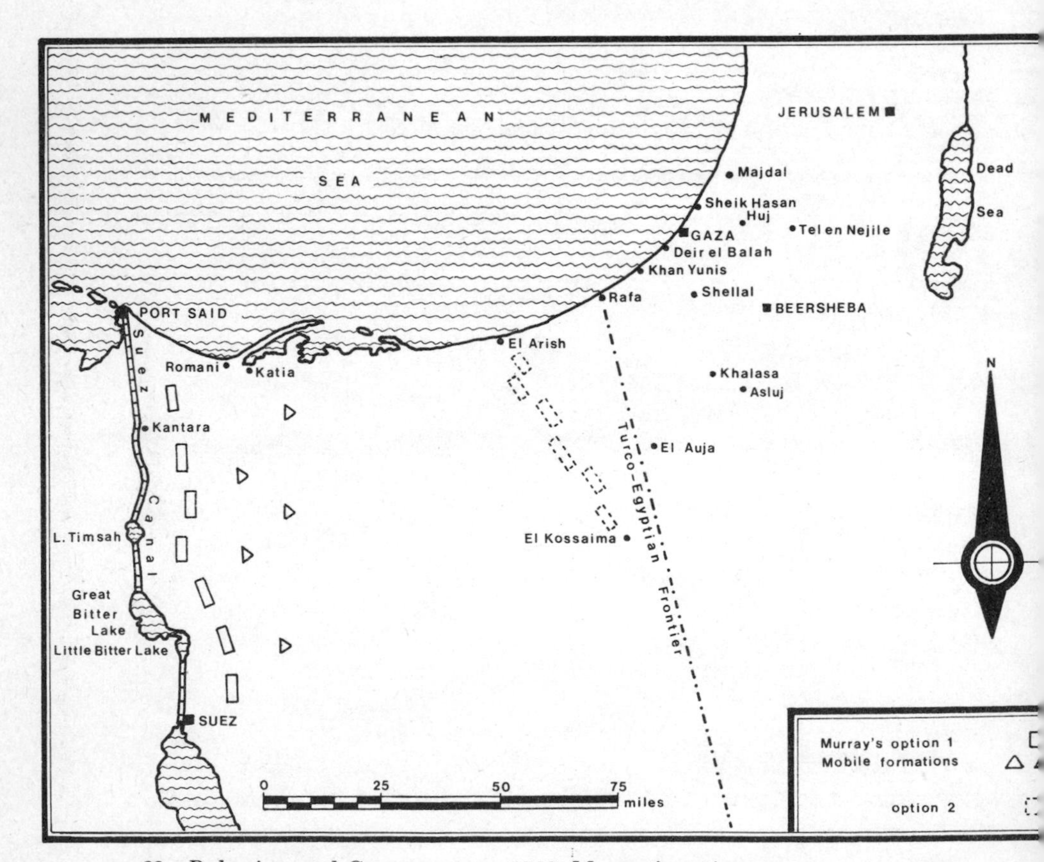

63 Palestine and Gaza: summer 1916, Murray's options

Murray's mandate was still a defensive one, and when he came to review the best method of implementing it *he was faced with two options:*

(1) To maintain the existing line of strong positions 11,000 yards east of the Canal with mobile formations operating in advance. This had the considerable advantage of forcing the enemy to launch his attack a long way from his base, with troops weary from crossing 150 miles of almost waterless desert.

(2) To bar the northern route across the Sinai peninsula by forming the strategical defensive base near to the Egyptian frontier on the line El Arish-El Kossaima. This would involve the construction of a standard-gauge railway from Kantara to Katia, and the laying of a pipeline for water; Murray estimated that to hold Egypt's eastern frontier he would need five divisions and four mounted brigades.

General Murray decided that the best defence of Egypt was to operate offensively and to advance his base to the El Arish line (option 2), and he accordingly submitted this appreciation to the new C.I.G.S., Sir William Robertson. The General Staff agreed with what was obviously the better decision (for apart from strategical considerations it removed the fighting further from an unsettled Egypt), and gave their consent for the additional requirements necessary for the operation.

The Turks were not slow to realize the danger of being denied the water supply in the Katia basin, and any threat to El Arish deprived them of their only base capable of supporting a large-scale offensive. In April 1916 Kress led a strong raiding force of 3,500 men, and gained a considerable tactical success at Katia, but this in no way interfered with Murray's strategic plan. Troops continued to leave Egypt for France, and Eastern Force, under the immediate command of Lieutenant-General Sir Charles Dobell, continued its advance; by the time Kress was ready to strike again in July it had nearly consolidated its position on the frontier.

This time, reinforced by strong German technical units, Kress's army mustered some 16,000 men and thirty guns; their first objective was to seize a line from which the Canal could be shelled and blocked to shipping. The opposing forces met near Romani, and on 4 August the Turks, who had attacked with resolution and fought most courageously, were decisively beaten and suffered heavy casualties. This action was sufficient to ensure the safety of the Canal, and from now on the initiative rested with the British. By the beginning of 1917 all Egyptian territory was free of the enemy.

On 11 January Murray was told that although for the time being no reinforcements could be spared him (indeed, one division was to be taken from him in March) and his role must therefore remain defensive, he should make preparations for an autumn offensive. By 21 March the railway had got beyond Rafa to Khan Yunis, and the pipeline was following closely. Kress had withdrawn to the Gaza–Beersheba line, and General Dobell felt that a limited offensive should be undertaken so as to form a firm base for the bigger autumn one, and to prevent Kress from slipping away. General Murray instructed him to take Gaza by a *coup de main*, and to ensure that the enemy was intercepted in his retreat. Thus was the scene set for the first battle of Gaza.

During the first three months of the year certain changes had been made in the composition of Eastern Force, and Dobell now had under his command the 52nd, 53rd and 54th Divisions, with the 74th being formed from dismounted yeomanry brigades and not yet with artillery or other divisional troops. The cavalry arm consisted of two divisions (the Anzac and Imperial), both of three brigades with a fighting strength of 900 rifles each; in addition there was the Imperial Camel Brigade, composed of four battalions with some 1,600 rifles, and two Light Car Patrols. The spearhead of the force was the Desert Column commanded by Lieutenant-General Sir Philip Chetwode, and comprising the two mounted divisions, the 53rd Division and the Light Car Patrols. The official history gives Dobell's total strength as 25,000 rifles, 8,500 sabres and ninety-two guns (mostly ten-pounders, although there were six sixty-pounders).

This was considerably in excess of what Kress could muster. He had strengthened the Gaza garrison to seven infantry battalions, five batteries and four machine-gun companies – about 3,500 rifles and twenty guns. There was a small garrison in Beersheba, and the remainder of his troops (12,000 rifles, 900 sabres and fifty-four guns) held positions some ten miles east and south-east of Gaza in the Ḥuj–Tel esh Sheria areas. He had realized the threat to Gaza and was in a position to reinforce the garrison with those of his troops (a fair proportion) that were in striking distance.

The gateways to Palestine are the two towns Gaza and Beersheba – lying a little short of thirty miles apart. Gaza stands on an eminence and commands the coast road; over the centuries it has many times been seared by the flame of war. The importance of Beersheba is its water supply – the last before the sun-baked, heat-sizzling hills that lie to the east and south. Gaza is well protected to the south by a natural barrier of dense cactus hedges, and to its west a belt of sand-dunes stretches down to the sea. This belt is in places quite wide – soft sand near the town, firmer ground near the sea – and from Rafa northwards there is a coastal strip, often ten miles wide, of good barley-growing land, which gradually gives way to the sterile brownish yellow dust of the desert. There were no roads immediately south of Gaza, but the cultivated strip was firm enough for everything but heavy motor traffic.

Two important features to the south of the town were the Wadi Ghazze and the Es Sire Ridge. The Wadi Ghazze, a dry watercourse (except when in spate during the rains) originates in the Judean hills and is fed by numerous smaller wadis; it reaches the sea about six miles south of Gaza. It was a formidable obstacle, being in places 100 yards wide with steep banks. The Es Sire Ridge rises from the north bank of the Ghazze, and continues in a chain of heights to Ali Muntar, which gives its name to the northern part of the ridge. To the south-east of Gaza there stretches an arid, and in parts undulating, plain towards Beersheba. There is a distinctive north-east, south-west running escarpment, the lower part of which was known as the Sheik Abbas Ridge, and below it is the fan-shaped Wadi Nukhabir. Sheik Abbas was an important feature in any attack on Gaza, for its possession by the British would effectively bar the garrison from receiving reinforcements from Beersheba.

Throughout the three battles for Gaza the supply of food, fodder, ammunition and above all water presented the commanders with many grave problems, and to a great extent dictated the course of operations. In Gaza and Beersheba there were numerous wells, a small quantity of brackish water fit for animals could be dug just above high-water mark along the beach, and the Wadi Ghazze had numerous springs capable of providing limited quantities of water; there were wells at Khan Yunis, Deir el Balah and Shellal, and east and south-east of Gaza in Turkish-occupied country numerous wells existed at Huj, Tel en Nejile and Sheria. But the main supply had to be piped to the water-head, and then transported chiefly by 1st Line Water Camels.

In making his plan to fulfil the Commander-in-Chief's instructions General Dobell had to rely on surprise and speed, because unless Gaza was taken by the end of the first day the mounted troops would have to return to base for water. *He had perhaps three options:*

(1) To screen Gaza with the mounted troops on the north, east and south-east sides in order to prevent the garrison escaping, or being reinforced from the Beersheba area, while the town was assaulted from the south-east by one division with a second buttressing the protective circle, and the third kept in reserve.

(2) To slip the mounted troops round the enemy's right flank to block the retreat, and make a one-divisional assault on Gaza, using another division to make a feint against the Hureira–Sheria detachments (thought to be a little over 5,000 men) to prevent them going to the assistance of the Gaza garrison, with the third division in reserve to reinforce either wing if necessary.

(3) To use the mounted troops to hold the ring (as in option 1), but mount a stronger attack in order to make sure that Gaza fell before Turkish reinforcements could arrive. This could be achieved by a two-divisional assault, with the protective screen strengthened by the inclusion of the Camel Brigade, enabling the mounted troops to attack Gaza from the north, if they were not needed to fend off Turkish reinforcements.

General Dobell's plan was for the Desert Column to hold the ring while one division attacked from the south-east, another buttressed the protective ring, and the third was in reserve (option 1). The Anzacs were directed on Beit Durdis (five miles east of Gaza), the Imperial Mounted Division on Khan el Reseim (seven miles south-east) and the Camel Brigade, which was attached to the Desert Column for the operation, completed the screen in the south (the El Mendur area). The 53rd Division was selected for the attack, with one brigade to advance up the Es Sire Ridge, another up the El Burjabye Ridge (just east of Es Sire) and the third in reserve. The 54th Division was to buttress the protective screen from the Sheik Abbas Ridge leaving one brigade near the Wadi Ghazze on call to 53rd Division. A small detachment under Colonel Money was to keep the enemy occupied along the coastal

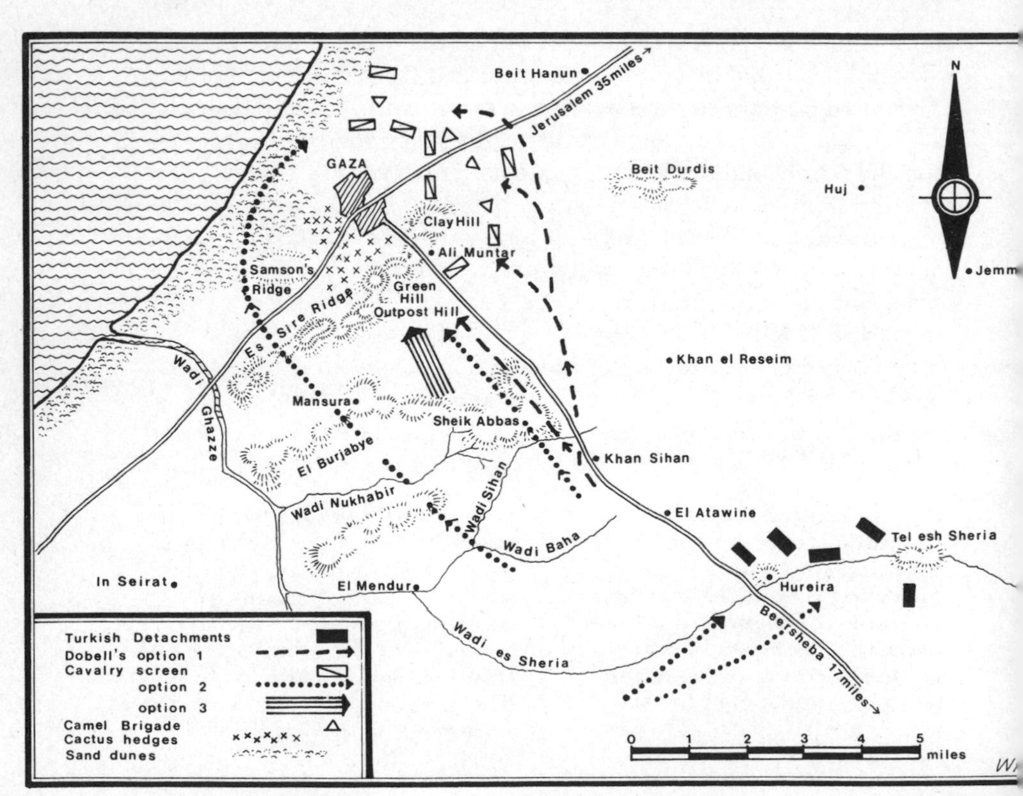

64 The First Battle of Gaza: March 1917, Dobell's options

sector. Headquarters Eastern Force was to be at In Seirat, and General Murray set up a command post in his train at El Arish.

In view of the outcome of the battle this plan has been criticized on the grounds that with time so short for a decision the punch should have been more heavily weighted (as suggested in option 3). It is true that neither the 52nd Division nor the Camel Brigade took any part in the fighting, but given the need to block enemy reinforcements by a screening force the front was fairly restricted, and while concentration is important congestion can be fatal.

As dawn broke on 26 March a dense sea fog swirled over the ground blotting out the few recognizable landmarks until nearly 7.30 a.m. This delayed General Dallas, commanding 53rd Division, in carrying out his reconnaissance and caused him to lose almost two hours of precious time; even so the two leading brigades reached their assembly positions by 8.30 a.m. The fact that the attack did not go in until nearly midday – and then only after urgent messages from General Chetwode – was mainly due to bad staff work resulting in orders not being transmitted promptly, the lateness of the reconnaissance and subsequent orders to brigades, and the fact that the brigade commander of 161st Brigade Group (on call from 54th Division) decided to use a different wadi to the one in his orders. This resulted in his entire force being 'lost' for most of the morning.

The mounted troops, however, through skilful navigation were in position by 10.30 a.m. Indeed the fog may have assisted them, and certainly enabled them to capture the commander of the 53rd Turkish Division and his staff as they were driving into Gaza. Some of these troops were now well placed to watch from their stern amphitheatre the opening stages of the battle.

The First Battle of Gaza, which should have ended with the capture of that town, was a sorry chapter of accidents. Bad communications, misunderstandings and incompetence (of the sort mentioned above) robbed the troops of their prize. When General Dallas was at last ready to begin the attack against troops holding the Ali Muntar position he did so on a two-brigade front, and when it was seen that these were making slow and painful progress in the face of a withering enfilade fire and the cactus hedges, he ordered his reserve brigade to reinforce the right of his attack. At about 3 p.m. the 'missing' 161st Brigade at last arrived, and an hour later (the delay seems quite unnecessary) Dallas directed it to advance between his left and centre brigades with orders to clear the enemy from a troublesome position later known as Green Hill. He thus had, by the late afternoon, four brigades directed on the Ali Muntar, Green Hill and Clay Hill positions south of the town.

Not long after 53rd Division's attack had begun General Chetwode realized that the appalling delays had thrown into jeopardy his whole plan to take Gaza that day, and at 1 p.m. he ordered General Chauvel to assault the north of the town with his Anzac division. At the same time he moved the Imperial Mounted Division and the Camel Brigade northwards to take over screening operations. This telegraphed order took an hour to reach Chauvel, who then had to call in his outposts, and it was not until 4 p.m. that the troops of this division hurled themselves against the almost impenetrable cactus hedges bristling with Turkish bayonets. But these formidable men never faltered, and by dusk (about 6 p.m.) they were in the northern and western outskirts of the town.

Meanwhile 53rd Division, assisted by two artillery brigade groups, had fought tenaciously throughout the afternoon, and by 6.30 p.m. had sent the Turks scuttling back to Gaza and completely cleared the Ali Muntar, Green Hill and Clay Hill positions. Their forward patrols had penetrated the town and in some instances made contact with the Anzac troops coming from the north. As the short twilight gave place to darkness Gaza was encircled save on the western side, where Colonel Money's small detachment had not been strong enough to break through.

The troops were in high spirits, the situation seemed eminently satisfactory, and come the dawn Gaza must surely surrender or be taken. But at command headquarters there were anxious times. At 4 p.m. messages had come in from Desert Column, later supported by air reconnaissance, of large Turkish reinforcements marching to the town's relief; indeed, troops of the Imperial Mounted Division had been attacked in the Beit Durdis area, and the engagement lasted until dark. Dusk came and the two generals at In Seirat were impatiently waiting for news of the capture of Ali Muntar; some progress had been reported, but it was after dark before the Turks were driven into the town from their outer defences. An agonizing decision then had to be taken about the mounted troops, and at 6.10 p.m. General Chetwode, with the approval of General Dobell, sent an order to General Chauvel for the mounted troops to disengage and retire behind the Wadi Ghazze.

It was a decision that has been criticized, but in fact Chetwode had no option. The infantry had not gained their objective; up to 4,000 Turkish reinforcements were reported closing in from the north; the horses were thought to be suffering from lack of water, and darkness was imminent. This withdrawal set off an unfortunate chain reaction.

The right flank of the 53rd Division was exposed by the withdrawal of the mounted troops, and 54th Division was ordered to close in to the Burjabye Ridge, but through an oversight this order was not communicated to General Dallas, the commander of 53rd Division. In consequence, when he was ordered to gain touch with 54th Division he protested, thinking that the division was still at Sheik Abbas; on being overruled, and in order to comply with what he thought was the position, he abandoned all the ground that had been so hardly won and fell back to a line south of Gaza. It was after midnight, and too late to countermand the order, before Dallas realized the true situation, and it was not until 5 a.m. on 27 March that Chetwode was fully appraised of this most damaging muddle. He at once ordered Dallas to reoccupy the Ali Muntar position, but it was too late; the Turks had taken fresh heart, and although at first successful the British were soon driven off by a strong Turkish counter-attack. Sheik Abbas had also been lost, and from this position the Turks could shell both British divisions now drawn up back to back on the Es Sire Ridge.

The First Battle of Gaza, so nearly a triumph for the perseverance and courage of the fighting soldier, had ended in failure. Ironically the three commanders threw up the sponge at almost exactly the same time. Kress, believing Gaza to be in British hands, and unwilling to risk his troops in night operations, ordered the relieving forces to halt at nightfall; Major Tiller, the German commander in Gaza, finding his Turks reluctant to continue the battle, had blown up his wireless station and resigned himself to captivity, and as we have seen General Chetwode ordered the withdrawal of his mounted troops. In war an ounce of optimism can sometimes win a battle.

Two important factors, had they been known at the time, would almost certainly have altered Chetwode's decision. Intercepted messages between Tiller and Kress indicating the former's hopeless plight were swiftly decoded and transmitted from Cairo, but for some unexplained reason did not reach forward headquarters until too late; and only a little while after this battle it was realized that horses could withstand desert conditions without water for more than twenty-four hours. The British casualties were 4,000 (523 killed) and the Turks 2,500 (300 killed).

General Murray's reports home concerning the battle spoke of the operation as being 'most successful . . . and just fell short of a complete disaster to the enemy'. Thus encouraged, and anxious for some spectacular success, the Prime Minister (Lloyd George) and the War Cabinet instructed Murray to clear the enemy out of south Palestine and take Jerusalem. General Dobell was accordingly ordered by his chief to proceed with this task at once.

Eastern Force had received new drafts. The mounted divisions got back their brigades from the Canal zone, bringing them up to four brigades each, the 74th Division now had three fully equipped brigades, although they lacked artillery, but the 53rd and 54th Divisions had their missing artillery brigades returned to them, and in general the artillery arm was stepped up to 170 pieces; however, only sixteen of these were medium calibre and over. Furthermore there were now eight tanks (experimental machines, but none the less useful), a French battleship and a supply of gas shells, all of which were new to this theatre. In the air the British were superior in numbers of machines, but the German Halberstadts could outmanoeuvre their Martynsides and BE-2s. Such a force could achieve much, but in all too short a time a lot of work was necessary to get the troops trained to fight with tanks and master siege warfare tactics; the staff had to work out naval co-operation plans, the proper use of gas shells had to be learned, and although the railway was now at Deir el Balah, and water arrangements were greatly improved, there would always be supply problems.

The Turks too had received reinforcements, though the British still outnumbered them greatly in mounted troops, and by about fifty per cent in infantry and guns. But there was no longer any thought of retiring from the Gaza–Beersheba line, and with three infantry divisions and a division of cavalry for his immediate front, together with two battalions and a battery at Beersheba, Kress now mustered 18,000 rifles and 101 guns. *In considering his defence he had two options:*

(1) He could use the bulk of his troops to hold a fortified front of some fourteen miles, and thereby ensure that Dobell would not be able to encircle Gaza again.

(2) He could favour mobility over a static position and keep, say, 10,000 of his troops as a mobile reserve ready to throw them against whichever position was most threatened by the British.

Kress decided to put his 3rd Division into Gaza (some 6,000 rifles and forty guns) and the 16th Division in the Abu Hureira–Tel esh Sheria area, while the 53rd Division plus a regiment formed a strong group between the 16th Division and the Gaza force. The 3rd Cavalry Division was kept in reserve at Huj. This was option 1. The Turkish line therefore stretched from the sea across the sandhills to Samson's Ridge (a feature two miles south-west of the town) through the maze of gardens and cactus hedges to Outpost Hill and then along a ridge between Outpost Hill and Ali Muntar, where there were numerous trench systems and *points d'appui*. East of Ali Muntar the defences comprised a number of redoubts strongly entrenched and capable of mutually supporting enfilade fire. These strongpoints were, from west to east, Beer trenches, Tank and Atawine Redoubts, and the Hureira–Sheria entrenchments.

It was an extremely strong defensive position, but without any mobile reserves Kress could be in difficulties if the British decided against making Gaza the principal objective. The extension of the railhead to Deir el Balah had greatly eased the British supply problem, and enabled Dobell to carry out a surprise concentration in any direction he desired, and perhaps gain a decisive decision before the Turks could bring reinforcements to the threatened sector.

Bearing in mind that his directive – to clear the enemy from southern Palestine and take Jerusalem – was wider than before his previous battle, *General Dobell had three options in planning his attack:*

(1) To pack a heavy punch with, say, four brigades along the coastal sector in order to clear a passage for the mounted troops to pass through. This would be comparatively easy, for the waterfront was defiladed from observation by the belt of sand-dunes. The troops would then get behind the Gaza force to pose a major threat to the enemy's communications, and quite possibly render the actual taking of the town unnecessary. One advantage of this plan was that with the naval flotilla able to co-operate, the enemy's strategic reserve (one division at Majdal) could be more easily turned. Moreover the water situation was favourable to a coastal push. Two divisions with suitable artillery support would hold the enemy's centre and right, ready to follow up when the line broke.

(2) To launch the main attack on Gaza in two stages, and on two days, operating on a three-division front – one advancing along the sand-dunes, one along the Es Sire Ridge, and a third swinging up on the right against the northern part of the town. The mounted troops would engage the enemy in the Hureira–Atawine areas, to act in protection of the infantry's right flank and to co-operate directly with the infantry in the northern attack. The first phase would be to secure a firm base from the sea to the Sheik Abbas Ridge, and the second to carry out the attack and right hook.

(3) To mount a strong attack north-eastwards from Sheik Abbas with two divisions in order to create a gap for the mounted troops to pour through, which would enable them to pounce on the Gaza force from the north-east, while the gap thus opened in the enemy line gave the shortest route to the Turkish railhead, and the ample water supplies at Huj and Jemmameh. The disadvantage to this plan was the water supply in the initial stages; because of this – as at First Gaza – the operation would have to be successful in one day.

General Dobell decided on the second option, but he made certain modifications when he received, on 10 April, more definite information on the enemy's dispositions. The first phase was to go in as planned, but the right hook appeared to be impracticable with the Turks so strongly entrenched. The right division was therefore to attack the enemy's centre, while the Desert Column engaged the Turkish position at Atawine with one division, and contained the 16th Turkish Division at Hureira with the second. Further orders to the troops on the right flank would depend upon the enemy's reaction.

There were those who thought that the best plan would have been to make the main thrust up the coast (option 1), but it appears that General Dobell dismissed it as offering too little scope for the mounted troops. The attack north-eastwards from Sheik Abbas (option 3) was given serious thought, but discarded on account of water difficulties. With hindsight we might agree with those who thought that Dobell would have been better attacking up the coast. But probably the best advice came from General Chetwode, who was against the battle being fought at all.

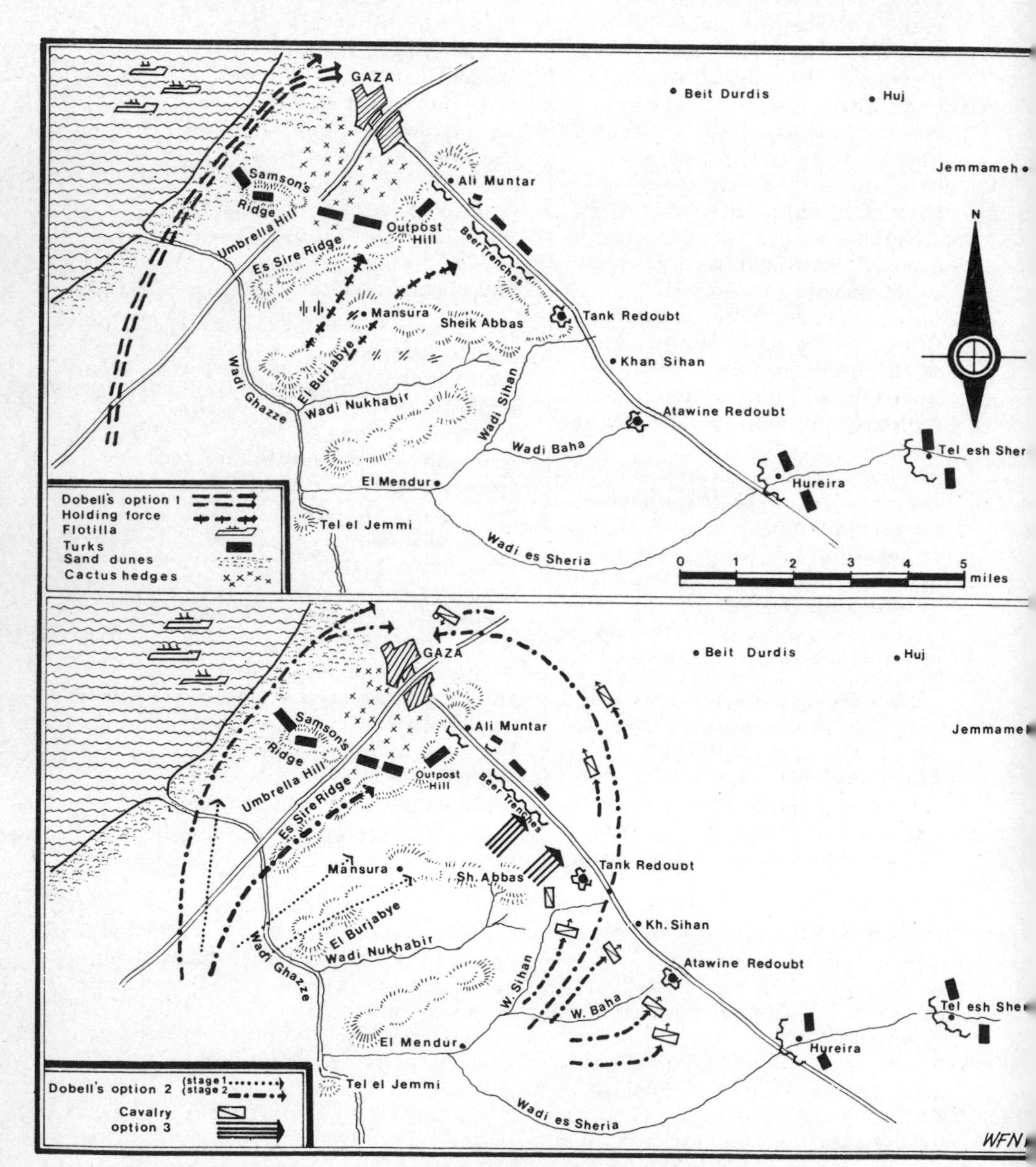

65 *The Second Battle of Gaza: April, Dobell's options*

In the first phase of the battle, which began early on 17 April, all went well, with Eastern Force advancing across ground well reconnoitred in the previous engagement, and occupying by evening the Es Sire Ridge–Mansura–Sheik Abbas Line. The opposition was comparatively light and casualties slight, although a tank, operating on the right with the 163rd Brigade of 54th Division, had been knocked out. On reaching their objectives the troops were subjected to heavy artillery fire, but this did not prevent them from consolidating the position.

During 18 April the enemy were pounded with a great weight of shells both from land and sea, but it does not appear that this softening up process had much effect on the well-prepared Turkish positions. The mounted troops on the right occupied themselves by driving in some outposts, but the chief activity was behind the line and concerned with bringing forward supplies of water and ammunition in readiness for the major offensive on the morrow.

At 5.30 a.m. on the 19th the naval and military bombardment was resumed, and at 7.15 the 53rd Division, shortly followed by the 52nd and 54th, went in to the attack. An hour later the mounted troops – fighting dismounted – were attacking the Atawine defences. During the course of this day more than 1,000 British and Anzac soldiers died for a few yards of blood-soaked sand, and a further 5,000 were wounded. Although not comparable to the slaughter on the Western Front, this was the hardest and most disappointing day's fighting that Eastern Force had so far encountered.

From left to right the attack was made by the 52nd, 53rd and 54th Divisions, with the Camel Brigade operating on the right of the 54th, and the Imperial Mounted Division taking care of the last three miles of the front. The 53rd Division managed to take Samson's Ridge, but the 52nd were bogged down for most of the day trying to gain Outpost Hill; the 54th Division made better progress, but being in advance of their neighbour were caught by enfilade fire from the Ali Muntar position. The Camel Brigade, helped by a tank, gained the Kh.Sihan position, but could not hold it for long, while on their right the Imperial Mounted Division – later assisted by the Anzac Division – made inroads on their whole front, but were too extended to press home any advantage.

Along the whole line the troops attacked with great gallantry over difficult ground and against unlocated positions. It was soon apparent that their breaching batteries were woefully inadequate, while the much vaunted gas shells showed little result; the greatly outnumbered but sternly embattled Turks were magnificent in defence and resolute in counter-attack. All this meant that initial gains, achieved at considerable cost, had mostly to be given up. By the late afternoon Dobell realized that to throw in the reserve would only cause further casualties, and that with the whole of the artillery committed and their ammunition stocks dwindling a breakthrough was unlikely. At 4 p.m. orders were accordingly given to adjust the position (which meant withdrawing the 54th Division, and the mounted troops to its right, on to the Sheik Abbas Ridge) and digging in on ground that had been won, preparatory to resuming the offensive at dawn the next day.

However, later that night, when the true position became clear, Dobell informed Murray that it was his and his divisional commanders' opinion that no further advance was possible. Thus ended the Second Battle of Gaza; if it had to be fought at all it should have been given a better appreciation, which must have resulted in more thorough planning.

The immediate result of this battle was the replacement on 21 April of General Dobell in command of Eastern Force by General Chetwode. Chetwode's command of the mounted troops was given to General Chauvel, and he in turn was replaced in command of the Anzac Mounted Division by General Chaytor. It was not long before the mounted troops, on receiving the 7th and 8th Mounted Brigades from Salonika, were reorganized into three divisions: Australian (late Imperial), Anzac and Yeomanry. During the months of May and June military activity was confined to minor raids, but Murray used the time to consolidate the existing front line, and to build up supply facilities by extending the railway from Rafa to Shellal.

The War Cabinet had been sadly disillusioned by events so far in Palestine, and now had to decide whether to fulfil the requirements demanded in order to continue an aggressive policy, or to forgo – at any rate for the time being – offensive operations. In deciding to keep up pressure against the Turks they were influenced by the effects of the Russian Revolution, which would release Turkish troops in the Caucasus for a renewed offensive in Mesopotamia. There was also the chance of providing a war-weary Britain with some glittering prize such as Jerusalem. The decision having been taken it was further decided that a change of command was desirable, and at the end of June General Sir Edmund Allenby (recently commanding the Third Army in France) replaced General Murray, whose achievements had been very considerable, in command of the Egyptian Expeditionary Force.

The arrival of Allenby, and his decision to command the army in person and move his operational headquarters from Cairo to Khan Yunis, acted as a tonic to the troops. He had the capacity to inspire absolute confidence, and the capability of building, through tireless industry and dynamic force, a large and disparate body of troops into a powerful fighting machine. He spent four days in thoroughly reconnoitring the front, and then submitted his requirements. For breaking the existing line and perhaps reaching Jerusalem, he wanted seven infantry and three cavalry divisions, with divisional artilleries at full strength and a large increase in heavy artillery. To go beyond the Jaffa–Jerusalem line he would need considerably more.

Allenby's requirements, save for a slight shortfall in his artillery demands, were met. He also got two extra air squadrons, which gave him complete air superiority. Eastern Force was abolished, and the army was now organized into two corps – XX and XXI, commanded by Generals Chetwode and Bulfin respectively – and the Desert Mounted Corps. The fighting strength was approximately 60,000 infantry (including the Camel Brigade), 12,000 cavalry and 475 guns.* Allenby realized that the time required to get his reinforcements acclimatized and properly trained meant that the proposed offensive would have to be postponed from September until the end of October, which was getting near to the time of the rains. But the delay enabled the railway to be double-lined for seventy miles beyond Kantara, and the branch line extended from Shellal to Kharm – only six miles from the Wadi Ghazze. The main water pipeline was extended to Shellal, and small-gauge pipes were also laid from various wells to the forward troops.

The Turkish successes in withstanding the British attacks on Gaza did not reflect the general well-being of their military machine. By the summer of 1917 this was in very poor shape, and the Germans – determined to keep their ally in the war – sent one of their finest soldiers, General Erich von Falkenhayn, to arrange with the Turks for a large German detachment of all arms to assist the war effort. Fortunately for the British Falkenhayn found unforeseen difficulties, and by the time it had been decided to use Army Group F (known as 'Yilderim', meaning, ironically, lightning) in Palestine rather than Mesopotamia, and Turkish delaying tactics had been overcome, 'Yilderim' was too late to be of assistance in the forthcoming battle.

Nevertheless the Turks had not been idle. Gaza had become a fortress held by troops of their XXII Corps, with XX Corps on their left. In all they held a front of almost thirty miles with some fifty battalions; every mile or so strongpoints were sited, except on the left between the Hureira and Beersheba groups, where there was a gap of about five miles. Their total strength was 33,000 rifles, 1,400 sabres and 260 guns. They were, therefore, outnumbered by almost two to one in rifles and guns, and seven times in cavalry; but a defending army can deploy greater firepower than one that is attacking.

The British line was also fully stretched, reaching from the sea to Sheik Abbas, and from thence back to Tel el Jemmi and along the Wadi Ghazze. On the left the line was in places no more than 400 yards from the Turks, but it had to bend back because the country between the Wadi Ghazze and Beersheba was virtually waterless. Thus any advance

* The official totals for infantry and cavalry are given as 80,000 and 15,000 respectively, but many of these were extra-regimentally employed.

in this area meant that all water consumed by the troops had to be carried, and animals could be watered only between bases. Water, always a major problem, would become a much greater one for any operations contemplated on the British right flank.

The enemy had to be dislodged by assault and stratagems. Allenby knew that the Turks had prepared positions in rear of the Gaza–Beersheba line, and that to push them back only a little way would merely ease the supply problem of their one-track railway; he himself was handicapped by lack of roads, for which sea transport was too problematical to compensate. No one knew better than the Commander-in-Chief the value of surprise to ensure success. In making his plan (which he did not long after assuming command) to achieve a complete breakthrough and pursuit, *Allenby had four options:*

(1) To make the main assault from Gaza to the Tank Redoubt – a frontage of from 10,000 to 12,000 yards. The principal disadvantage here was the need to attack the enemy at his strongest point. It would be a slow, laborious process needing a great weight of artillery, and posed no threat to the enemy's left flank. On the other hand it was the nearest point to the British line, the water problem was comparatively easy, and a breakthrough by superior numbers could drive the enemy at best (for them) on to their fairly weak switch line Deir Sineid–Tel en Nejile, from which it would not be difficult to dislodge them, or at worst into the mountains, where food and water could not be found.

(2) To put in the main attack against the enemy's left flank, with a holding operation against Gaza. This had some disadvantages, the chief of which was lack of water. It would be possible to advance the pipeline towards Bir Ifteis, and there was water behind the British line at Shellal, Qamle, Fara and Esani, but the operation would stand or fall on capturing the Beersheba wells intact. Speed and surprise would therefore be as vital as they were difficult to achieve, for a large force would have to be moved some twenty miles across an almost waterless desert, and the necessary preliminary work on extending the railway and pipelines might alert the enemy. To carry this operation out successfully would require a calculated dispersion of the force for a concentrated attack, which although in principle sound might on this occasion weaken the centre and invite a successful counter-attack.

The advantage of this option was the chance it offered for penetrating the enemy's flank, and from the high ground thus obtained for rolling up his left and centre by attacking north-westwards, eventually pivoting on the Sheik Abbas position to swing the right and prise the enemy out of Gaza.

(3) To make the main attack against the enemy's right, but to mask Gaza and pass a large

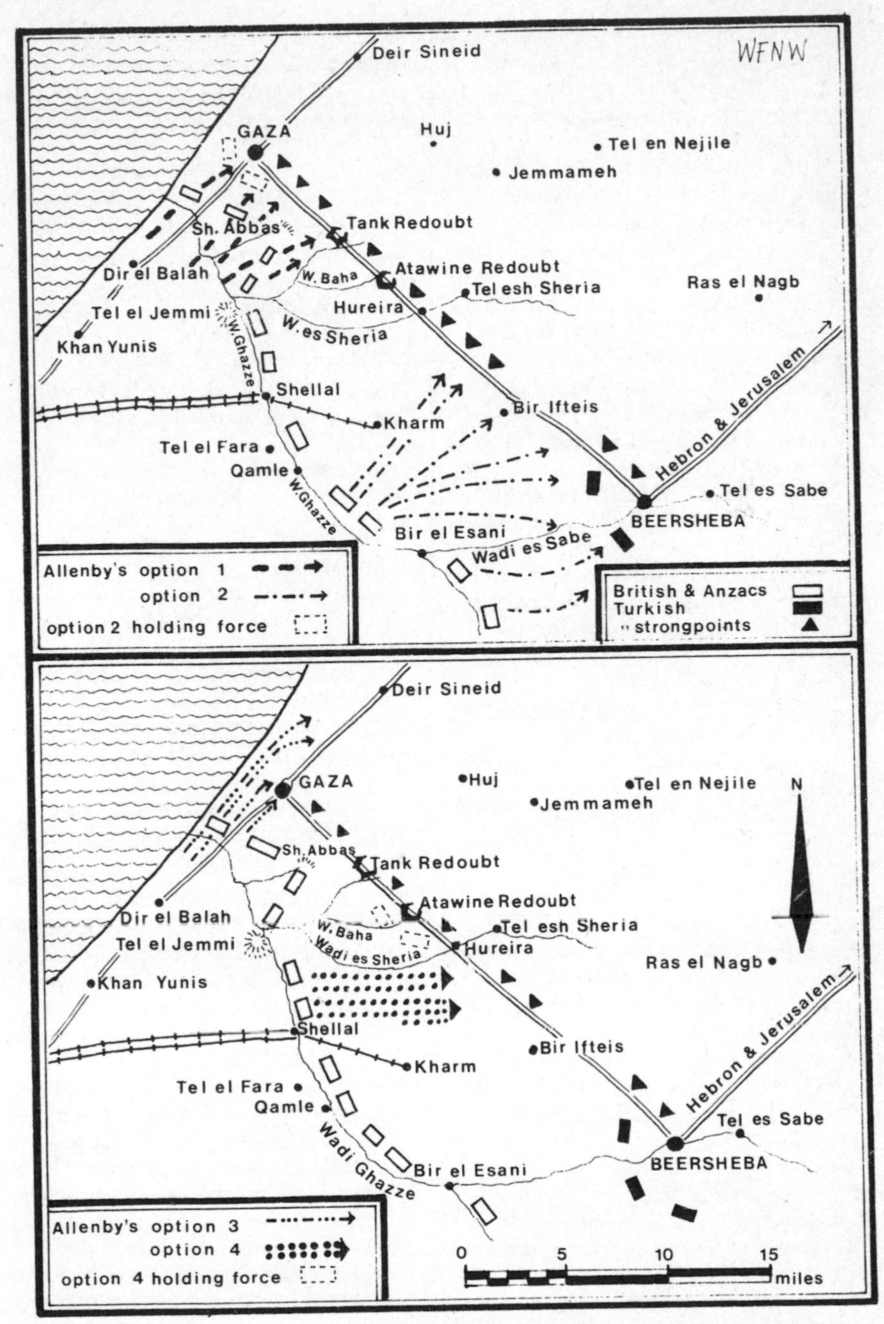

66 The Third Battle of Gaza: 31 October, Allenby's options

force through the defiladed sand-dunes to intercept retreat and attack the town from the south-west and west, rather than waste lives, ammunition and time trying tc storm the town from the south-east as previously. This would also need dispersion in order to get concentration, but the distance involved here was much less than in option 2, and two divisions for exploitation could be transferred to the left flank during the night of D-1. The water situation was also quite favourable. It could be argued further that with the enemy's front not being at right angles to his line of withdrawal and supply bases, a breakthrough on his left (as envisaged in option 2) might fail to catch the Gaza garrison's retreat, while a breakthrough on his right should catch the withdrawal of the left – unless the enemy took to the inhospitable mountains.

(4) To engage the Atawine group of defences in a holding operation, and to attack the Hureira group from the west. These defences were not so well fortified as the Gaza–Atawine line, and it might be possible to put a large enough force through the gap and to get behind the Beersheba troops, and then pivot the right against Gaza. The major obstacle to such a plan was that from the British side the ground was unfavourable to artillery observation, and there was also a shortage of water in the central area.

To attack the centre from approximately the Tank Redoubt – Atawine – Wadi Baha line was considered, but quickly dismissed as not being a viable option. The enemy's lateral communications were good; he could reinforce the centre from his left or right, and his positions were mutually supporting in artillery and small-arms fire.

General Allenby decided to attack the enemy's left flank (option 2). XX Corps and the Desert Mounted Corps were detailed for the attack, and would take up a position south of the Wadi es Sabe on the night of D-1. The first phase would be the capture of Beersheba. To accomplish this XX Corps' objective would be the outer defences and entrenchments, while the Desert Mounted Corps (less one division holding the ground between XX and XXI Corps) would co-operate by attacking Beersheba from the south-east and taking the town. Theirs would also be the task of dealing with the escaping garrison.

In the second phase XX Corps would attack north and north-westwards, rolling up the Sheria and Hureira trenches, while the Mounted Corps (having collected the detached division) would operate against the enemy's left flank from the Tel el Nejile area. The Gaza front would be subjected to heavy bombardment from land and sea for three or four days prior to the opening of the attack; XXI Corps would attack the Gaza defences from the south-west after Beersheba had fallen, but before the rolling up of the Sheria position commenced.

The plan succeeded and has therefore been acclaimed as the best, but at the time and after there was a strong 'Gaza School', who thought that the main thrust should have been up the coast (option 3). They argued that despite every precaution surprise, which was achieved, should not have been, and that had the Turks blown up the wells (and it is still a mystery as to why they failed to do this, for the charges were all in position) the whole right wing would have had to fall back to base for water.

Allenby's plan was formed largely from a very good appreciation made by Chetwode after Second Gaza, and presented to Allenby on his assuming command. Undoubtedly to Chetwode Gaza represented failure, and since the last battle the place had been made even stronger. Moreover, both Allenby and Chetwode were cavalrymen, and to use cavalry to turn a flank must have had great appeal – and to Allenby, fresh from France, a flank was something new and exciting. The official reason given for discarding the Gaza option was that when the fortress was overwhelmed the enemy would pivot on his Ḥureira strongpoint, swinging his right back to partly prepared positions on the Deir Sineid–Jemmameh line, thereby maintaining his communications and still blocking the way to Jerusalem. The argument against this is that that line was far too weak to hold for long in the face of a resolute foe flushed with victory.

Two eminently distinguished field-marshals (Chetwode and Wavell), when, at a later date, reviewing Allenby's decision, refused to admit that there was a better plan than the one adopted. Nevertheless, it is tempting to think that the coastal one would also have succeeded, and even perhaps entailed fewer anxious moments.

As already explained, the need for surprise, and therefore deception, was a *sine qua non* for success, and the Intelligence Branch spared no pains in this direction. Elaborate measures were taken to strengthen the enemy's belief that Gaza was the principal objective. Naval activity – soundings, etc. – off the coast north of the town, and cipher messages and dummy camps in Cyprus, were designed to make the Turks believe that a landing behind Gaza was to take place. On the enemy's left a game of double bluff was risked through sending strong reconnaissance patrols against the Beersheba defences in the hopes that the Turks would think this merely a design to lure his reserves away from Gaza.

Evidence collected after the battle tended to show that most of these measures had little effect, and that the enemy had a fairly accurate idea of the number of troops being concentrated opposite his left flank. But one exceptionally well-conceived and executed ruse does seem to have made its mark. A G.H.Q. staff officer on a reconnaissance feigned being hit when fired on and dropped his haversack (previously smeared with fresh blood), which contained a number of cleverly framed papers, signals, ciphers and letters, many purporting to show that the attack would fall on Gaza. These were brought to the notice of Kress (now commanding VIII Army), who from the Intelligence reports submitted to him *had three options:*

(1) To assume the documents were genuine and leave his reserves at Gaza.

(2) To treat them as yet another ruse and accept the build-up on his left flank as

evidence that the attack would come in there, and transfer at least a part of his reserves to bolster his comparatively weak left.

(3) To assume (as in option 2) that his left flank was to be attacked and abandon Beersheba, bringing his line back to rest on Khuweilfeh in the hills. This would make a strong position with no flank to be turned.

Kress, although not entirely convinced by the ruse, nevertheless acted as though it was genuine. He said later that he simply could not risk his right being turned (a favourable point for the 'Gaza School'), which he thought must be the British objective, and so he adopted option 1. As a matter of interest he had previously sought permission to withdraw to the virtually unassailable line in the Judean hills, but first Enver Pasha, the Turkish Commander-in-Chief, and then Falkenhayn forbade it. The arable land round Beersheba was too valuable to abandon, and Falkenhayn wanted Beersheba as a jumping-off base when he passed to the offensive on the arrival of 'Yilderim'.

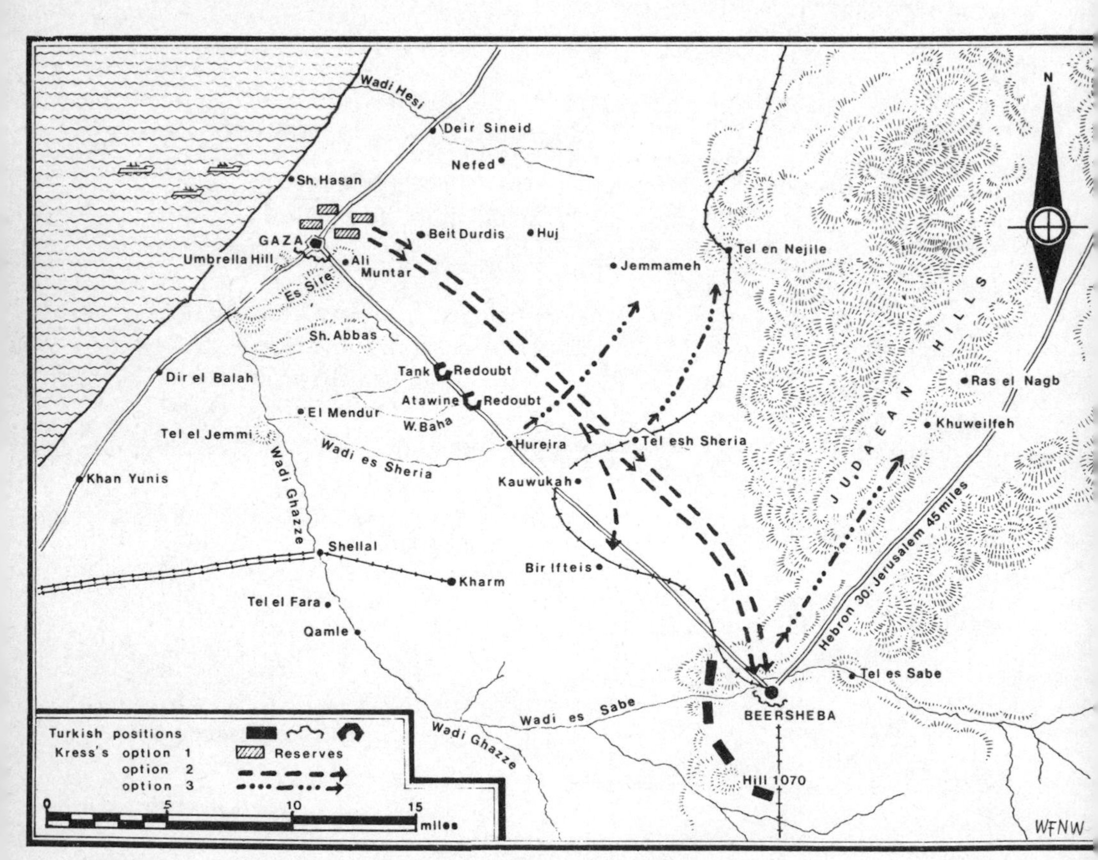

67 The Gaza–Beersheba Line: October, Kress's options

The concentration of the attacking force started on 21 October and took ten days, for the troops (especially the mounted ones) had to make a long circuitous march, mostly by night, to Khalasa and Asluj, and supply dumps and water storage had to be arranged. While this was in progress outposts of the Yeomanry Division (which was the one left between XX and XXI Corps) were heavily attacked, and a spirited action took place before the Turks were beaten off. XX Corps completed its approach march on the night of 30–31 October.

The Turkish defences of Beersheba lay to the south and south-west in a semi-circle stretching for almost ten miles from the railway line north of the Wadi Sabe, south and south-east back to the railway about four miles south of the town. The centre sector only was to be attacked (by the 60th and 74th Divisions), the north and south sectors being closely watched. The Anzac Division was to move round to the north-east of the town, with the Australian Division behind it. General Chauvel had orders that his attack was to go in before the enemy realized the full extent of the infantry attack on the perimeter defences.

At 5.55 a.m. on 31 October the bombardment of the outer defences began. The first objective was Hill 1070, an outlying strongpoint; by 8.30 a.m. 181st Brigade of the 60th Division had taken the position, and by this time the Anzacs had halted within attacking distance of the town. By 1.30 p.m. the whole sector being attacked (some three miles of strongly defended positions) had been carried by the infantry, and the troops watching the other sectors now moved into action. By 7 p.m. all the entrenchments were clear of the enemy, and XX Corps' work was completed for a loss of not much over 1,000 men. Five hundred Turks and six field guns had been captured.

The Anzac Division went into the attack north and east of the town astride the Hebron road at 9 a.m., and met with very stiff resistance in the area of Tel es Sabe. Severe fighting (mounted and dismounted) took place during the morning, with the New Zealanders, now helped by the Australians, relentlessly closing in on the enemy. Not until 3 p.m. was Tel es Sabe finally taken, and by this time General Chauvel, conscious of his orders to capture Beersheba before dark, felt that more drastic measures were needed. He therefore ordered Brigadier Grant to launch a mounted attack straight into the town. In charged the 4th Australian Light Horse Brigade, each regiment in three successive lines, each man extended to three yards, the sunlight flashing on their drawn bayonets (they carried no sabres); the horrified Turks manning the trenches were confronted by a solid frieze of galloping horses, the men upon them bent on killing. Some dismounted and cleared the trenches, others raced on and saved the wells. It was all in keeping with the great tradition of these troops, and with the fall of Beersheba the tide had at last turned for the Egyptian Expeditionary Force.

On the Gaza front no less fruitful work was being done. On learning of the success against Beersheba, the Commander-in-Chief decided that the Gaza attack would go in on the night of 1–2 November; there was a moon and it was felt that a night attack would give better protection against Turkish machine-gunners. The attack was to be carried out in four phases on a front of 5,000 yards extending from Umbrella Hill in the south to Sheik Hasan on the coast to the north. All three brigades of the 54th Division, and one attached from 53rd Division, were committed, supported by heavy artillery and naval gunfire.

A preliminary bombardment, begun four days earlier, undoubtedly had good effect – particularly in the sandhill areas – but the divisional attack, which began at 11 p.m., was met with fierce opposition, and in the centre sector the troops had to be content with a final penetration that was somewhat less than planned. Nevertheless, by 6.30 a.m. on 2 November when 162nd Brigade had captured Sheik Hasan – its last objective – the tasks given to 54th Division had been mainly accomplished, for it was never intended that the town should be captured. During the next four days there were no important engagements on this front, and on the night of 6–7 November the Turks, whose left was by then in disarray, evacuated Gaza. British losses on this front were 2,697 killed, wounded and missing. Almost 1,000 Turks were later buried on the field.

The operations and pursuit that followed the taking of Beersheba lasted well into November, and saw some of the hardest fighting of the campaign. The rolling up of what was now the Turkish left, stretching from Kauwukah through Sheria to Hureira, was to be done by XX Corps with the Camel Brigade protecting its right flank, while the Desert Mounted Corps, as soon as resistance was broken, were to make for Huj and Jemmameh, where there was a good supply of water. But before the flank attack could go in the Turks (masterminded by the Germans), being determined to regain Beersheba, put in a strong counter-attack. In spite of the severity of this attack, which the enemy kept up for four days in the El Nagb–Khuweilfeh area, Allenby refused to oblige them by pulling out troops from other sectors. This decision to hold fast to the original strategic plan was a battle-winning one.

At 5 a.m. on 6 November the 74th Division, which was to bear the brunt of the fighting, was at last able to open the attack on the strong Turkish positions in the Kauwukah area. There was bitter fighting all that day and the next, both against the Turkish left and for the Mounted Corps in the Sheria–Khuweilfeh area, but by the evening of the 7th the Turks began to give way all along the Gaza–Beersheba line,* which they had defended so tenaciously for the past nine months.

*Save for a small force which stubbornly held on to the Atawine–Tank Redoubt trenches and, allowing the British to lap round them, considerably delayed the pursuit.

The pursuit of the retreating Turks, which lasted from 7 to 16 November, was no procession, for the enemy's withdrawal was slow, stubborn and very dangerous. It was General Allenby's hope that the mounted troops would get behind the retreating enemy and cut off large numbers, but this was not to be. Although the going was mostly good, so were fields of fire, and the enemy machine-guns took a fearful toll. By 9 November the right wing was level with Sheik Hasan, and the men of XXI Corps, who had been fighting their way up the coast, joined hands with those of XX Corps. Across the Plain of Philistia for some fifty miles the army moved forward remorselessly, throwing an iron ring around the Turks and penning them back upon Jerusalem. Turkish losses were enormous, 10,000 prisoners and 100 guns being taken, but they managed to escape complete destruction.

Three weeks later Mr Lloyd George got the Christmas present he wanted for the people of Britain. On 9 December Jerusalem surrendered – a fitting reward for the courage and patient endurance of those soldiers who had struggled through the wavering heat bars of the desert, and the cold, rain-soaked hills of Judea.

14 Anzio

Background to Beachhead

AFTER the German surrender in Tunis in May 1943, the Allies turned their attention to Sicily, which was taken – after some very stiff opposition – by the end of July, and from there the invasion of Italy was launched. General Montgomery's Eighth Army crossed the Straits of Messina on 3 September, and General Clark's Fifth Army became engaged in an opposed landing at Salerno on 9 September.

But before the troops made their landings stirring political events had been taking place in Italy. On 25 July the King of Italy had replaced Mussolini with Marshal Badoglio as head of government, and although the latter at first declared his intention of continuing the war, it was not long before he was negotiating with the Allies an enormous piece of treachery, whereby he would surrender the Italian army unconditionally, and later even become a co-belligerent with his former enemies. Negotiations were protracted, for at first Badoglio demanded greater safeguards against German vengeance than the Allies were able to give; later he discovered that as no proper plans had been made for dealing with his recent German colleagues they had stolen a march on him, and were ready to put into operation an immediate plan to disarm the entire Italian army. Nevertheless, on the eve of the Salerno landing Badoglio did broadcast to the nation the surrender of the Italian army.

All this had a considerable effect on the German Führer's military thinking. Hitler's was the deciding word in all important matters of strategy, and although he would sometimes take the advice of his generals he needed much persuasion to alter course once his mind was made up. He now decided that there was little point in riveting German armour on to the soft bellies of all that remained of the Italian legions, and that southern Italy should be abandoned. Field-Marshal Kesselring, who commanded there, was ordered to withdraw his troops to the north, fighting delaying actions and disarming the Italians as they went; he was then to come under the overall command of Field-Marshal Rommel.

Kesselring argued fiercely against this strategy, for he felt that any Allied breakthrough south of Rome would be less disastrous than one in the northern Appenines and the Po Valley; moreover, a strong force south of Rome might prevent any Allied attempt to disrupt the source of raw materials in the Balkans, and keep enemy air bases at arm's length from southern Germany. These arguments, combined with Kesselring's very successful delaying tactics once the Allies had landed, persuaded Hitler to hold out south of Rome. Rommel's opinion that troops holding a line too far south might be cut off by an amphibious landing in the north (and he was nearly proved right) was disregarded, and early in October Kesselring was told to hold the Cassino line in strength. In November Rommel was transferred to France and Kesselring was given the new unified High Command Army Group C.

This muddled German strategy may well have saved General Clark's army from being hurled back into the sea at Salerno, for at the time of the landing the O.K.W.*, in compliance with the Führer's orders, had no thoughts for reinforcing the south. Nevertheless, Kesselring made full use of the troops available, and had brought 16th Panzer Division across from the Adriatic coast in time to disarm the Italian garrison at Salerno, and be ready to resist along part of the thirty-six mile stretch of coast selected for the Fifth Army's landing. Soon other German units were rushed to the scene, and Kesselring's build-up outmatched that of the Allies. His troops took up a commanding position on the hills overlooking the bay, and for nine long, perilous days, Fifth Army (helped towards the end by the Eighth) struggled to retain the foothold it had so precariously gained on the European mainland. It was not until 19 September that General Alexander could report to his Prime Minister that the battle of Salerno had been won.

Both armies were now firmly established, and began to inch their way up the Italian leg. Fifth Army broke through the ring of hills and were in Naples on 1 October; the Eighth, separated from them by the central mass of the Abruzzi, moved up the Adriatic coast and through the valleys of central Italy. Both had to contend with fierce opposition and the eternal force of many rivers. The autumn weather held, progress while not spectacular was steady, and spirits rose with the thought of reaching Rome before winter set in.

But Allied confidence was confounded when, in the darker days of December, the very formidable Gustav Line, stretching the breadth of Italy with strong defensive positions many miles deep, was reached. It did not need Allied Intelligence to confirm that German High Command

* The O.K.W. (Oberkommando der Wehrmacht – Armed Forces High Command) was responsible through its chief, Field-Marshal Keitel, to Hitler in his capacity (at this time) as Commander-in-Chief of the army. It was the superior organization of the three services high commands.

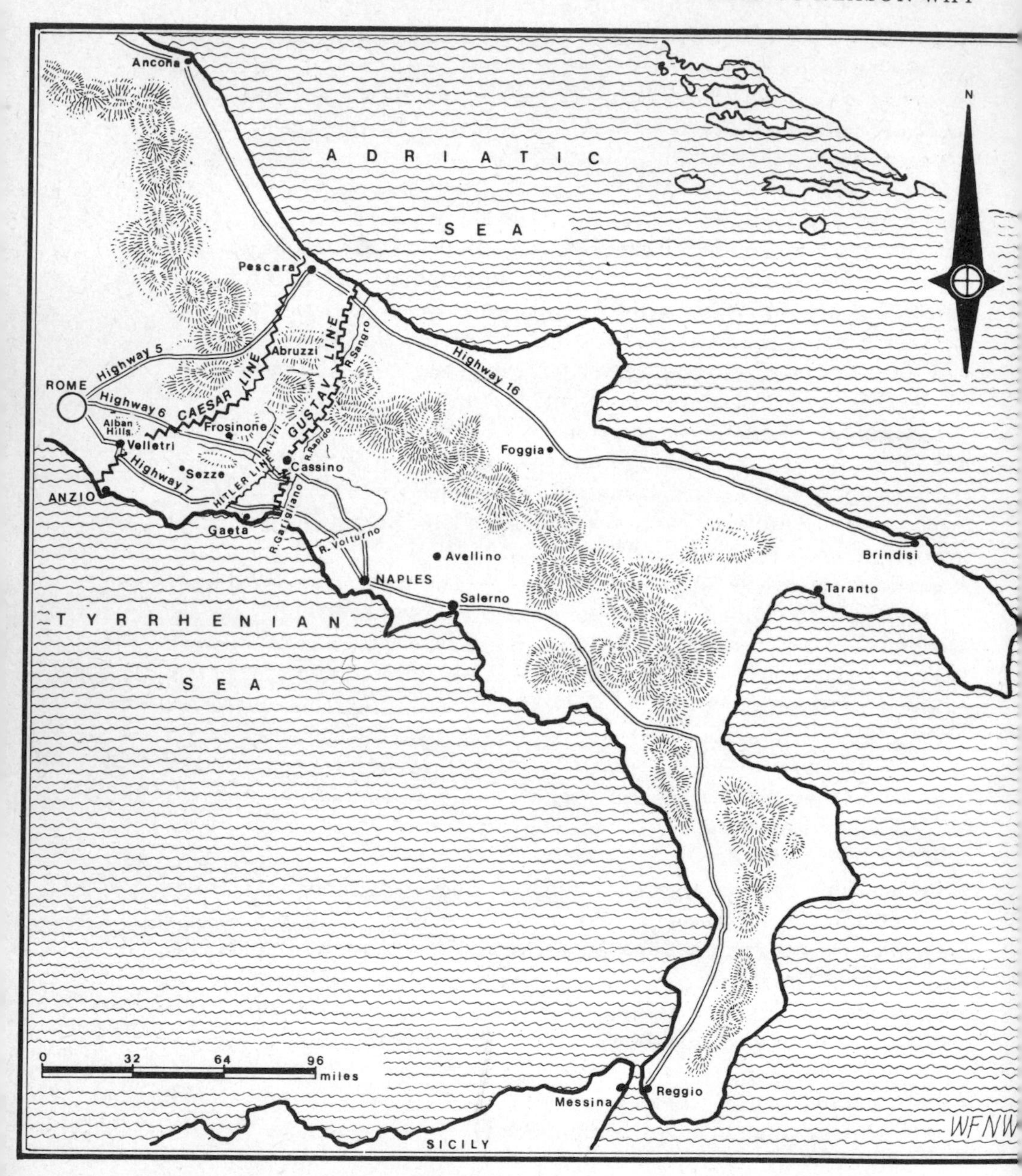

68 Anzio: background to beachhead

policy was now to stand firm south of Rome. Winter came with frozen intensity, and the great mountain ranges – beautiful, yet savage beyond belief – now presented a cheerless, snow-covered panorama. In the valleys the wind and sleet swept like a curtain, turning all to mud and slush, the mountain streams became torrents, and the rivers broke their banks. The troops became disillusioned; a general weariness set in. Somehow a way had to be found to break the deadlock, pierce the Gustav Line and capture Rome.

There were three roads to Rome. Highway 6 ran through the centre of the leg; Highway 7 ran close to the west coast, and Highway 16 hugged the Adriatic to Pescara, where Highway 5 branched off to Rome. Massive mountain ranges and three great rivers (the Garigliano, Rapido and Sangro) lay astride these roads. The excellent natural defences had been reinforced most thoroughly, so that any break-through could be achieved only with overwhelming superiority on land and in the air, and then at great cost. Except in the air, where the weather often nullified the advantage, the Allies did not have this superiority. In the immediate battle area the Germans could muster fifteen divisions to the Allies' thirteen, which even allowing for the slightly fewer men in a German division still gave them the edge in manpower, and in heavy weapons there was little difference. The German 88-mm was an outstanding dual-purpose gun and their *Nebelwerfer* was a most demoralizing mortar.

There was a fourth route to Rome – by sea, or by sea and air combined. Some weeks before his troops reached the impenetrable barrier of the Gustav Line, Alexander had seen the obvious advantages of an amphibious operation. But although the Allied navies had command of the sea there was a problem over landing craft, most of which were shortly to be sent to England for Operation Overlord – the invasion of France. The original plan was simple, and extremely optimistic. The Eighth Army was to take Pescara and cut the road to Rome, while the Fifth Army erupted into the Liri Valley and reached Frozinone. When these two targets had been achieved a landing south of Rome by one division and paratroops would complete the discomfort of the enemy. By the middle of December the Fifth Army was still thirty miles and the Gustav Line from Frosinone, and the plan was abandoned.

But Alexander was hopeful that Operation Shingle – as it was called – would not be dropped altogether, and in this he had the wholehearted support of Mr Churchill. It was the sort of rapier thrust that had immense appeal to the British Prime Minister, and he had the clear-sightedness to realize the prize and the sagacity to measure the consequences. Overlord would remain the top priority and must not be endangered, but with the advantage given to a convalescent* Churchill was prepared to use his immense influence with President Roosevelt to change American Mediterranean policy in favour of greater action in Italy, and to delay the return of vital landing craft. At least 88 L.S.T.s –

* Winston Churchill was recovering at this time from a severe bout of pneumonia.

Landing Ship Tanks – out of the 105 then in the Mediterranean would be required for a two-divisional lift, which was considered the absolute minimum force necessary to survive a landing. These Churchill obtained for a limited period, and a new plan was at once contrived for hurling the Prime Minister's 'wild cat' ashore behind the German lines.

This plan was basically the same as that proposed by Alexander in early November, differing in the place of landing and the size of force to be landed. Now it was intended to put ashore an Anglo-American force of two divisions, Ranger and Commando battalions, and a Tank and Parachute Regiment (all part of Fifth Army's VI Corps), with follow-up troops as shipping became available. The Anzio–Nettuno area was selected, and any idea of a subsidiary landing north of Rome was ruled out, on logistic grounds and because of a lack of troops and landing craft. It had always been Alexander's intention, and it was confirmed in his orders, that this force should reach out to the Alban Hills (Colli Laziali), and by cutting the German communications along Highways 6 and 7 force them to withdraw from the Gustav Line under pressure, simultaneously applied, from the troops facing that line. If successful this operation must lead to the fall of Rome, and would probably trap a number of German divisions.

However, in the paper he prepared for discussion with his commanders,* Alexander sounded a more cautious note, and General Clark, who was to be in charge of the operation and who was not at the Carthage conference when the full plans were discussed with Churchill, was given to understand that his troops were first to secure the bridgehead and exert steady pressure on the enemy. The main thrust (to be made by armour from a third division landed on D + 3) would then be passed through as the spearhead of an advance towards fulfilling the main objective of closing the trapdoor. No mention was made in Alexander's notes of holding the Alban Hills, which would not have been possible for a small force unless the enemy withdrew from the Gustav Line under the threat of encirclement (which was extremely unlikely), or their front at Cassino was pierced. This had an important bearing on what followed, and the controversy it engendered.

The date fixed for Operation Shingle was 22 January 1944. Between the 17th and 20th the British X Corps, the Free French Expeditionary Force and the American II Corps were to open a major offensive against the Gustav Line, for the whole plan was based upon a successful body punch and left hook. Major-General John Lucas commanded VI Corps, and the two divisions earmarked for the original assault were the 3rd U.S. Division commanded by Major-General Lucian Truscott, and the 1st British Division commanded by Major-General W. R. C. Penney. The

*See *Alex* by Nigel Nicolson, pp. 230, 231.

United States Rangers (three battalions) were deputed to seize the port in the first wave with a follow-up of two British commandos – No. 9 Commando and 43 Royal Marine Commando. The subsequent landing of Combat Command A (a part of the American 1st Armored Division) would depend upon the turn-round time of the landing craft, and that depended upon the speed of unloading and the weather – but three days was a hoped-for target. The 45th U.S. Infantry Division was also to be landed at a later date.

The four men most closely connected with the detailed direction of the landing, and subsequent battle, were the army commander, General Mark W. Clark, and Generals Lucas, Penney and Truscott. Clark was a tall, spare man of great energy, marked capacity and undoubted courage. Spurred by ambition his rise up the military ladder had been spectacular; now, at forty-eight, he was younger than most of those holding similar positions. Like his fellow commander on the Adriatic coast he appreciated the value of publicity and was not averse to personal limelight, but beneath his flamboyance there was plenty of drive and efficiency. He was a very good commander and a popular one.

His corps commander, General Lucas, cannot be given such a high encomium, although he came in for harsher criticism than was perhaps deserved. He had been given command of the corps during the battle of Salerno, where his predecessor had proved inadequate, and he had fought many a hard battle up the leg of Italy. He arrived at the Garigliano a tired man; on his fifty-fourth birthday he wrote in his diary, 'I am afraid I feel every year of it.' He did not inspire confidence in his appearance or his manner; he was a cautious and indecisive commander, and totally unsuited to an enterprise such as Operation Shingle. Furthermore, he was never in full accord with his British allies and there was a mutual feeling of mistrust. Inevitably he was the scapegoat of initial failure, although Clark stood by him most loyally.

Truscott, Lucas's fellow-American divisional commander, was of an entirely different stamp. He had the panache and élan of an English Cavalier or an American Confederate general, and like so many of them, he was a born leader of men. He had worked in Combined Operations in London, and had gained there a valuable insight into British ways, which was of value to him now. He possessed charm, courage and intelligence, and was one of the outstanding American generals of the war.

Truscott's opposite number, General Penney, was commissioned into the Royal Engineers and later transferred to the Signals. It came as something of a surprise to him – and some others – that he should be given command of the 1st Division, but he was to justify the trust most admirably. He possessed enormous industry, and in attempting (successfully) to make up for lack of experience he spared neither himself nor his staff. He did not suffer fools gladly, and at times became impatient with the way matters went, but he was a kind-hearted man well liked and admired by those who served under him.

Operation Shingle did not get off to a good start. There was time only for one rehearsal, which took place off a beach near Naples said to resemble the one at Anzio. It resulted in a complete shambles – especially for the American 3rd Division. A serious error in navigation stopped the transports too far out to sea, DUKWS (amphibious troop-carrying trucks) were launched too soon, men were drowned, guns and other valuable equipment were lost.

While these troops were going down to the seabed in DUKWS, their comrades of the X British Corps, using similar vehicles, were struggling to cross the Garigliano, and further inland men of the American II Corps were hurling themselves at the Rapido. Both these river crossings were fiercely contested. The attacks were pressed home with great courage, but although a bridgehead was secured across the Garigliano, penetration along the whole line was very limited. The natural features of mountain and river, so strongly defended, remained a barrier to the sought-after Liri valley, and the already scarred and corpse-strewn Italian landscape claimed a further 1,000 men. This meant that any quick link-up with a successful Anzio landing could be achieved only if the threat to his rear made Kesselring pull back from the Gustav Line.

The actual landing met with unexpected success. During the night of 20–21 January ships crammed with troops and equipment sailed from Naples, Salerno and two other ports. During the 22nd some 374 craft were making their slow way up the Italian coast unhindered and unseen, thanks to the Allied air forces and some fortuitous fog. Shortly before midnight the ships had reached the rendezvous. H-hour was timed for 2 a.m. on the 23rd. The L.S.T.s anchored three miles off-shore, and the larger vessels some two miles farther out. Between midnight and 2 a.m. the minesweepers cleared a channel, and marker buoys for the various beaches were set out. It is easy to imagine the tension on board the landing craft as men gazed across the dark foam-flecked sea towards the unseen outline of a hostile coast. Their appointment with destiny was at hand. All was uncannily, unbelievably quiet.

This time the navy made no mistake, and the landings went like clockwork. Three battalions of Rangers and 509th Parachute Infantry Battalion were landed near the harbour to gain immediate control of the port, for much depended upon a quick and efficient unloading. Six miles north-west of Anzio the British 2nd Special Service Brigade (9 and 43 Commandos) were landed with instructions to block the highway above the town. By 8 a.m. the 2nd Brigade of the 1st Division and a few Sherman tanks were ashore in the same area, and on the extreme right – some four miles east of Anzio – General Truscott's 3rd Division had got ashore untroubled even by mines. All the troops in these early landings had orders to move inland and consolidate the beachhead.

Surprise was complete. The Rangers encountered some Germans manning the port, whose batteries discharged a few rounds before the men were captured or killed, and one or two staff officers had hastily left the town. As daylight came the whole harbour presented a busy appearance of ships unloading with the precision of a well-executed exercise; barrage balloons were up and guns of the fleet ready, but unwanted. One enemy bombing raid had managed to avoid the many Allied air sorties, but little damage was done and by mid-morning more than half the 40,000 men and 5,000 vehicles that had come with the first convoys had been landed, and by midnight on D-day it was estimated that 36,000 troops and 3,200 vehicles were ashore. VI Corps had lost thirteen killed and eighty-seven wounded and taken 227 prisoners.

Generals Alexander and Clark were early arrivals at the beachhead, and pleased to see that everything was going so exceedingly well. It is unlikely that they gave any fresh orders to General Lucas, who noted in his diary that Alexander thought he had done 'a splendid piece of work'. But now, with his troops streaming ashore and pushing ahead unopposed to form a beachhead seven miles deep and sixteen miles long, Lucas had a difficult decision to take. His Intelligence had told him that there were two German armoured divisions south of Rome, but General Alexander felt sure that these had been rushed to strengthen the Cassino front. Lucas, a cautious man, was not so sure, and his own armoured division had not yet arrived. *There were three options open to him:*

(1) To go right forward while there was no opposition, and try to establish himself on the Alban Hills, twenty miles distant, perhaps even making a dash for Rome.

(2) To order a limited advance to seize Campoleone and Cisterna, thus holding the enemy's certain counter-attack at arm's length, and giving himself a larger beachhead in which to manoeuvre.

(3) To be content with carrying out his primary task, to dig in and consolidate the beachhead along the approximate line already reached, and make no further thrust before his build-up (particularly Combat Command A) was complete.

General Lucas decided on option 3, a decision for which he has been frequently, and somewhat unfairly, criticized. If anyone was to blame for the fact that VI Corps may have missed an opportunity it was the Commander-in-Chief, and through him General Clark, for the orders concerning the advance 'on' the Alban Hills were too vague, whereas those that Lucas received from Clark concerning his primary task of securing the beachhead were most emphatic. Moreover, when Alexander and Clark had seen for

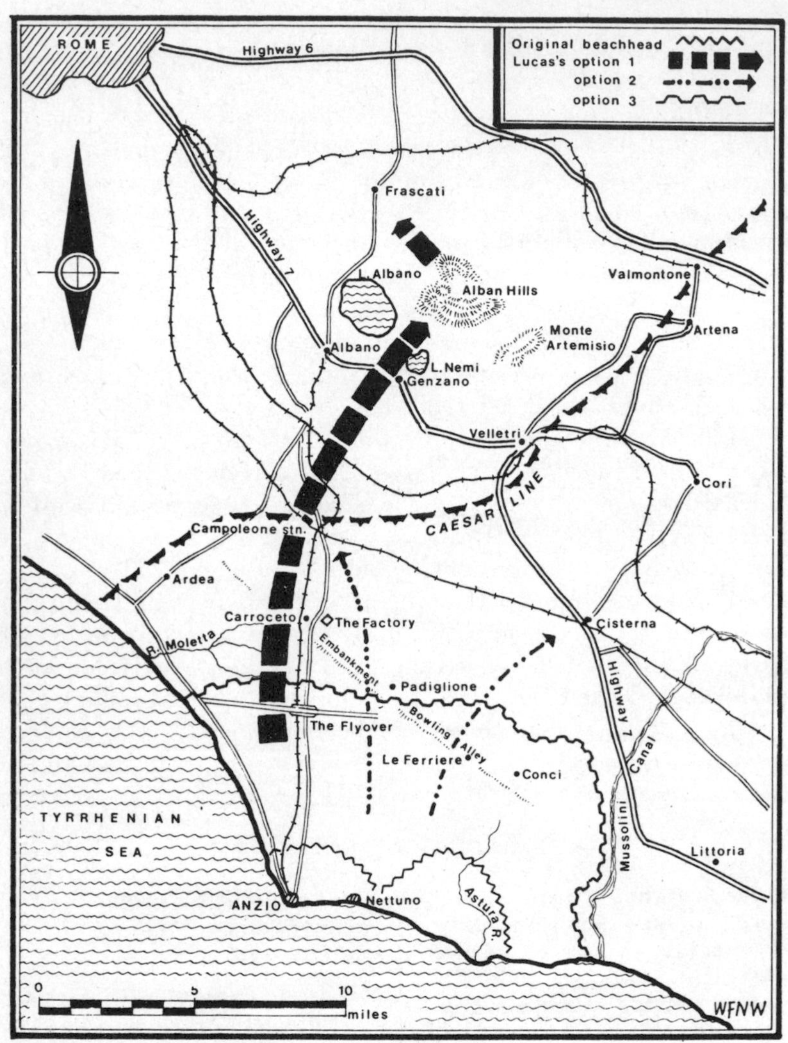

69 Anzio: 23–24 January 1944, Lucas's options

themselves, on the morning of D-day, that the landings had been virtually unopposed and the troops were becoming well established, they were in a position to order Lucas to do something more daring than his gentle probing.

A more thrusting commander might have ordered his troops forward to seek the enemy on the Alban Hills or even in Rome. The enemy Commander-in-Chief wrote in his memoirs, 'Yet as I traversed the front [on the afternoon of D-day] I had the constant feeling that the Allies had missed a uniquely favourable chance of capturing Rome and of opening the door on the Garigliano front',* but it seems that

* *The Memoirs of Field-Marshal Kesselring*, p. 194.

only one of the Allied commanders agreed with him. Alexander, Clark, Truscott and Templer (Commander of 56th British Infantry Division), writing or talking of the battle afterwards, were all of the opinion that to thrust out to the Alban Hills, let alone Rome, before the beachhead was completely secure would have enabled the Germans to clamp a vice of steel round the long communication line, cut off the corps and retake the port.*

However, Field-Marshal Lord Harding, † in a recent conversation with the author, said that he always felt that a great opportunity was missed. In his opinion the immediate occupation of the Alban Hills (which was a perfectly feasible operation directly after landing, and there were back-up troops on the way) would have so disrupted Kesselring's build-up that a withdrawal from the Gustav Line would have been a distinct possibility.

If we give Lucas the benefit of the doubt over the controversial first option, there remains the second. Not very much has ever been said about this, although considerable space has been devoted to the fierce battles that were soon to rage for the capture (or rather the failure to capture) of Campoleone and Cisterna. After the third day Lucas lost Time as an ally to Kesselring. Had he been more resolute he could have taken those towns without the fearful losses he suffered a little later. It might have been the best solution, and he could perhaps be criticized for not attempting it.

Field-Marshal Kesselring, in command of the German armies in Italy, was not in a particularly enviable position at the time the Allies were preparing their double thrust against the Gustav Line. He was short of troops, and many of those he did have lacked high-altitude winter clothing and were extremely battle-weary; he was short of ammunition and inferior in the air, he had a long communication line to guard, and being without command of the sea he was in constant fear of an amphibious landing.

But Kesselring was not a man to despair, nor one to be stampeded into panic measures. He had fought as an artillery officer in the First World War, being transferred to the Luftwaffe in the 1930s. Created a field-marshal in 1940, he was soon to show outstanding qualities of leadership and technical skill. He never despaired under the repeated blows of fortune, and proved a dour opponent throughout the Italian campaign. Like many of his senior colleagues he was at sea in politics, but he was sufficiently cultured to preserve the glories of Rome from the ravages of war. For this alone he deserved a better fate than a sentence of death.‡

When making his strong representations to the O.K.W. that Italy should be defended south of Rome, Kesselring was well aware of the danger threatening the long Italian coastline. Landings might be expected in several places from Genoa southwards, but Kesselring

* Nigel Nicolson, *Alex*, pp. 231, 232.
† General Harding (as he then was) was Alexander's Chief-of-Staff from January 1944.
‡ Remitted to imprisonment.

always considered Rome to be the most sensitive area, for he was fairly sure that Alexander would attempt to break the Gustav Line with a double punch. In Italy at the beginning of 1944 – broadly speaking – General von Mackensen's Fourteenth Army in the north had the task of maintaining order, while von Veitinghoff's Tenth Army held the Cassino front with eight divisions. But it was not as simple as that, because units of the Tenth Army were in the process of being pulled back to the Rome area to rest, or transferred to the Adriatic front. The situation on the German right shortly before the Allied offensive of 17 January could certainly be described as fluid, if not muddled.

Elaborate arrangements had been made to counter any invasion from the sea, under the codeword Case Richard, but Kesselring was not content to rely solely on reinforcements from the north: he felt that he must have at least two divisions nearer at hand. In consequence, despite von Veitinghoff's protests, the 29th and 90th Panzer Grenadier Divisions were withdrawn from the Tenth Army to the Rome area, where Kesselring also had I Parachute Corps headquarters, which could be activated to take charge of any defensive operations, and the 4th Parachute Division in the process of being re-formed.

The British attack across the Garigliano on 17 January, and the establishment of a bridgehead, followed by the success of the Free French and Americans on the Rapido, created a serious problem. Any further penetration could outflank Cassino, and pose a threat to the whole Gustav Line. General von Veitinghoff clamoured urgently for the return of his two divisions, with a promise that with them a successful counter-attack could be made, which would enable him to restore the line and return the divisions in a matter of a few days at most. *Kesselring had to decide:*

(1) Whether to agree to denude troops from his reserve in order to shore up the weakening Tenth Army front, and enable von Veitinghoff to launch a counter-attack to restore the situation, or

(2) Whether to back his hunch that a landing was imminent, and retain the only troops he had that could drive the enemy back into the sea.

On 18 January Kesselring sent the two panzer divisions, and H.Q. I Parachute Corps, back to the southern front. The difficulty of this decision is emphasized by the fact that the two men most closely concerned disagreed over the wisdom of it. Siegfried Westphal, Kesselring's chief of staff, was to write later, 'Looking back, this was clearly a mistake, for the attack and landing on the Tenth Army's front were only diversionary manoeuvres, designed to tie down our forces, if possible, to lay bare the defences of Rome.' But Kesselring was to argue that to allow the Tenth Army's right wing to be driven in, with a consequent uncontrolled retirement, would also lay bare the defences of Rome. He contended that the threat to the Gustav Line was too dangerous to be disregarded, and that the Allies were not attacking in the south merely to form a screen for a landing, but that success there was to make possible the encircling movement of a landing. He felt it essential to clear up the existing mess before meeting any fresh challenge.

Later, as has been noted, Kesselring was to say that he thought VI Corps had missed a unique opportunity of taking Rome after the initial landing. In this he may well have been wrong, but he was almost certainly right to send von Veitinghoff his two divisions, for at all costs he had to hold the Gustav Line.

General Westphal tells us that when the two divisions had been sent south the only troops available south of Rome, apart from coastal batteries, were just two battalions.* Kesselring was understandably nervous about his rear, but could not activate Case Richard until he knew exactly where any landing would take place. He did put all troops in the Rome area on an emergency alert, but when his staff complained that this was tiring the men unnecessarily he stood them down – on the night of the landing. He may have been influenced in this decision by the chief of German counter-espionage, Admiral Canaris, who, while visiting his headquarters on 21 January, made the incredibly rash pronouncement that there was no evidence that there would be any enemy landing in the near future. A few hours after his departure the Allies were ashore at Anzio!

Undoubtedly the Anzio landing found Kesselring off balance. The Fifth Army's offensive had not only dented his line, but had forced him to commit his reserves and leave the coast bare. However, such was the speed and resolution with which he reacted to this new threat that he was able to seal off the beachhead before General Lucas felt himself strong enough to begin a deep penetration.

As soon as he realized that the landing was an invasion and not just a raid, Kesselring went into action. Case Richard was activated, and troops from northern Italy and as far afield as France, Germany and Yugoslavia prepared to move towards Anzio. But most impressive of all was the speed with which units in the battle area were concentrated. Within hours the lower slopes of the Alban Hills were bristling with General von Pohl's dual-purpose anti-aircraft batteries, and by midday they were ready to bar the way to any armour.

The immediate defence of the area was entrusted to General Schlemmer, the Commandant of Rome, but during D-day General Schlemm,

*In the actual Anzio area the only combat troops were one infantry company.

commanding I Parachute Corps, was recalled from the Garigliano front; by that evening he had set up a tactical headquarters, as had 3rd Panzer Grenadier Division and the Hermann Göring Panzer Division. They could now muster three battalions of the 4th Parachute Division, and one battalion each from 90th Panzer Grenadier, 3rd Panzer Grenadier and the Hermann Göring Panzer Divisions.

Kesselring described this rapid build-up as 'a higgledy-piggledy jumble'; certainly during the early days battalions came tumbling in from all directions, but it could not be helped that in the interests of immediate defence organized formations were badly split up, and command and communication problems were appalling. By first light on 24 January the Germans had assembled the approximate equivalent of two divisions, made up from bits and pieces of no fewer than nine divisions, and on that day General von Mackensen, commander of the Fourteenth Army, was ordered to take over from General Schlemm.

Many of the soldiers (especially those from 4th Parachute Division) were inexperienced and only partly trained, and regimental commanders and staff were new to each other, but they sufficed for Kesselring to tell von Mackensen on the 24th that he considered the position secure from any reverse.* Von Mackensen was now ordered 'to strengthen the defence ring and institute measures to narrow and remove the bridgehead'.

These were bold words, for when they were uttered the situation was still fraught with danger. Von Veitinghoff, whose line had been dangerously thinned, was looking over his shoulder with the utmost anxiety, and suggesting withdrawal to meet the new threat; but Kesselring was adamant that the line must be held. He possessed that invaluable ounce of optimism which the commanders at First Gaza had lacked.

* By the end of the month, when H.Q. LXXVI Parachute Corps was under orders to move from the Adriatic, and units of 114th Light and 715th Infantry Divisions were arriving from the Balkans and France respectively, Kesselring felt confident enough to withdraw units of 29th and 90th Panzer Grenadier Divisions in order to reconstitute his mobile reserve.

15 Anzio
Battle for Survival

THE beachhead, where by 24 January VI Corps had established itself, was a narrow coastal plain partly reclaimed by Mussolini's government from that treacherous quagmire called the Pontine Marshes. The principal irrigation conduit was the Mussolini Canal, along which General Lucas rested his right flank; the protection this deep waterway gave was further heightened by the Germans having flooded the area to its east and south, with the result that the ground had virtually reverted to its former impassable condition.

North of Anzio town for about five miles there was a belt of woodland called the Bosco di Padiglione; the principal metal road, the Via Anziate, ran through this woodland due north through Carroceto and Campoleone to Albano and Highway 7. About seven miles north of Anzio the road passed under a flyover – soon to become notorious – which carried the Lateral Road (San Lorenzo–Padiglione) over it, and just short of Carroceto it went under an embankment. A secondary road ran from Nettuno through Le Ferriere to Cisterna and eventually to Velletri. Beyond these two roads (and the Lateral Road) there was little but cart tracks in the beachhead. The two major roads led eventually to the Alban Hills, a volcanic basin rising some 3,000 feet and providing the most perfect grandstand view from which the enemy could observe almost the whole beachhead.

On the western side of the beachhead was the Moletta river, and northwest of the Flyover the deceptively flat farmland was dominated by the Buonriposo ridge. Another ridge – the Vallelata – lay to the west of the Via Anziate between Carroceto and Campoleone. In this area a pleasant landscape of farms and meadows assumed a much more sinister appearance at close quarters, for much of the ground was interlaced with very deep and wide ditches, whose crumbling banks were usually covered with brambles. These wadis (as the desert-hardened troops came to call them) proved an obstacle to tanks, and although they occasionally offered protection they soon became water-filled trenches in which soldiers had to live and die, courageously coping with the highly competent German infiltration tactics.

The right of the beachhead from Fosso di Spaccasassi to the Mussolini Canal was to be held by the 3rd U.S. Division under General Truscott, while General Penney's 1st British Division, which had been taken out of reserve on 24 January, stretched from the American left across to the Moletta river. By now 179th Regimental Combat Team* of 45th U.S. Division had arrived at the beachhead, and with the promise of more troops to come Lucas felt strong enough to order probing reconnaissance patrols northwards towards Albano and Cisterna. Troops from the 24th Guards Brigade of the 1st Division reached the Flyover without encountering opposition, but Aprilia (the Factory †) was held by troops from the 3rd Panzer Grenadier Division, and the patrol returned with five prisoners. On the right troops from the 3rd U.S. Division came into contact with the Hermann Göring Panzer Division, and the 15th and 30th Infantry Regiments were firmly held a mile beyond Le Ferriere. At this stage the beachhead was some twenty-six miles long by eight deep.

It was only after Lucas had ordered forward these probing missions that Intelligence informed him that by now Kesselring had pushed into the beachhead some 40,000 troops, while many more were hurrying to the front. It was clear to him therefore that no major offensive could be contemplated until his build-up had progressed. On the evening of 23 January the Luftwaffe bombed Allied shipping in the harbour, and one of their radio-guided bombs sank H.M.S. *Janus*; there were other losses to follow from air bombardment, but the weather and Allied air superiority kept them within bounds. More serious at this time was the effect the weather had on Allied shipping. The 1st Reconnaissance Regiment was not fully ashore until 25 January, which inhibited any early reconnaissance in depth, and Combat Command A was held up by weather until 28 January.

In the early days all that Lucas felt he could do was to expand the beachhead so as to form a firm base from which to launch his main attack towards the end of January. The Grenadiers took the Factory, and the two other battalions of the Guards Brigade (Scots and Irish) pushed on to the area of the Embankment and Carroceto, where they were strongly engaged. Between the British and American sectors of the battlefield the Rangers moved forward in concert with the Grenadiers, and by 26 January they had cleared the Fosso di Spaccasassi of enemy to a point north-east of the Factory. The 3rd Division's attempt to reach Cisterna had not been successful, however; Truscott halted them three

* An American R.C.T. is about equivalent to a British brigade.

† The village of Aprilia contained a number of brick buildings designed, under the Fascist regime, as a model farm settlement. The whole complex closely resembled a factory, hence its name.

miles short of the town, pending the full-scale attack that the corps commander was planning.

Lucas's plan to break out of the beachhead at the end of January was ambitious. It was to be on a two-division front with naval, air and artillery support, and he envisaged the cutting and control of Highways 6 and 7. The main thrust was to be undertaken by 1st British Division, with Combat Command A of 1st U.S. Armored Division operating on its left; 3rd U.S. Division, with the Ranger battalions and 504th Parachute Infantry Regiment in support, was to carry out a preparatory operation on the right, starting twenty-four hours earlier.

The British were to advance up the Via Anziate from the Factory and seize the road-railway junction at Campoleone, which would give them a springboard for a further advance on the Alban mass; meanwhile General Harmon's armour was to swing round the left of the 1st Division and Campoleone, and come on the Alban Hills from the west. On the right Truscott's men were to take Cisterna, cut Highway 7, seize the high ground at Valletri and be ready to move towards Valmontone and eventually control Highway 6. Because the forces engaged in this two-pronged attack would diverge, Lucas wisely kept a tight control from his headquarters in order to forestall any attempt by the Germans to drive a wedge between the two divisions.

Lucas was not a particularly good general, nor a lucky one. He depended very much on air support, and at this time (and later as well) the weather went sour on him. Even though much of the German concentration was done at night, the fact that the Allied air forces were badly restricted considerably helped Kesselring's quick build-up, and the weather again closed in on 30 January for several days.

The Germans were able to oppose VI Corps' attack with four large battle groups under command of Generals Pfeiffer, Gräser, Raapke and Conrath, who commanded elements of (reading from left to right of the Allied front) 65th Infantry Division, 3rd Panzer Grenadier Division, 71st Infantry Division and the Hermann Göring Panzer Division respectively. In addition 26th Panzer Division had just become operational and, with parts of 715th Infantry Division, was in support of the Hermann Göring Panzer Division. There were 238 field guns and thirty-two *nebelwerfers* at the front by this time.

The attack was opened on the night of 29–30 January; it got off to a bad start. To spearhead the 3rd Division's assault on Cisterna two battalions of the Rangers were to infiltrate along a dry irrigation ditch which seemed to offer an ideal covered approach to the town, while the third battalion, with 15th Infantry Regiment, were given the task of clearing the road of tanks ready for the full-scale drive on the town at dawn. Unfortunately the importance of the Pantano ditch had not escaped the Germans, and they ambushed the Rangers with superior numbers, and after some desperate fighting virtually annihilated them. Out of 767 men who had formed the two infiltrating battalions only six escaped.

Elsewhere the attack that night made little progress and the losses of the 15th Infantry and 504th Parachute Infantry, in gaining the line Ponte Rotto-Isola Bella, had been very heavy. On 31 January Truscott renewed the attack using all his artillery, a tank battalion and a tank destroyer battalion, but the opposition was too strong. With the division exhausted after three days' hard fighting (and a casualty list of 3,000 since the landing), he ordered them to dig in some 1,500 yards short of Cisterna, and be ready for the counter-attack he felt sure would soon be coming.

On the left the British and Americans fared a little better. General Penney had decided to attack with his hitherto uncommitted 3rd Infantry Brigade, supported by 46th Royal Tank Regiment, and three Field and one Medium regiments R.A. Their first objective was to be the railway–road crossing at Campoleone, and the second the town of Osteriaccia. General Harmon had decided to advance his armour along the embankment running north-west from Carroceto. The attack was scheduled to start at midday on 30 January. But first it was necessary to gain suitable start lines. The 24th Guards Brigade was ordered to seize a lateral track about a mile south of Campoleone for this purpose, and units from 1st U.S. Armored Division a part of the disused railway track.

This operation, which took place on 29–30 January, did not run as smoothly as was hoped. The Americans on the left of the road ran into heavy opposition from the Buonriposo ridge, while the Grenadiers of the 24th Guards Brigade had the misfortune to lose four officers (three of them company commanders) when an order group took the wrong turning. This left the Scots and Irish Guards to carry out the attack, and although after very stiff fighting they reached their objectives, they were temporarily driven back in the early hours of 30 January. However, the advance of 3rd Infantry Brigade was only delayed three hours, and by 6 p.m. they had fought their way through to just below the railway line in the vicinity of Campoleone station, and had consolidated their objectives.

This was very satisfactory, except that the forward troops of 1st Division were now stuck out on a narrow salient; General Penney must have been anxiously awaiting news of the armoured thrust to his left, which was to sweep round Campoleone. But this was not to be, for Harmon found that what might appear to be tank country from air photographs was in fact impassable to his vehicles; tanks were soon floundering in the mud, and being brought to a halt by waterlogged ground and deep ditches. Lucas, in view of this lack of progress, decided to split the armour; half were to persevere across country, while the remainder stood by to push up the Via Anziate as soon as Osteriaccia – the second objective – had been taken.

The 2nd Battalion of the Sherwood Foresters, who had spent a dangerous and uncomfortable night in reserve, was to pass through the two forward battalions of the 3rd Brigade to take the second objective. As soon as they left their slit trenches at 10.30 a.m. on the 31st they came under intense and accurate fire from strong enemy posts holding Campoleone and the station. Doggedly they advanced, leaving behind swathes of broken men; by midday they had become bogged down, but under cover of a new fire plan, and supported by British and American armour, they renewed the attack that afternoon. However, neither they nor the armour (wallowing in the mud north-west of Carroceto) could make any further progress. Buildings along the railway line bristled with machine guns and tank emplacements, and the line itself was a veritable death trap. The Foresters suffered enormous casualties, and General Lucas's attack had spent itself short of Campoleone.

Thus the brave attempt to break out of the beachhead had failed. The towns of Campoleone and Cisterna were strongly held by the enemy; VI Corps was penned up in a very restricted area, for only at two points did their line stretch inland by as much as fifteen miles. The Germans held the initiative, and could be expected to counter-attack at any moment. Lucas had ordered the 1st U.S. Armored Division back to the Padiglione Woods and into corps reserve. The men of the 1st Division must have watched the tanks pass through with deep concern, for their position at that moment was most unenviable. The 3rd Brigade was at the tip of a salient, some four miles long by two miles at the base, which thrust itself towards the enemy, crying out to be chopped.

On the very day that the attack had been finally halted General Alexander arrived on the beachhead. His urbane, unruffled appearance, coupled with his clear judgment and cool, incisive mind, always acted as a tonic to those he met, be they private soldier or general. During his visit, which lasted until 2 February, he toured the front and met the commanders; as a result an interesting situation developed. Alexander's role in the Italian campaign called for considerable tact. Not liking to override his American army commander (although Churchill made it plain that he could do so), he avoided interfering with Fifth Army's strategy as much as possible. He was perhaps too obsessed by the fear of treading on American toes. *Having toured the battlefield he had two options:*

(1) To instruct General Clark to remain on the defensive in the present positions, to strengthen the defence and prepare to receive the German counter-attack.

(2) To gain more elbow room and secure a firmer base for further operations by resuming a limited offensive with which to take Campoleone and Cisterna, and widen the beachhead to the west.

General Alexander decided that VI Corps must take Cisterna and Campoleone before any real progress could be made. He appreciated that the corps was not yet ready for a major offensive, but in expanding the beachhead and taking the two towns room would be made for building up reserves, and a good springboard would then be available for launching the main attack at a later date (option 2). After consultation with Clark, he instructed him accordingly. However, before Clark had given the necessary orders Lucas had already set in motion a thorough defensive organization of the present position. Clark then had second thoughts about Alexander's instruction and agreed with what Lucas was doing. Alexander did not overrule Clark.

It would seem that only in theory was the Commander-in-Chief right, for in practice, as Lucas must have known, the troops were in no state to capture the two towns that they had so recently failed to take, even though von Mackensen, who had only two divisions at full strength, was fully expecting them to make the attempt.

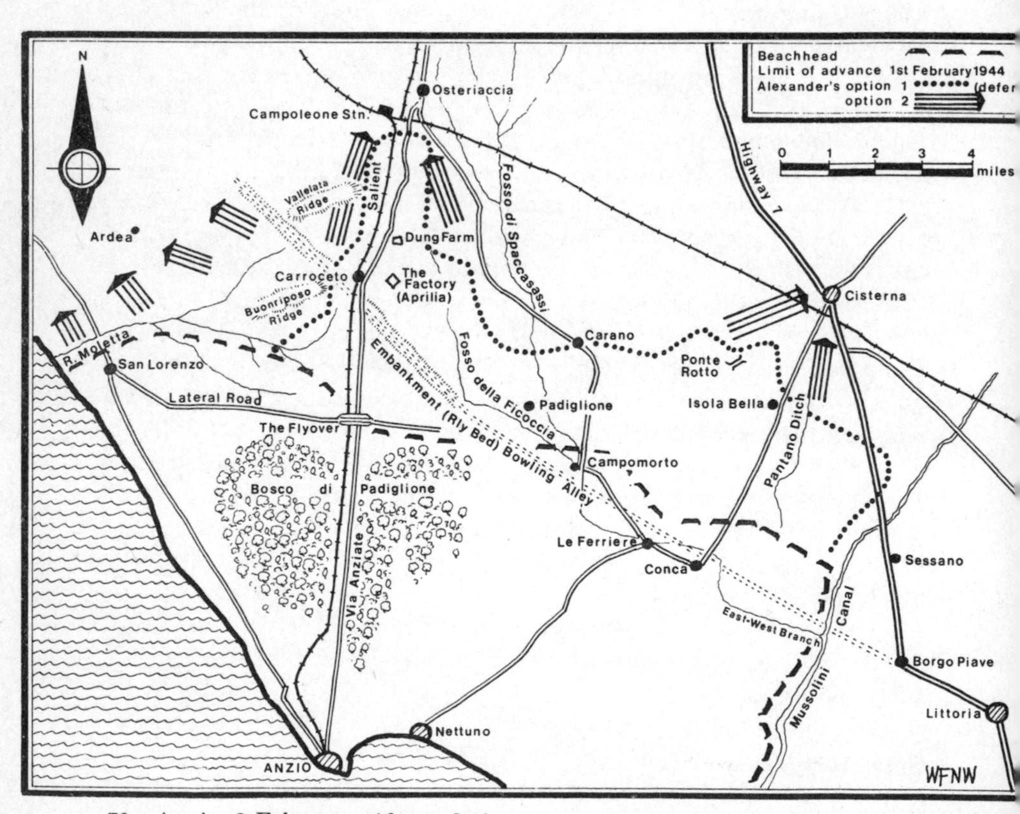

70 Anzio: 2 February, Alexander's options

In preparing his defensive position against the inevitable German counter-attack, Lucas had been helped by the arrival of substantial reinforcements. The whole of 45th U.S. Division was now on the beachhead, 168th Infantry Brigade had been detached from 56th British Infantry Division, and also from the Garigliano front came three regiments of 1st Special Service Force. These latter were sent to the extreme east of the beachhead to hold the line of the Mussolini Canal with 504th Parachute Infantry. On their left was the 3rd U.S. Division, then the 1st British Division, and on the extreme left the 157th Infantry Brigade from the 45th U.S. Division. The remainder of that division and the 1st Armored were in corps reserve.

General Penney's division was in a precariously strung out position, and if this did not cause the corps commander anxiety, it most certainly did the divisional commander. As General Penney awaited instructions from corps, *Lucas had two options:*

(1) To allow Penney to withdraw the 3rd Brigade to the base of the salient in the area of Dung Farm, which would give a better defensive line and avoid the grievous harm that could arise from a German envelopment.

(2) To order Penney to hold firm in the salient, so as to use it as an advance base through which to pass troops and fan out when the time came to go over to the attack.

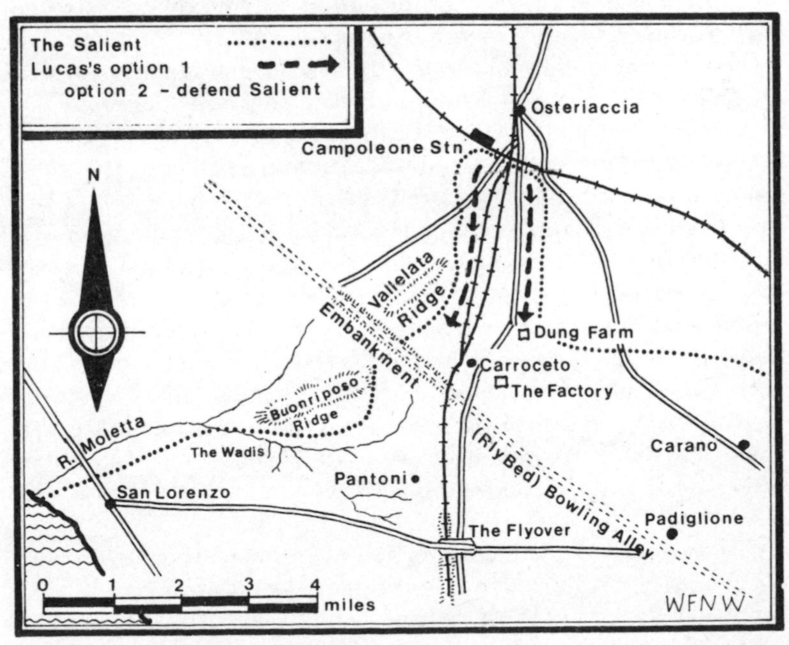

71 Anzio: 2 February, Lucas's options

General Lucas very seldom visited the forward troops. It appears that he still did not fully appreciate that armour could not manoeuvre to the west of the main road, and he retained hopes of being able to hold the salient long enough to withstand a counter-attack, and resume the armoured-infantry thrust from this advanced base. One does not need the benefit of hindsight to see that it must have been the wrong decision.

Lucas's attack had somewhat shaken the fourteenth Army, and had disorganized von Mackensen's offensive planning, which was further disrupted by a warning from Kesselring that he would require some troops for the Cassino front, which had once again reached a critical stage. Von Mackensen had decided that the line of the Via Anziate offered the best chance of success for his intended powerful thrust. This was divided broadly into three phases; pinching out the 1st Division's salient, capturing Carroceto and the Factory, and using these as a firm base from which to launch the third phase of the plan, the all-out drive to the sea.

When Lucas determined that the salient must be held, General Penney strengthened its flanks as best he could by putting 24th Guards Brigade about 1,000 yards behind 3rd Brigade's battalions on the left flank facing west, while 2nd Infantry Brigade was given a similar protective task on the right flank. After a preliminary bombardment and an attempt to dent the nose of the salient during the afternoon of 3 February, the main business was put in hand at 11 p.m. that night when Battle Group Pfeiffer closed in to the attack on the western side of the salient, and Battle Group Gräser dealt with the right flank.

Almost all the fighting in this difficult, trappy Anzio country, with its wadis, bogs and rivulets, was conducted if not on a platoon, certainly on a company level; with battalion commanders often quite out of touch with the situation, it was some of the toughest, most dangerous and beastliest fighting of the whole war. To describe any individual actions would be invidious. For long stretches every man in the forward areas was called upon to face indescribable discomfort, living in rain-filled ditches with enemy shell and shot ploughing wide lanes of dead and dying all around him. Men imbued with the offensive spirit, the outcome of perfect training, retained their sanity; others less fortunate would lose hope and begin to disintegrate. But through the bruised and battered ranks of each battalion the spirit of companionship flowed like a silver thread.

And so, in its most simple form, it can be said that after fearful battling the British 3rd Brigade was able to make a skilful withdrawal on the night of 4 February to the base line of the salient. They had lost some three miles of ground and suffered about 1,400 casualties in carrying out the order to stand firm in that exposed position. But von Mackensen had not got the Factory, and 1st Division now held a line from Buonriposo ridge curving round the Embankment and running just north of Carroceto and Aprilia.

All day on 7 February it poured with rain, turning the ground into a worse morass than usual. 1st Division had reorganized its line with three brigades forward, and as the darkness deepened the German shelling of their positions intensified. Von Mackensen's men were about to make another, and greater, effort to take the Factory. Much the same tactics were employed: a bombardment, deep infiltration behind the lines, followed by the main push, which once again would be a pincer movement to left and right. And much the same dogged determination was shown by the soldiers of both armies. The fire never slackened; the deep bass rumble of the heavy guns, the constant crackle of rifle and machine gun; jagged metal, jamming bren-guns; the wadis blocked with bodies of the fallen. Such was the pattern of those three days (and many before and after).

There was much heroism: Major Sidney of the Grenadiers won his Victoria Cross holding a wadi against a mass of infiltrating enemy, in much the same style as Horatius Cocles had once done with a bridge thirty miles up country. But it was in vain; inch by inch the Germans edged their way forward and, despite the efforts of eighty-four medium bombers who showered high explosive on them, by 10 February von Mackensen had the Factory, Carroceto and the ground he wanted from which to advance for the third and main phase of his attack.

The very serious situation now facing the Allies was fully realized by the commanders; General Penney was constantly calling for troops to relieve his battle-weary men, for fresher men better suited to mount a counter-attack against the Factory before it was too late. 1st Division was operating at about half its effective strength, and General Lucas sent two regiments from 45th U.S. Division, which was in reserve, to take over some of the front. For the first time since the landing he visited 24th Guards Brigade headquarters, on the morning of 10 February, for a conference.

The main purpose of this conference was to plan the attack to regain the Factory; what followed must rank as one of the worst examples of decision-taking in the annals of war. General Penney gave the assembled division and brigade commanders his appreciation of the situation, and recommended a strong and immediate counter-attack. Little was said for a minute or two after he had spoken, then the Corps Commander turned to General Eagles – commanding 45th Division – with the words, 'O.K. Bill, you give 'em the works.'* Soon afterwards

* Wynford Vaughan-Thomas, *Anzio*, p. 127.

Lucas departed, leaving the generals with no appreciation, plan, orders or even guidance.

Eventually out of the vacuum came an order for one regiment of 45th Division to retake the Factory. The regimental commander passed it to the battalion commander, who delegated it to two rifle companies supported by two companies of tanks. The artillery programme was not properly co-ordinated between the American and British gunners, and the attack could not be mounted before dawn on the 11th, which Penney thought would be too late. And so it proved to be. The men of 179th Infantry and 191st Tank Battalion fought most gallantly, and at 2 a.m. on the 12th two fresh companies were thrown into the fray, but they – like the first wave – were driven out by the strongly posted Germans, and the attack was abandoned. It was a case of too little and too late. There were no options here; a straightforward decision was called for, and it was badly taken.

There was now to be a very short respite during which both armies carried out some reorganization ready for the next trial of strength. General Alexander had decided on 10 February to transfer the remainder of 56th British Division from the southern front (168th Brigade had already arrived). They did not replace 1st Division, but enabled the latter to be pulled back for a short while. The 2nd Brigade of 1st Division was to man the rear defensive line just south of the Flyover, from which there was to be no withdrawal; it was in fact the initial line taken up after the landing. The front was now held, from right to left, by 3rd U.S. Division, which stretched from the Mussolini Canal to Carano; 45th U.S. Division, with three brigades up and reaching to just west of the Via Anziate; the 56th British Division and, on the extreme left of the Allied line, the 36th U.S. Engineer Regiment.

The heavy fighting of the past week had undoubtedly stretched the Allies to the uttermost, but it had not left the Germans unscathed. The toughness of the opposition and the fact that the Allied airmen had been constantly pulverizing them had shaken morale among some of the troops. Nevertheless, optimism ran high on the whole, especially among the senior commanders, with the exception of von Mackensen, who was a cautious, painstaking commander not given to over-confidence, and never anxious to rush in before all was absolutely ready. However, his pleas for delay were now overruled by Kesselring, who was anxious to finish off the beachhead before any further reinforcements could arrive.

Kesselring estimated that the Allies had a numerical superiority in troops and as much as 130 more tanks, as well as undoubted artillery and air superiority. He was certainly correct in regard to the air, and probably the tanks, but the Germans now had parity in guns, although their shortage of ammunition meant that their artillery programme had to be carefully orchestrated. The LXXVI Panzer Corps had five divisions

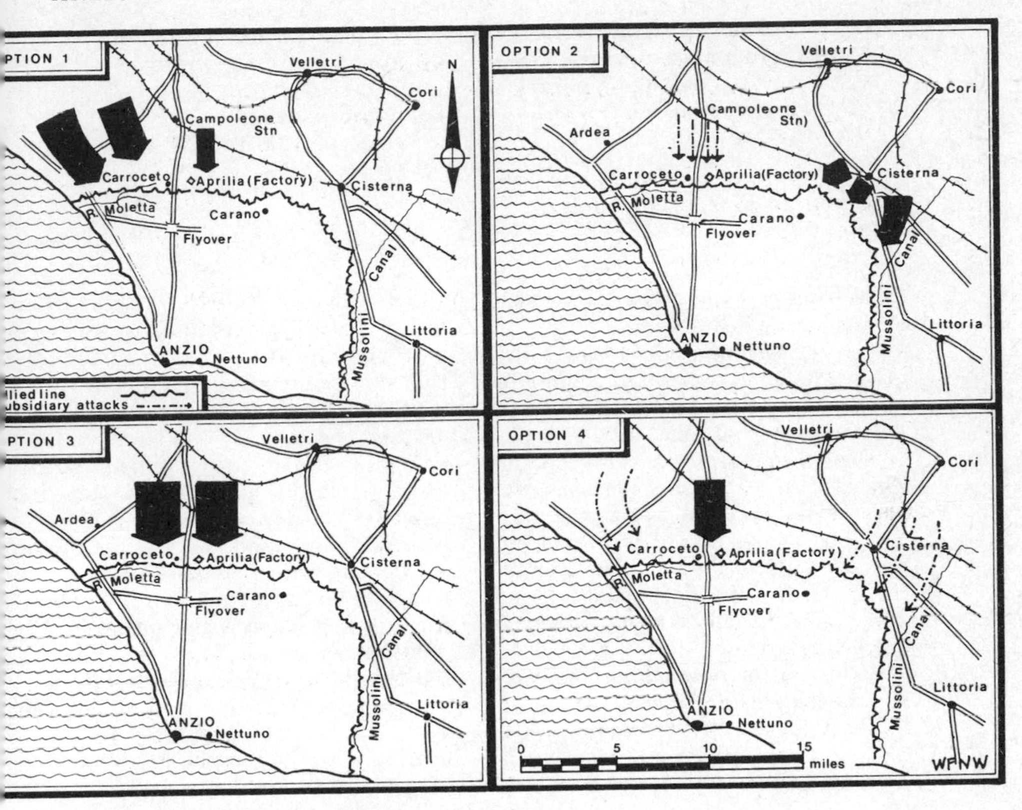

72 *Anzio: 16 February, Kesselring's options*

totalling thirty battalions, and I Parachute Corps had fifteen battalions; it is therefore possible that German troops slightly outnumbered the Allies, with both armies having around 100,000 men. Von Mackensen had also been sent two new 'weapons'. The Goliath was a small tracked vehicle filled with TNT, steered and detonated by remote control, designed to operate ahead of the troops and clear minefields and other obstacles. Hitler had also insisted on him having the crack Infantry Lehr Regiment to spearhead the next assault. This was a demonstration regiment composed of élite soldiers, most of whom had so far never been in action, but from whom great things were expected. Both the Goliath and the invincible demonstration regiment were to prove disappointing failures.

In planning the attack which was to drive the Allies off the beachhead, *Kesselring and von Mackensen had four options:*

(1) To unhinge the beachhead with a flank attack, simultaneously supported by a frontal push. The left of the Allied line was the only 'open' flank.

(2) To launch the main attack from Cisterna against the 3rd U.S. Division, with subsidiary attacks east and west of the Via Anziate.

(3) To make the Via Anziate the main axis, and attack on a broad front.

(4) To use the same axis, but to concentrate the main thrust on a narrow front, with perhaps some subsidiary attacks on a flank.

Kesselring and von Mackensen very quickly dismissed a flank attack (option 1), for although in theory such an envelopment movement could work, in practice it would be broken up by Allied naval guns, and, if these failed, by the wooded and mine-sown country. The eastern sector of the beachhead was not the quickest way to the sea, and the obvious line of advance was the one already used by both armies – straight down the main road, for that alone offered the best scope for exploiting armour.

Hitler was by now taking an active part in the Anzio planning, but it was not easy for anyone, however good a strategist (and Hitler was no tyro), to plan a detailed attack from hundreds of miles away. Nevertheless, that is what he did, insisting that the attack should be concentrated on a narrow front, and that the Infantry Lehr Regiment should lead the assault. What the Führer decreed few commanders were bold enough to decry, but Kesselring wrote later, 'We had to pay for both these mistakes.' Hitler had argued that a narrow front would enable the German artillery to pulverize the enemy position, but that could cut both ways, for it gave a much better target for the Allied gunners; von Mackensen also felt that he would be engaging far too few enemy troops. The generals on the spot were probably right, but in broad outline option 4 was the one adopted.

The attack was to be bounded on the east by the Fosso di Spaccasassi and on the west by the Buonriposo ridge, and was to be in two phases. First would come the breaching of the Allied defences, to be followed by armoured penetration driving on Nettuno, or if necessary mopping up any flank resistance left intact through the narrowness of the front. A diversionary attack on the Cisterna front was to be made by the Hermann Göring Panzer Division. LXXVI Panzer Corps was allotted the primary role of accomplishing the breakthrough, and would attack east of the road (against 45th U.S. Division) using 3rd Panzer Grenadier, 114th and 715th Divisions, and the Infantry Lehr Regiment as the first wave, with 29th Panzer Grenadier Division and 26th Panzer Division to exploit the penetration. The role of I Parachute Corps was to protect the LXXVI's flank in the difficult wadi country.

Operation *Fischfang** commenced, as planned, on 16 February. The attack had to go in by day, for with so many troops new to the ground a

* A curious name, perhaps inspired by the hopes of catching fish when the coast was reached!

night battle would have been hopeless. The diversion on VI Corps' right flank gained about 1,000 yards at one point (Ponte Rotto), but was not a success elsewhere. In the centre the fighting was as furious as any that had gone before. The Infantry Lehr Regiment was found wanting under fire and very soon broke,* but the 3rd Panzer Grenadiers and the 715th Infantry Division broke through the forward positions of 45th U.S. Division and penetrated about a mile. West of the road troops of I Parachute Corps managed to infiltrate into 168th Brigade's position, but could not break through.

The sixteenth of February was not a very satisfactory day for the Germans. The fierce resistance they had encountered, the heavy air bombardment (174 tons were dropped that day) and the artillery pounding to which they had been subjected had made grievous inroads into their numbers. The recent hard frost had broken, and their armour could make little headway, while Lucas had not yet committed his 1st U.S. Armored Division. Von Mackensen ordered his commanders to pursue their punishing infiltration tactics that night to create what gaps they could, while he himself had to decide on the employment of his reserves for the next day. *He had two options:*

(1) Should he commit the two panzer divisions (26th and 29th) which were earmarked for the second wave? Kesselring had been urging him to employ them during the 16th, for he felt that the weight of their attack would have turned the scale.

(2) Should he continue to hold them back, for he still had some reserves of the first wave to rely on, and it was questionable whether the time had yet come for the ultimate push.

Von Mackensen knew that the battle had still to reach its culmination, and he judged it wrong to throw in his powerful reserves; he expected a breakthrough within twenty-four hours, and the panzer divisions would be needed then to complete the business. He felt that more preparatory work was needed that night (16–17 February) by infiltration tactics to create a gap that could be penetrated later by armour. Kesselring was not prepared to overrule his army commander, and reluctantly agreed to the postponement (option 2).

It is difficult to say who was right, Kesselring or von Mackensen. It is possible that neither was. The infiltration on the night of 16–17 February was so successful that a gap was prized open between 157th and 179th Infantry Regiments down the line of the Via Anziate. At dawn on the 17th the Germans began to widen the salient thus formed in 45th Division's ground, strongly aided by the Luftwaffe, who managed to carry out one of their rare successful air strafes. That day the German attack made considerable progress, as will be shown later, and Lucas was moving up reserves to strengthen the last line of defence on

* Later they were to fight with considerable distinction.

either side of the Flyover. On the 16th and 17th German casualties were said to have been more than 2,600, and their battalion fighting strength was now down on average to 150 men. To have thrown in the reserve divisions on the 16th as Kesselring had urged would have been unwise, but on the 17th they could just conceivably have achieved a breakthrough. By the 18th, when they were sent in, the momentum of attack had passed its zenith; 2nd Brigade on the final line was ready, and, fighting with incredible doggedness and courage, proved too great a barrier for the panzers, who were perhaps a day too late.

The situation on the morning of 17 February was critical on 45th Division's front. The gap made during the night was quickly widened by elements of three German divisions (the 65th, 114th and 715th), aided by more then fifty tanks; by the end of the morning it had stretched to some two miles wide and one deep. The commander of 179th Infantry Regiment was forced to pull back and shorten his line, and as this had to be done in daylight – though under cover of smoke – the regiment suffered heavy casualties. With much of 45th Division pinned back upon his final line of defence, Lucas hurried reinforcements from reserve including artillery, armour and dual-purpose anti-aircraft guns.

The 1st Division was sent from reserve as back-up troops between 56th British and 45th U.S. Divisions. Both British divisions now came under command of General Templer, for Penney had just been wounded by a shell that struck his caravan in the Padiglione Woods. Penney (who was soon to come back) had fought his division with distinction during the fearful weeks since the landing, but for the next most critical days the British troops under Templer could not have been in better hands than those of this resolute, clear-sighted, forward fighting general.

On the afternoon of the 17th von Mackensen broadened his attack, throwing in the reserves of his first wave, but still he held back the two panzer divisions. The gap bulged dangerously under pressure from fourteen enemy battalions supported by tanks, but Lucas got considerable air support, and part of his 1st Armored Division was now in action. And more important still was the courage and sheer grit shown by the American G.I.s; although back almost to the last line of defence, they refused to break.

The eighteenth of February was probably the most decisive day for the beachhead. The Germans could feel well pleased by the progress made on the 17th, although von Mackensen, who was in a position to know more than Kesselring, was subject to his usual cautious pessimism and argued that with battalions at such low strength it might be as well to call off the attack. But apart from Kesselring's optimism, the Führer had decreed that the Anzio 'abscess' was to be lanced, and the

O.K.W. felt certain that this was within an ace of being achieved. On the Allied side there was considerable gloom at headquarters, but farther forward confidence was not lacking.

By the evening of the 17th the Germans had forged a salient not so narrow as VI Corps' at Campoleone, but too narrow for comfort. They were less than 1,000 yards from the Flyover, but crammed into a space which gave the massive Allied artillery an opportunity for a maximum massacre. Every attempt was made to broaden the salient; German armour rumbled down Bowling Alley and pushed back 180th Regimental Combat Team, while to the west of the Via Anziate 157th R.C.T. gave ground very grudgingly. Both sides were throwing in all that they had, and VI Corps was fighting for its very existence.

In every big battle those closest to the firing can see only a small part of the whole; mutilated and mystified, often surrounded, sometimes overwhelmed, but somehow managing to rise again with courage and resilience to resume the fight, battalions like the 1st Loyals (and many others) learned only afterwards of the vital part they had played in saving the situation. Only 1,000 yards to go for what must be final victory; clouds of exhausted, struggling field-grey figures lapped against the Lateral Road, but never established themselves on it. Then almost abruptly the tide turned; the German soldier had had enough, his will to win had gone. The attack petered out, the ragged ranks withdrew; the farthermost extent of penetration had been reached. This was the first rash of defeat.

Bitter fighting went on throughout the night of the 18–19th, during the 19th, and even for a short time on the 20th, but Kesselring knew by the 18th that the attack could not succeed. Lucas mounted a counter-attack by two task forces in the early hours of the 19th, which took the Germans by surprise, for the heavy artillery concentration on the previous day had disrupted their signal system. Moreover, by now their ability to organize any positive response had greatly diminished. Harmon's armour and 2nd Infantry Brigade reached their objectives, and advanced a mile up the Via Anziate. On that same afternoon Hitler sanctioned a temporary ending to the attack.

The fighting between 16 and 20 February had cost the Germans 5,389 casualties, and the Allies scarcely less (the 45th Division alone had suffered 2,400 casualties). This last German offensive was a battle within a battle, and was an undoubted Allied victory, for the beachhead still held, and it was becoming very obvious to many Germans that it would continue to hold. But Hitler did not see it that way, and if Kesselring did he was careful not to show it, for he knew very well that he could not possibly continue to fight on two fronts.

The beachhead had to be obliterated, and Kesselring ordered the hapless von Mackensen to make plans to do this. For the next few days there was a welcome surcease to the savage fighting that had become the hallmark of this hell on earth. But for the soldiers, squatting in their miserable slits, there could be no respite; the thunder of the heavy guns, the thumping of mortars, the crackle of tracer sparkling along the line, and always the incessant rain, mud and stench of death. Such was their rest before the greater agony began again. But everyone was now aware that a change of command at the top was on the way, and it must surely herald a brighter future for VI Corps.

16 Anzio
Break-Out

SINCE early February, at least, General Alexander had become dissatisfied with Lucas as a corps commander. He had communicated his anxiety on this score to General Clark on more than one occasion, but Clark was not anxious to do anything that might hurt Lucas, for he thought that he had commanded the corps well from Salerno up to the Garigliano. However, in the end he too felt that more thrust was needed at the top, and on 17 February General Truscott was sent to corps H.Q. as Lucas's deputy, and the 3rd U.S. Division was given to General John W. O'Daniel.

Lucas rightly saw this move as the beginning of the end of his command, and on 22 February he was relieved by Truscott and appointed as Clark's deputy. He left with considerable bitterness, because he honestly believed that he was doing well, and he felt sure that he was the victim of British intrigue, and that Clark had been pressurized by Alexander and Churchill; he was even angry with Clark for submitting to such pressure. He was relieved not for any failure to take the Alban Hills or Campoleone and Cisterna after the initial landing, but because he was clearly worn out. The fact that he was absolutely the wrong man for the job became quickly apparent when Truscott took over. Here was a man who commanded from the front, who exuded confidence and was constantly to be seen by the forward troops; a capable man in a crisis and a general who could take decisions. In other words he possessed all the qualities that poor Lucas lacked.

The new general was not given much time to play himself in, for it was very obvious that Kesselring would launch a further attack. It seems probable, however, that both the German Commander-in-Chief and his Fourteenth Army commander knew that the failure to break the Allied line on 19 February represented the farthest limits of expansion of which the German troops in the beachhead were capable. But Hitler was not prepared to admit failure short of the Flyover, and ordered Kesselring to complete the business of annihilation. Kesselring therefore made ready for one more push, which he hoped might at least so constrict the beachhead as to enable his guns to make its continued tenure impossible. For once he was not optimistic. *In making his plans he had two options:*

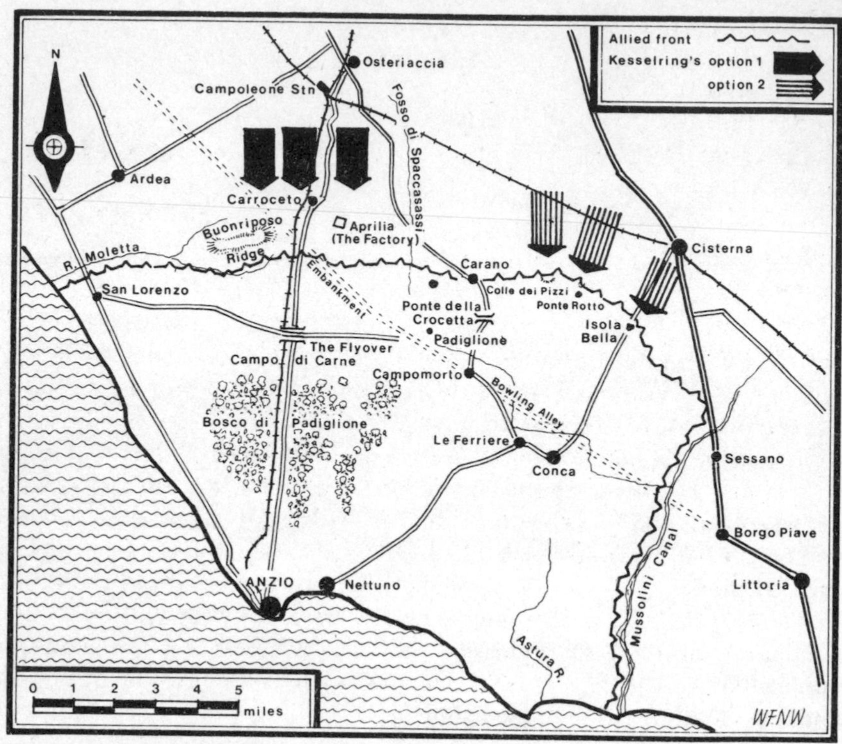

73 Anzio: 29 February, Kesselring's options

(1) To resume his offensive down the central axis of the beachhead, where a considerable dent had already been made.

(2) To switch to the right flank of the Allied line, which had remained more or less static during the previous battle.

Although the 3rd U.S. Division had not received the punishment that had been the lot of the troops holding the line either side of the Via Anziate, and were therefore not quite so battle-weary, Kesselring, in deciding to switch his attack (option 2), made the right decision; it is a well-known military maxim never to bang one's head against a wall that refuses to give. Moreover the tanks had not had an easy passage (once they were off the Via Anziate) in the previous battle, and it was hoped that the ground on the Allied right might prove slightly better for armour.

Neither Kesselring nor von Mackensen was going to risk another attack on a narrow front; instead they planned to break the Allied line and push through between Isola Bella and Ponte della Crocetta. But once again Hitler intervened tactically, and insisted that the line of attack should be shifted farther towards the Mussolini Canal, for he considered the chosen line too close to the main axis of the previous battle; he also thought that the better tank country (as seen from Berchtesgaden!) to the east would facilitate the advance on Nettuno. In theory the two generals on the spot bowed to the superior wisdom of their Führer, but for various reasons the attack went in along almost the exact lines planned by von Mackensen and agreed to by Kesselring.

Von Mackensen's plan in outline was to attack on a five-mile front with three divisions of LXXVI Panzer Corps. The Hermann Göring Panzer Division was to pierce the Allied line in the Isola Bella area, the 26th Panzer Division was to go in west of Ponte Rotto, and the 362nd Infantry Division had Colle dei Pizzi and Ponte della Crocetta as its targets. As soon as the breakthrough had been achieved the divisions were to head for the Astura river in the area of Conca as the first bound towards Nettuno. I Parachute Corps was to create a diversion before the main attack went in against the British 1st and 56th Divisions in the Via Anziate–Buonriposo ridge area.

Obviously to achieve this new axis of attack von Mackensen had to regroup his regiments; the appalling weather between 22 and 28 February not only caused the attack to be postponed for one day (from 28 to 29 February), but so restricted the movement of armour in an area of sodden ground and few tracks that it played little part in the battle. Kesselring made a number of internal switches with divisions and regiments, the most important of which was the transfer of 362nd Infantry Division from coast defence in the Rome area to the LXXVI Panzer Corps, which thus gained a fresh division for the attack.

Allied air reconnaissance during this bad weather was confined to artillery spotter light aircraft flying below cloud base, but this, combined with active patrolling and intercepts, was sufficient to give General Truscott a clear idea of where the attack was to come in, and negated I Parachute Corps' deception role. He had sufficient time to readjust the 3rd Division's line, which before the Germans attacked was held west of Carano to Isola Bella by 30th, 7th and 15th Infantry Regiments in that order, with 504th Parachute Infantry Regiment and the Special Service Force on the line of the Mussolini Canal. In the early hours of 29 February likely enemy forming-up positions and approach tracks were subjected to a very intense bombardment – a counter-preparation programme to which the German gunners replied with an unusually heavy stonk before the first waves of infantry went into the assault at 4.30 a.m.

For two days the tired, battle-weary German soldiers dragged themselves forward and with great gallantry tried to pierce the Allied line. Here and there they succeeded, but with equal gallantry the

Americans counter-attacked and threw them back. The freezing rain came down in torrents, and it must have seemed to these young men, who had given so much in the previous fighting, that they and their comrades were dying in their thousands for a purpose which for them had long since sunk into obscurity. Be this as it may, they had already given of their best and this time their assault lacked punch.

By the evening of 1 March, with his infantry exhausted, his tanks bogged down almost everywhere, and I Parachute Corps' private war on the western sector being held amid some of the most bitter of that dreadful wadi fighting, Kesselring realized that all hope had gone. At 6.40 p.m. he ordered the attack to cease. The next day the skies cleared, and to compound the German misery there was a fearful drumming in the air, and the ground trembled beneath an avalanche of 566 tons of bombs dropped during the day from hundreds of Allied bombers and fighter-bombers. The Germans had made their last attempt to drive the Allies into the sea; it had cost them nearly 3,000 casualties and gained them nothing.

Between D-day and the end of this last German offensive the Allies had suffered 20,943 battle casualties; the German numbers are uncertain, because Fourteenth Army figures were incomplete, but they were probably about the same. There now ensued a period of stalemate. Both sides contemplated renewing the offensive, but neither was in a position to do so without undertaking a considerable amount of reorganization. For the soldiers, however, the misery, the wounding and the dying continued unabated. Patrolling, mining and wiring were a daily and nightly routine; everyone was digging like moles, for this was like the trench warfare of former years.

The beachhead, now restricted to some sixteen miles in breadth, seven in depth at the centre and nine from Nettuno to the Cisterna front, was a horribly crowded place. At Anzio there were no back areas. Soldiers 'resting', doctors, nurses and orderlies were all in the front line and along with everyone else they suffered casualties from the German guns, especially the big railway ones – Anzio Annie and her mates – who were shunted out of tunnels in the Alban Hills to belch forth destruction and then disappeared again.

It became abundantly clear that until the Germans gave way along the Gustav Line there could be no solution to this stalemate, and the big attack launched against Cassino in the middle of March had, after the now familiar swaying back and forth, failed to pierce the defences. Hence the beachhead had to be sustained, supported and strengthened. The supply system had worked well and smoothly despite the hazards of weather (undoubtedly some credit for this should go to Lucas); during February a daily average of 2,000 tons of food, ammunition and equipment had been landed, and in March this had been stepped up to

5,000 tons. The quartermasters – those gallant, unsung heroes of 'B' echelon – were the troops' life-line and the conduit along which these vital supplies ran. But the landing craft, which – protected by the maritime power of the Allied navies – carried most of these stores, were at full stretch, and L.S.T.s were urgently wanted for Overlord. Furthermore, Operation Anvil – the invasion of southern France – was imminent.

Alexander needed to increase the force in the beachhead preparatory to a break-out, and this could not possibly be done if the existing supply services were cut back. In modern warfare there has to be a master strategist (soldier or statesman) capable of instinctively balancing, at any given time, the requirements of one campaign against another in many different and often widely separated theatres. It needed, therefore, some persuasive pleading before the Allied chiefs-of-staff succeeded in retaining sufficient L.S.T.s for a further period, and obtaining priority for the capture of Rome over the invasion of southern France. While these delicate negotiations were in progress – and indeed for some time previously – Generals Alexander and Clark were regrouping the beachhead force.

On 7 March the 24th Guards Brigade – or what remained of it – left the beachhead, and on the same day units of the 56th Division, which were drastically under strength, were relieved by the British 5th Division (Major-General Gregson-Ellis), transferred from the Garigliano front. This division was undoubtedly stronger than the 56th, but its three brigades (17th, 15th and 13th Infantry Brigades) and its machine gun battalion (7th Battalion Cheshire Regiment) were well below full strength. At the end of March the 3rd U.S. Division, after an almost unprecedented sixty-seven consecutive days in the line, was pulled back to Nettuno and its place was taken by the 34th U.S. Division. At the same time 504th Parachute Infantry Regiment and 509th Parachute Infantry Battalion left the beachhead. The sorely tried 1st British Division (with General Penney back in command) remained much as before with 18th Infantry Brigade replacing the Guards Brigade.

General Truscott now had five infantry divisions and one armoured division (less one combat command) in a beachhead that had once held just two divisions. As before, the British held the left – the 5th Division in the Moletta area, and the 1st around the Flyover – while the American 3rd Division was on the right of the line with the 34th and 1st Armored in support and reserve.

The Germans, too, did a considerable amount of reorganizing, and in the process reduced the quality, but not the quantity, of their army. I Parachute Corps remained much as before, with 4th Parachute Division, and those two stalwarts, 65th Infantry and 3rd Panzer Grenadier Divisions. But LXXVI Panzer Corps lost 114th Jaeger Division to the

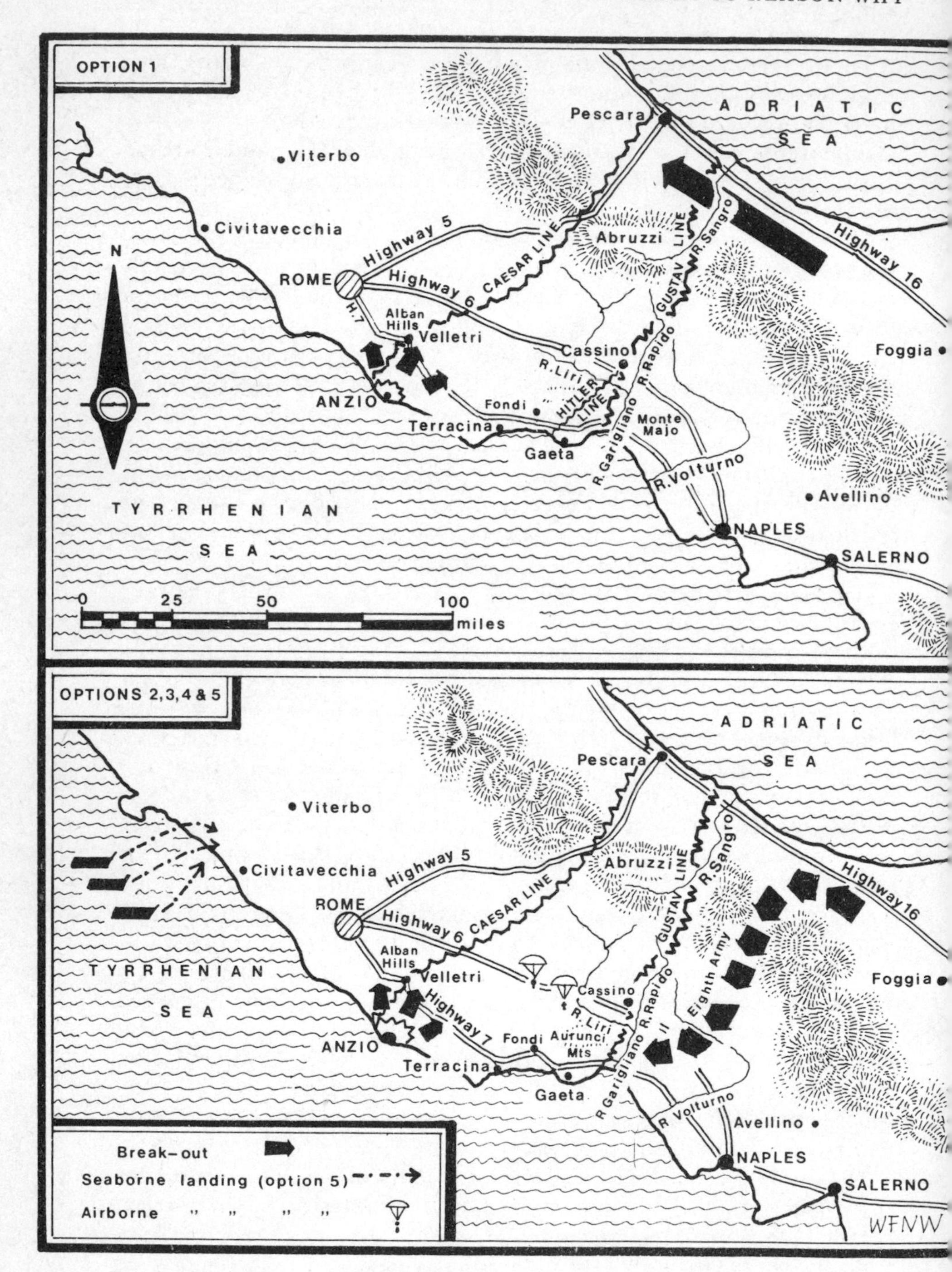

74 Anzio: 9 May, Alexander's options

Balkans, the Hermann Göring and 16th S.S. Panzer Grenadier Divisions to north Italy, and the Infantry Lehr Regiment to Germany. The 26th Panzer and 29th Panzer Grenadier Divisions were pulled back into general reserve south of Rome. These troops were replaced by a variety of units including two Italian battalions, none of which was of much worth.

All this reflected Kesselring's uneasiness as to where the blow would fall, for he knew full well that Alexander was now getting the reinforcements and supplies he wanted, many of which were being shovelled into the beachhead. What he was unable to discover was the place and timing of the Allied strike. There were many possibilities, such as a further landing (perhaps in the Civitavecchia area), an attack with its centre on the Rapido, or on the Garigliano, and obviously at some time a break-out from the beachhead.

To guard against as many eventualities as he could, Kesselring had brought back his two armoured divisions ready to go south or to the beachhead as required; thousands of Italian soldiers and civilians were impressed into repairing roads and railways, and constructing a new line in depth across the southern slopes of the Alban Hills – to be called the Caesar Line. But, uneasy as he was, with the lull continuing into the month of May, and seemingly little sign of immediate action, he allowed some senior officers (including von Veitinghoff) to go back to Germany on leave, and he was without Westphal, his Chief-of-Staff, who was on sick leave.

Kesselring would have given much to know what had been going through General Harding's (Alexander's Chief-of-Staff since January) mind for some months past as he carefully weighed the factors on which the forthcoming Allied offensive depended. The strategic objective must be the destruction of a large part of Kesselring's armies, for the purpose of forcing the Germans to send more troops into Italy (and therefore away from France) in order to safeguard the runways of northern Italy from which southern Germany could be so easily hammered. A breakthrough and encirclement were therefore necessary, for which Alexander had to have numerical superiority at a given point, and the prospect of good weather for air support.

These factors dictated the lull and build-up period, but with the coming of summer weather in May there could be no further delay if success was to be achieved before Overlord. *There were several options open to General Alexander:*

(1) To break through on the right across the Sangro, where there was no German switch line (Hitler Line), and simultaneously break out of the beachhead.

(2) To bring the Eighth Army across from the Adriatic to

make a forceful punch on the left (across the Garigliano and into the Liri valley), but after the break-out from the Anzio beachhead.

(3) As above, but with the southern front opening the battle, and the break-out following later.

(4) As above, but with a simultaneous break-out from the beachhead.

(5) To carry out these operations, but supplemented by a further landing up the coast in the form of a large-scale raid to draw off enemy troops. Alternatively, an airborne landing in the Liri valley.

A breakthrough on the Adriatic coast could be quickly dismissed, because there was little future to it. To attack along the whole line was not even an option, for it would have been impossible to obtain superiority of numbers at any one point. Clearly the best course was to force a passage between the Cassino and Majo mountains and burst into the Liri valley – as had always been the target. It was equally clear that VI Corps had to be ready to break out to complete the encirclement, but the problem was to time the double offensive so as to keep Kesselring guessing as to where and when to commit his reserves. General Alexander did consider breaking out from Anzio first in order to draw German reserves from the Garigliano front, but he decided against this because he had reason to believe that Kesselring was expecting it, and because any setback on the beachhead would be difficult to correct.

The plan decided upon was to bring the Eighth Army (now commanded by General Oliver Leese) across from the Adriatic to form the main part of the punch against the Gustav Line, with that part of the Fifth Army in the south supporting the attack between the Liri river and the Tyrrhenian coast. VI Corps would not break out of the beachhead until the Commander-in-Chief deemed the moment right. This was option 3, but in addition landing craft were to be assembled in the Bay of Naples to give the appearance of a pending amphibious assault.

Operation Diadem, as the attack against the Gustav Line was called, was timed to begin on 11 May. The broad plan was for the Eighth Army to spearhead the attack at Cassino with four corps, while the Fifth Army's II Corps would thrust up Highway 7 in support. General Juin's Free French Expeditionary Force, which contained some highly skilled mountaineering troops, was to attack up the Ausente valley and take Monte Majo, thereby giving support to the southern flank of the Eighth Army and to the northern part of Fifth Army's II Corps.

Shortly before midnight on 11 May the offensive began, with hundreds of guns thundering destruction between Cassino and the sea. At first, in spite of gaining surprise and having certain other advantages,* the attack did not progress well on any part of the front, but by nightfall on 13 May, after two days of some very fierce and clever fighting, Juin's Frenchmen had taken Monte Majo, and two days later they were in Ausonia and amid the trackless Aurunci mountains. This threatened the flank of those troops opposing the II U.S. Corps and forced them to fall back. By 18 May the Poles had Cassino and the XIII British Corps had cut the Via Casilina (Highway 6) behind the town. On the 19th the French and Americans were across the Pico-Formia road and threatening the southern sector of the Hitler Line between Fondi and Terracina; and on that same day the Canadian Corps were pressing up the Liri valley. After more than a week of savage fighting on land, supported by heavy pounding from the air, the troops of six nations were leaping upon the shell-shattered defenders of the Hitler Line, and along its whole length the Germans were bending and breaking before the storm.

Undoubtedly Kesselring had been outgeneralled. With the meagre Intelligence resources available to him he had been unable to gauge exactly where or when the Allied attack would strike. At the time of Diadem he was still thinking in terms of another landing, and his reserves were poised accordingly. Not until two days after the attack had begun was he satisfied from aerial reconnaissance that no invasion ships were at sea. He had then to decide whether to reinforce the south with the 26th Panzer Division and the 29th Panzer Grenadiers.

A decision in favour of strengthening the southern front was not a difficult one to take. It is interesting that in the official O.K.W. report after the war it was emphasized that the capture of Rome resulted not from the battles on the beachhead, but rather from the effects of the Allied offensive against the southern flank of the Tenth Army, which caused a collapse of the German defences south of Cassino – although this same report states that it was the linking up of the main force with the beachhead that gave the Allies a flanking position with which to roll up the front of the Tenth Army.

In order to win the battle on the Hitler Line Kesselring therefore brought two divisions across from the Adriatic front, redeployed the 90th Panzer Grenadiers, and brought south from their stand-by position the two panzer divisions. But his orders were not in all cases obeyed promptly or efficiently. Von Mackensen in particular was opposed to the transfer of his reserve division (29th Panzer Grenadiers) and was so slow in releasing it that by the time it eventually reached the Terracina area (20 May) the Americans were in a position to prevent its most effective

*General von Senger, whose XIV Panzer Corps was holding the important Majo-Garigliano sector of the front, was one of those senior commanders away in Germany, and his deputy was far less proficient.

use. There were to be other disagreements with von Mackensen, which were soon to lead to his replacement by General Lemelsen.

Decisions on a higher level were being taken at this time, at O.K.W. headquarters, where it was being suggested to Hitler either that Kesselring should abandon his front south of Rome, or else that much greater air strength should be employed in order to stabilize the present line. But Hitler was not prepared to commit the fully stretched Luftwaffe to assist Kesselring, and ordered him to continue his resistance. All this quickly came to Alexander through deciphered Enigma messages, and strengthened his belief that the time had now come for General Truscott's troops to play their part in this great battle. The twenty-third of May was the day Alexander decided that VI Corps would commence their break-out from the beachhead.

Plans for this had, of course, been formulated some weeks past. *There were four possible options, which were considered under somewhat strange code names:*

(1) To strike south-east in the direction of Littoria-Sezze in order to join forces with troops pushing up from the south, and threaten the right flank of the Tenth Army. This was given the code name Grasshopper.

(2) Crawdad was a plan to break out in the opposite direction, north-west through Ardea, and more or less parallel to the coast. The roads were few and the country difficult, but it was the shortest route to Rome.

(3) Buffalo envisaged a north-eastward advance through Cisterna, Cori, Artena and Valmontone. This advance would create a dangerous salient, but it would cut off the Tenth Army's line of retreat via Highway 6, although there were other possible escape routes beyond Valmontone.

(4) Turtle was an attack astride the well-trodden path of the Via Anziate, northward through Carroceto and Campoleone to a junction with Highway 7. Although not the shortest this was probably the quickest way to Rome, but it would hit the Caesar Line where the defences were strong, just south of the Alban Hills.

As soon as General Alexander had decided that the optimum time for the beachhead break-out was when the southern offensive had got thoroughly under way, Grasshopper was viable only if the main attack became bogged down. Such an attack by VI Corps would be pointing away from Rome, and offered little scope for trapping the Tenth Army. Crawdad was virtually ruled out by the going, which greatly favoured defensive tactics. Turtle would do little towards Alexander's main objective, the destruction of the Tenth Army, and might involve VI Corps in the heaviest fighting of the four options. This left Buffalo (option 3), which was the plan most closely aligned to the original concept of Shingle; Alexander had no hesitation

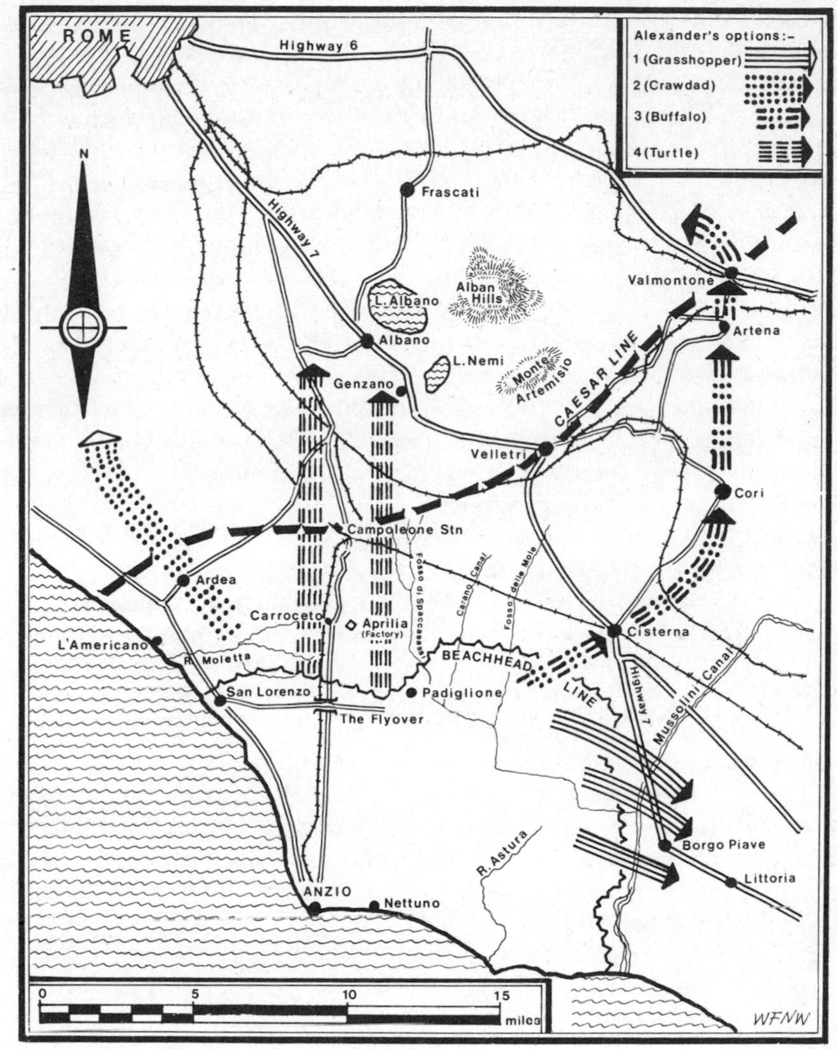

75 Anzio: 22 May, Alexander's options

in telling Truscott, when he presented the choices, that it was the only one of the four plans that interested him.

It was not, however, the only one that interested Clark. He wished to be free to move in any direction that the course of the battle might dictate – and he had his eye on Rome as the legitimate prize for his Fifth Army. Already incensed that Alexander was giving orders over his head direct to the corps commander, he telephoned the Commander-in-Chief from the beachhead to express his irritation at such interference, and to complain against what he called 'preconceived ideas'. Alexander did not argue, and thereby gave his subordinate to think that he could, if he wished, alter and invalidate the most important part of the plan.

General Truscott's plan for the break-out envisaged initially a powerful punch on the right sector by the 1st U.S. Armored and 3rd Infantry Divisions. General Harmon was very averse to using armour as a spearhead, and would have preferred the more conventional role of exploiting a breakthrough, but Truscott felt that the strong enemy position could best be broken by armour. Harmon was to advance from south-west of Cisterna along a line parallel to the Fosso delle Mole, cut the railway north-west of Cisterna and then advance to Highway 7.

General O'Daniels 3rd Division was first to isolate and then to capture Cisterna; his right flank would be protected by 1st Special Service Force, who were to advance from the Mussolini Canal to cut Highway 7 south-east of Cisterna. The 45th U.S. Division (General Eagles) was to cover the left flank of the offensive, advancing in the area bounded by the Carano and Spaccasassi canals. The 34th U.S. Division remained in rear, ready to assist either 1st Armored or 1st Special Service Force if necessary after they had completed their first phase. The 36th U.S. Division was on its way to Anzio; the plan was for it to pass through the 3rd Division after the capture of Cisterna.

Intelligence reports indicated that the Germans were expecting the break-out to be made along the Via Anziate (Turtle), and to encourage this belief General Truscott mounted a deception plan known as Operation Hippo. This was to be carried out by troops of the 1st and 5th British Divisions a few hours before the main attack went in. The more ambitious deception, known as Operation Wolf, fell to the 5th Division, who were to force the Moletta river in a battalion strength and attack enemy positions in L'Americano area. The 1st Division's diversion was on a smaller scale along the Lateral Road. Both these deception measures had the desired effect of keeping large numbers of enemy in the area until it was too late for them to influence the main battle to the east.

To meet the Buffalo charge the Germans held the line with five divisions; these were – from left to right facing the Allied troops – 4th Parachute, the 65th, 3rd Panzer Grenadiers (that staunchest of divisions), 362nd and 715th. Behind them was virtually nothing, except the Hermann Göring Panzer Division in army reserve, which was to appear after the fall of Cisterna. General Greiner, commanding 362nd Division, complained in his memoirs that his division and the 715th, which bore the brunt of the initial thrust, were far too extended at the expense of a greater concentration astride the Via Anziate.

Shortly before 6 a.m. on 23 May somewhere around 500 guns opened up on enemy positions partly obscured by a morning mist. After 123 days of close and very dangerous confinement the troops of the Anzio beachhead were about to break out. Progress on the first day was a little disappointing; Harmon (as he had forecast) lost many tanks and tank destroyers, and could get no farther than the Rome–Cisterna railway. The 3rd Division had their hardest battle of the whole war, suffering just under 1,000 casualties and being denied Cisterna. The next day (24 May) was altogether more encouraging. The 3rd Division had still not completely eradicated the tough German resistance in Cisterna (the 362nd Division), but the armour had got to Highway 7 north of the town, and the 1st Special Service Force had got beyond the road south-east of it.

The twenty-fifth of May was a day of destiny in the struggle and torments of all those engaged in the bitter battles of the Liri valley and the Anzio beachhead. On this day, at a place called Borgo Grappa, troops from the southern front made contact with those on the beachhead. Anzio was no longer an 'island'. Cisterna fell to the 3rd Division, the Eighth Army thrust through the Hitler Line in three places, and General Clark came to a momentous and controversial decision.

By now the position in the beachhead was as follows. The Americans had suffered 2,872 casualties, and the British divisions 469. The 1st U.S. Armored and 3rd U.S. Infantry Divisions were advancing steadily towards Highway 6. Cori had been taken, and the infantry was heading for Artena. The armour had met some very stiff opposition in the Velletri area, but was about to thrust up the Valmontone corridor to Giulianello. The 45th Division had gained ground on the Carano sector, and the British troops were playing a valuable, if less spectacular role by holding the I Parachute Corps in place. Von Mackensen had transferred forty-eight anti-tank and eight 88-mm guns from this corps to the LXXVI Panzer Corps, but would send nothing more, for he still thought that a major attack would erupt against his right wing.

The German situation was becoming critical. The 1st Armored Division's thrust had driven a wedge between the 362nd and 715th Divisions, and Allied air attacks had seriously hampered the movement of troops – although the Herman Göring Panzer Division was now coming into the line, and an infantry regiment from the 92nd Division had got through to support the 715th Division. Despite these reinforcements, the LXXVII Panzer Corps was faltering, and at midday on the 25th Hitler agreed that both the Tenth and the Fourteenth Armies should fall back to the Caesar Line. The O.K.W. recognized that the vital sector would be directly north of the Alban Hills, astride Highway 6 – which was in fact Alexander's immediate objective. It was not long before British Intelligence had passed this information to the Allied command.

This was the situation facing Clark after the third day's fighting. There were before him two options, and in trying to decide which to adopt there were many imponderables to be overcome. *His options were:*

(1) He could carry out the Buffalo operation as originally planned, keeping Truscott's main striking force on the Valmontone–Highway 6 axis. The object of this line of advance was to annihilate the Tenth German Army, and it would not be too optimistic to anticipate that troops of VI Corps would be astride the German line of withdrawal in about twenty-four hours. But had the Eighth Army broken through the Hitler Line in sufficient strength to push von Veitinghoff into the trap? Were there other escape routes open to the enemy? Would this line of advance expose the corps to a strong flank attack from the Alban Hills?

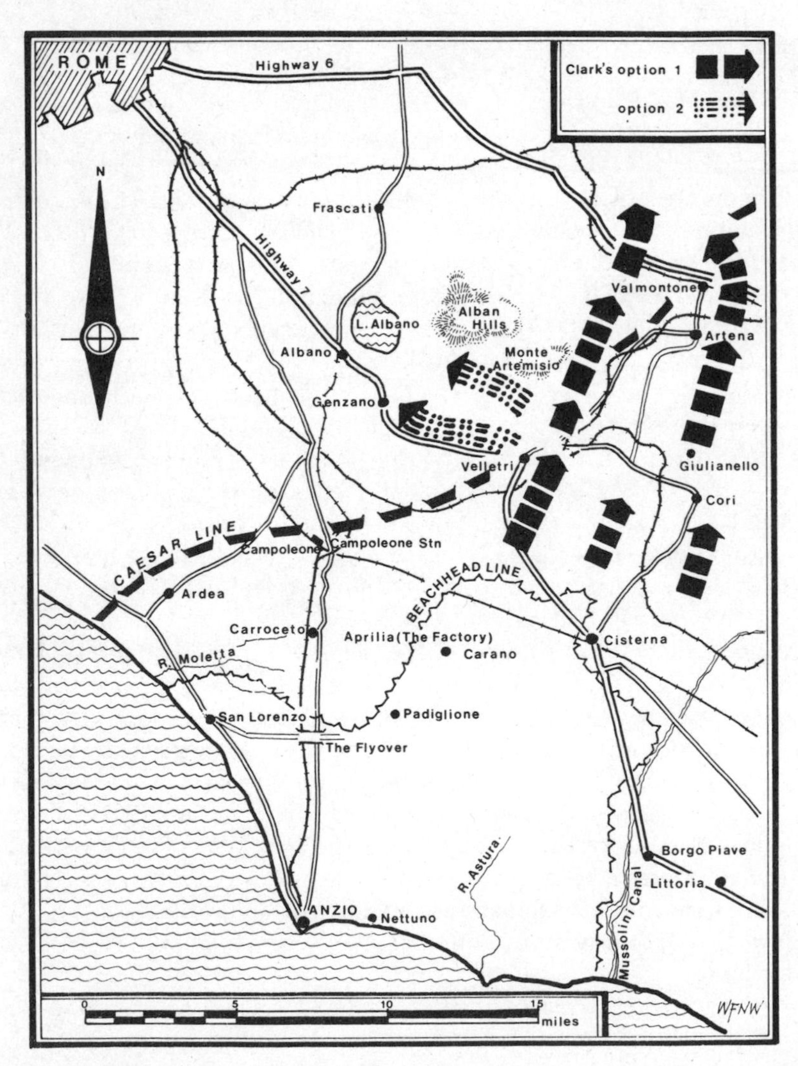

76 *Anzio: 25 May, Clark's options*

(2) He could switch all or part of his force away from the original axis and advance north-west, skirting the Alban Hills and going up Highway 7. This would remove any chance of destroying the German army, but it offered a quicker route to Rome for the Fifth Army, especially as the Alban Hills were reported to be strongly held. On the other hand it was known that I Parachute Corps, which barred the north-west route, had hardly been dented in the recent fighting.

Clark decided that Buffalo should be all but buried. To the surprise, and indeed consternation of his VI Corps' commander he ordered him on 26 May to swing the axis of attack north-west, with the exception of the 3rd Division and 1st Special Service Force which were to continue along the original axis to Valmontone and Highway 6, and in due course come under command of II Corps.

In his memoirs Clark makes little attempt to defend this decision which, apart from being a complete obfuscation of the Commander-in-Chief's orders (which admittedly, for political reasons, left much to the discretion of the Fifth Army commander), was absolutely wrong. Clark states that he was mindful of three considerations: the desire to take Rome before the invasion of France, the need to spare the city from destruction, and the importance of destroying the German army. Through the change of plan the first of these was nearly forfeited, because the new line of attack involved some desperate fighting against the comparatively fresh I Parachute Corps in the strongest sector of the Caesar Line; the second was achieved under orders of the enemy Commander-in-Chief; and the third was always unobtainable with the use of an emasculated force, whereas for several days the German strength in the Valmontone area and westward to the Alban Hills would not have been capable of resisting the main effort of VI Corps.

There is a fourth consideration, or confession, in his memoirs, which was a determination to capture Rome with the Fifth Army, and a gnawing anxiety that some other force 'was trying to get in on the act'. A very understandable aspiration, for the Fifth Army had fought its way up the leg of Italy with courage and determination and endured to the full the hardships, horrors and sacrifices of war; but the irony is that had its commander not taken this controversial, and wrong, decision it would have been in Rome sooner, and possibly – although not certainly – would have added to its laurels by the annihilation of large numbers of German troops.*

The Germans continued the fight with their customary stubborn ferocity. General Schlemm's corps, holding the Ardea sector of the Caesar Line, inflicted heavy casualties on the British 1st and 5th and the American 45th Divisions, while the 3rd U.S. Division and 1st Special

*Lord Harding, in conversation with the author, said that the Commander-in-Chief only learned of General Clark's change of plan after he had made it, when it was too late to interfere. In his (Harding's) opinion had Clark carried out the plan as originally conceived, the Tenth Army's withdrawal must have been seriously impeded, and possibly some (although certainly not all) of it destroyed.

Service Force kept up their hammering of Valmontone. The Germans also took punishment, but Kesselring could consider this worth while, for it gave him valuable time in which to bring back the Tenth Army and attempt to stabilize the Caesar Line around the Alban Hills. Indeed it seemed as though he would succeed, for the country was as unsuitable for the Allied armour as it was suitable for German rearguard fighting, at which they were adept. But the break came on 30 May through a stratagem of rare brilliance on the part of the 36th U.S. Division, which had recently come into the line in relief of 1st Armored Division below Velletri.

The key to unhinging the Caesar Line lay in Monte Artemisio, and because of its rugged steepness and also his shortage of men, von Mackensen occupied it only with a small observation post. Men of General Walker's 36th Division discovered a trail up the mountain, and just before midnight on 30 May, by the dim light of a young moon, the 142nd Infantry Regiment, marching in column of battalions, picked their way up the path; by dawn they had gained the summit and captured the artillery observers – one in his bath! They were followed by the 143rd Regiment. The capture of this commanding height was soon exploited, and Clark could now commence a powerful thrust with II Corps to divide the 4th Parachute and Hermann Göring Divisions, while VI Corps continued its bitter struggle with General Schlemm's tough paratroopers.

By 2 June, with II Corps on Highway 6 beyond Valmontone and VI Corps on Highway 7 at Albano, Kesselring knew that Rome was lost. That night the Fourteenth Army fell back through and beyond Rome at the beginning of what was to be a long and expertly managed withdrawal. On 4 June leading elements of the Fifth Army entered Rome, and on the next day their commander entered the city to claim the prize. The battle for Rome was over, and the triumph in Italy duly acclaimed a day before France became the cynosure of all eyes. The toll had been heavy, and temporarily obscured by the thrill of Overlord. From the beginning of Operation Diadem the total Allied loss had been 40,205; the Germans had suffered slightly less with a casualty list of 38,024, and Kesselring and his men were far from being finally defeated.

Was Operation Shingle justified? At the Gustav Line the Allies were holding the shortest front they ever held in Italy – only eighty-five miles – and therefore a flank attack would seem to have been the obvious answer. The principal objective was not Rome, and only in part to get Kesselring to withdraw (which would have had the advantage of forcing him to hold a longer and harder line), but rather to destroy as many German divisions as possible by catching them in a trap between two forces.

The principal objective was therefore not accomplished – although, as we have seen, it might well have been. However, there was a very important secondary objective, which was to hold down German formations in Italy and therefore away from France. This was definitely achieved, and it was probably better to engage these troops in the beachhead rather than on the southern front, where facilities for defence were much better. The best solution would have been to land in greater strength, but this was not possible with the number of L.S.T.s available. General Westphal thought that the losses would have been fewer if there had been a second landing in May near Leghorn; General Templer could never understand how the Anzio landing could ever work.*

The debate will continue for as long as men are interested in the business of war. Two things, however, are certain: Anzio was one of the toughest, the cruellest and the most unpleasant battles of the whole war; but for the most part it brought forth a shining spirit of concord and comradeship between the commanders and soldiers of the two great allies who fought and won it.

* *The German Army in The West*, p. 166, and *Alex*, p. 232.

Orders of Battle

Appendix I: Saratoga Campaign

Effective strength and composition of the British Army under the command of Lieutenant-General John Burgoyne prior to the Battles of Saratoga, 19 September and 7 October* 1777

BRITISH GENERAL STAFF . . . 26

RIGHT WING – Brigadier-General Fraser
24th Regiment (411 men) – Major Robert Grant
Grenadier Companies (495 men) – Major Acland
Light Infantry Companies (490 men) – The Earl of Balcarres
Rangers (110 men) – Captain Alexander Fraser
Total 1,506
British Artillery – 71 men, 10 cannon
Auxiliaries (Canadians 123 men, Tories approx. 682, Indians approx. 90)
Total . . . 895

Lieutenant-Colonel Breymann's Corps
Brunswick Grenadier Regiment (328 men) – Lieutenant-Colonel Breymann
Brunswick Chasseur Regiment (132 men) – Major Ferdinand von Bärner
Brunswick Jaeger Company (40 men) – Captain von Geyso
Total . . . 500
Hessian Artillery – 31 men, 2 cannon

CENTRE COLUMN – Brigadier-General James Hamilton
9th Regiment (400 men) – Lieutenant-Colonel Hill
20th Regiment (404 men) – Lieutenant-Colonel Lynd
21st Regiment (413 men) – Brigadier-General Hamilton
62nd Regiment (417 men) – Lieutenant-Colonel Anstruther
Total . . . 1,634
Artillery – 61 men, 4 cannon

LEFT WING – Major-Generals Riedesel and Phillips

GERMAN GENERAL STAFF . . . 43

Brunswick Dragoons (38 men)
Brunswick Chasseurs (100 men, taken from Major von Bärner's regiment in the Right Wing)

* Figures taken from a report on the strength of the British Army by Charles W. Snell, February 1951. Published by Saratoga National Historical Park, Stillwater, New York, 28 February 1951.

Hessian Artillery (56 men)
Total . . . 194

First Brigade – Brigadier-General Johann Specht
Brunswick Regiment von Rhetz (458 men) – Lieutenant-Colonel von Ehrenkrook
Brunswick Regiment von Riedesel (490 men) – Lieutenant-Colonel von Spaeth
Brunswick Regiment von Specht (441 men) – Brigadier-General Specht
Total . . . 1,389 men

Second Brigade – Brigadier-General W. R. von Gall
Hesse Hanau Regiment Crown Prince (594 men) – Brigadier-General von Gall

Major-General William Phillips' Brigade
British 47th Regiment (289 men) – Lieutenant-Colonel Sutherland
British Artillery – 231 men, 15 cannon
Total . . . 520
Auxiliaries: British seamen, bateaux men, etc. – 238 men

Total at the commencement of the first Saratoga battle (Freeman's Farm):
British Regulars – 3,818
German Regulars – 2,751
Auxiliaries – 1,133
Total . . . 7,702 men, 35 cannon

Total at the commencement of the second Saratoga battle (Bemis Heights):
British Regulars – 3,618
German Regulars – 2,749
Auxiliaries – 816
Total . . . 7,183 men, 35 cannon

Effective strength and composition of the American Army under the command of Major-General Horatio Gates

RIGHT WING – Major-General Horatio Gates
Brigadier-General John Patterson's Brigade (976 men)
10th Massachusetts Regiment – Colonel Thomas Marshall
11th Massachusetts Regiment – Lieutenant-Colonel Benjamin Tupper
12th Massachusetts Regiment – Colonel Samuel Brewers
14th Massachusetts Regiment – Colonel Gamaliel Bradford

Brigadier-General John Glover's Brigade (1,685 men)
1st Massachusetts Regiment – Colonel Joseph Vose
4th Massachusetts Regiment – Colonel William Shepard
13th Massachusetts Regiment – Colonel Edward Wigglesworth
15th Massachusetts Regiment – Colonel Timothy Bigelow
2nd Albany County Regiment N.Y. Militia – Colonel Abraham Wemple
17th Albany County Regiment N.Y. Militia – Colonel William B. Whiting
Dutchess and Ulster County Militia N.Y. – Colonel Morris Graham

Brigadier-General John Nixon's Brigade (1,126 men)
3rd Massachusetts Regiment – Colonel John Greaton
5th Massachusetts Regiment – Colonel Rufus Putnam
6th Massachusetts Regiment – Colonel Thomas Nixon
7th Massachusetts Regiment – Colonel Ichabod Alden

LEFT WING – Major-General Benedict Arnold
Brigadier-General Enoch Poor's Brigade (2,116 men)
1st New Hampshire Regiment – Colonel Joseph Cilley
2nd N.H. Regiment – Lieutenant-Colonel Winborn Adam
3rd N.H. Regiment – Colonel Alexander Scammel
2nd N.Y. Regiment – Colonel Philip Van Cortlandt
4th N.Y. Regiment – Colonel Henry Beckman Livingstone
Connecticut Militia – two regiments commanded by Colonels Cooke and Lattimore

Brigadier-General Ebenezer Learned's Brigade (1,243 men)
2nd Massachusetts Regiment – Colonel John Bailey
8th Massachusetts Regiment – Colonel Michael Jackson
9th Massachusetts Regiment – Colonel James Wesson
N.Y. Regiment – Colonel James Livingstone

Colonel Daniel Morgan's Brigade (674 men)
Colonel Morgan's Regiment of Virginia Riflemen
Major Henry Dearborn's Light Infantry

Cavalry: Connecticut Light Horse (200) – Major Elijah Hyde
Artillery: 248 men, 22 cannon – Major Ebenezer Stevens
Engineers and artificers: 71 men – Colonel Thaddeus Kosciuszko

Total *effective* strength for the second battle of Saratoga (Bemis Heights), by which time General Gates had been considerably reinforced by Major-General Lincoln's Massachusetts Militia and Brigadier-General Abraham Ten Broeck's N.Y. Militia . . . 13,065 men

Appendix II: Waterloo Campaign

Effective strength and composition of the Anglo-Allied Army, under the command of Field-Marshal the Duke of Wellington*

	Strength at commencement of campaign
I CORPS – His Royal Highness the Prince of Orange	
1st Division – Major-General Cooke	
1st British Brigade – Major-General Maitland	1,997
2nd British Brigade – Major-General Sir John Byng	2,064
Artillery – Lieutenant-Colonel Adye	

* Strengths taken from D. Gardner & Dorsay's *Quatre Bras, Ligny and Waterloo*, London 1882, W. Siborne's *War in France and Belgium* as corrected by Colonel Charles C. Chesney's *Waterloo Lectures: a Study of the Campaign of 1815*, London 1868, and Colonel Jean-Baptiste Charras's *Histoire de la Campagne de 1815: Waterloo*, Brussels, 1851.

3rd Division – Lieutenant-General Sir Charles Alten	
5th British Brigade – Major-General Sir Colin Halkett	2,254
2nd Brigade King's German Legion – Colonel von Ompteda	1,527
1st Hanoverian Brigade – Major-General Count Kielmansegge	3,189
Artillery – Lieutenant-Colonel Williamson	
2nd Dutch-Belgian Division – Lieutenant-General Baron de Perponcher	
1st Brigade – Major-General Count de Bylandt	3,233
2nd Brigade – H.S.H. The Prince Bernhard of Saxe-Weimar	4,300
Artillery – Major von Opstal	
3rd Dutch-Belgian Division – Lieutenant-General Baron Chassé	
1st Brigade – Major-General Ditmers	3,088
2nd Brigade – Major-General D'Aubremé	3,581
Artillery – Major van der Smissen	

TOTAL I CORPS . . . 25,233 and 48 guns

II Corps – Lieutenant-General Lord Hill	
2nd Division – Lieutenant-General Sir H. Clinton	
3rd British Brigade – Major-General Adam	2,625
1st Brigade King's German Legion – Colonel du Plat	1,758
3rd Hanoverian Brigade – Colonel Halkett	2,454
Artillery – Lieutenant-Colonel Gold	
4th Division – Lieutenant-General Sir Charles Colville	
4th British Brigade – Colonel Mitchell	1,767
6th British Brigade – Major-General Johnstone	2,396
6th Hanoverian Brigade – Major-General Sir James Lyon	3,049
Artillery – Lieutenant-Colonel Hawker	
1st Dutch-Belgian Division – Lieutenant-General Stedmann	
1st Brigade – Major-General Hauw	
2nd Brigade – Major-General Eerens	6,389
	Strength at commencement of campaign
Dutch-Belgian Indian Brigade – Lieutenant-General Anthing	3,583

TOTAL II CORPS . . . 24,021 and 40 guns

Reserve	
5th Division – Lieutenant-General Sir Thomas Picton	
8th British Brigade – Major-General Sir James Kempt	2,471
9th British Brigade – Major-General Sir Denis Pack	2,173
5th Hanoverian Brigade – Colonel von Vincke	2,514
Artillery – Major Heisse	
6th Division – Lieutenant-General Hon. Sir L. Cole	
10th British Brigade – Major-General Sir John Lambert	2,567
4th Hanoverian Brigade – Colonel Best	2,582
Artillery – Lieutenant-Colonel Brückmann	

7th Division	
7th British Brigade	1,216
British Garrison Troops	2,017
Brunswick Corps – H.S.H. The Duke of Brunswick	
Advanced Guard Battalion – Major von Rauschenplatt	672
Light Brigade – Lieutenant-Colonel von Buttlar	2,688
Line Brigade – Lieutenant-Colonel von Specht	2,016
Artillery – Major Mahn	
Hanoverian Reserve Corps – Lieutenant-General von der Decken	
1st Brigade – Lieutenant-Colonel von Bennigsen	9,000
2nd Brigade – Lieutenant-Colonel von Beaulieu	
3rd Brigade – Lieutenant-Colonel Bodecker	
4th Brigade – Lieutenant-Colonel Wissel	
Nassau Contingent – General von Kruse	2,880

TOTAL RESERVE . . . 32,796 and 64 guns

CAVALRY – Lieutenant-General the Earl of Uxbridge

British, and King's German Legion –	
1st Brigade – Major-General Lord E. Somerset	1,286
2nd Brigade – Major-General Sir W. Ponsonby	1,181
3rd Brigade – Major-General Sir W. Dörnberg	1,268
4th Brigade – Major-General Sir J. Vandeleur	1,171
	Strength at commencement of campaign
5th Brigade – Major-General Sir Colq. Grant	1,336
6th Brigade – Major-General Sir H. Vivian	1,279
7th Brigade – Colonel Sir F. von Arentsschildt	1,012
(6 British Horse batteries attached to cavalry)	
Hanoverian	
1st Brigade – Colonel von Estorff	1,682
Brunswick Cavalry –	922
Dutch-Belgian –	
1st Brigade – Major-General Trip	1,237
2nd Brigade – Major-General de Ghigny	1,086
3rd Brigade – Major-General van Merlen	1,082

TOTAL CAVALRY . . .	14,542 men and 44 guns
TOTAL INFANTRY . . .	82,050 men
TOTAL ARTILLERY . . .	8,166 men and 196 guns
TOTAL SAPPERS, MINERS, WAGONERS, STAFF CORPS . . .	1,240 men
GRAND TOTAL . . .	105,998 men and 196 guns

Effective strength and composition of the Prussian Army under the command of Field Marshal Prince Blücher von Wahlstadt

	Strength at commencement of campaign
I Corps – Lieutenant-General von Zieten	
1st Brigade – General von Steinmetz	8,647
2nd Brigade – General von Pirch II	7,666
3rd Brigade – General von Jagow	6,853
4th Brigade – General von Henkel	4,721
Reserve Cavalry of I Corps – Lieutenant-General von Röder	
Brigade of General von Treskow Brigade of Lieutenant-Colonel von Lützow	1,925
Reserve Artillery – Colonel von Lehmann	1,019
TOTAL I CORPS . . . 30,831 men and 96 guns	
II Corps – General von Pirch I	
5th Brigade – General von Tippelskirchen	6,851
6th Brigade – General von Krafft	6,469
7th Brigade – General von Brause	6,224
8th Brigade – Colonel von Langen	6,292
Reserve Cavalry of II Corps – General von Jürgass	
Brigade of Colonel von Thümen Brigade of Colonel Count Schulenburg Brigade of Lieutenant-Colonel von Sohr	4,468
Reserve Artillery – Colonel von Röhl	1,454
TOTAL II CORPS . . . 31,758 men and 80 guns	
III Corps – Lieutenant-General von Thielemann	
9th Brigade – General von Borcke	6,752
10th Brigade – Colonel von Kämpfen	4,045
11th Brigade – Colonel von Luck	3,634
12th Brigade – Colonel von Stülpnagel	6,180
Reserve Cavalry of III Corps – General von Hobe	
Brigade of Colonel von de Marwitz Brigade of Colonel Count Lottum	2,405
Reserve Artillery – Colonel von Mohnhaupt	964
TOTAL III CORPS . . . 23,980 men and 48 guns	
IV Corps – General Count Bülow von Dennewitz	
13th Brigade – Lieutenant-General von Hacke	6,385
14th Brigade – General von Ryssel	6,953
15th Brigade – General von Losthin	5,881
16th Brigade – Colonel von Hiller	6,162

Reserve Cavalry of the IV Corps – General, Prince William of Prussia

Brigade of General von Sydow Brigade of Colonel Count Schwerin Brigade of Lieutenant-Colonel von Watzdorf	3,081
Reserve Artillery – Lieutenant-Colonel von Bardelezen	1,866

TOTAL IV Corps . . . 30,328 men and 88 guns

GRAND TOTAL . . . 116,897 men and 312 guns

Effective strength and composition of the French Army, under the command of Napoleon Bonaparte

IMPERIAL GUARD – Marshal Mortier, Duke of Treviso
1st and 2nd Regiments of Grenadiers – Lieutenant-General Friant
3rd and 4th Regiments of Grenadiers – Lieutenant-General Roguet
1st and 2nd Regiments of Chasseurs – Lieutenant-General Morand
3rd and 4th Regiments of Chasseurs – Lieutenant-General Michel
1st and 3rd Regiments of Tirailleurs – Lieutenant-General Duhesme
1st and 3rd Regiments of Voltigeurs – Lieutenant-General Barrois
Lancers and Chasseurs à Cheval – Lieutenant-General Lefebvre-Desnouettes
Dragoons and Grenadiers à Cheval – Lieutenant-General Guyot
Artillery – Lieutenant-General Desvaux de St Meurice

Total infantry	12,554
Total cavalry	3,590
Total artillery	3,175
Total sappers	109

GRAND TOTAL IMPERIAL GUARD . . . 19,428 men and 96 guns

I CORPS D'ARMÉE – Lieutenant-General Count D'Erlon (on 10 June)
1st Division – Lieutenant-General Alix
2nd Division – Lieutenant-General Baron Donzelot
3rd Division – Lieutenant-General Baron Marcognet
4th Division – Lieutenant-General Count Durutte
1st Cavalry Division – Lieutenant-General Baron Jaquinot

Total infantry	16,200
Total cavalry	1,400
Total artillery	1,066
Total sappers	330

GRAND TOTAL I CORPS D'ARMÉE . . . 18,996 men and 46 guns

II CORPS D'ARMÉE – Lieutenant-General Count Reille (on 10 June)
5th Division – Lieutenant-General Baron Bachelu
6th Division – Prince Jerome Napoleon
7th Division – Lieutenant-General Count Girard
9th Division – Lieutenant-General Count Foy

2nd Cavalry Division – Lieutenant-General Baron Piré

Total infantry	19,750
Total cavalry	1,729
Total artillery	1,385
Total sappers	409

GRAND TOTAL II CORPS D'ARMÉE . . . 23,273 and 46 guns

III Corps d'Armée – Lieutenant-General Count Vandamme (on 10 June)
8th Division – Lieutenant-General Baron Lefol
10th Division – Lieutenant-General Baron Hubert
11th Division – Lieutenant-General Berhezène
3rd Cavalry Division – Lieutenant-General Baron Domon

Total infantry	14,508
Total cavalry	932
Total artillery	936
Total sappers	146

GRAND TOTAL III CORPS D'ARMÉE . . . 16,522 men and 38 guns

IV Corps d'Armée – Lieutenant-General Count Gérard (on 31 May)
12th Division – Lieutenant-General Baron Pecheux
13th Division – Lieutenant-General Baron Vichery
14th Division – Lieutenant-General Hulot (this officer succeeded Lieutenant-General de Bourmont after the latter had deserted)
7th Cavalry Division – Lieutenant-General Maurin
Reserve Cavalry Division – Lieutenant-General Baron Jaquinot (returned as commanding the light cavalry division of I Corps on 10 June)

Total infantry	12,589
Total cavalry	2,366
Total artillery	1,538
Total sappers	201

GRAND TOTAL IV CORPS D'ARMÉE . . . 16,694 men and 38 guns

VI Corps d'Armée – Lieutenant-General Count Lobau (on 10 June)
19th Division – Lieutenant-General Baron Simmer
20th Division – Lieutenant-General Baron Jeannin
21st Division – Lieutenant-General Baron Teste

Total infantry	8,152
Total artillery	743
Total sappers	189

GRAND TOTAL VI CORPS D'ARMÉE . . . 9,084 men and 38 guns

Reserve Cavalry – Marshal Grouchy
I Corps – Lieutenant-General Count Pajol (in June)

4th Cavalry Division – Lieutenant-General Baron Soult 5th Cavalry Division – Lieutenant-General Baron Subervie	2,324
Artillery	317

II Corps – Lieutenant-General Count Excelmans (in June)	
9th Cavalry Division – Lieutenant-General Strolz 10th Cavalry Division – Lieutenant-General Baron Chastel	2,817
Artillery	246
III Corps – Lieutenant-General Kellermann (Count de Valmy) (in June)	
11th Cavalry Division – Lieutenant-General Baron L'Heritier 12th Cavalry Division – Lieutenant-General Roussel d'Hurbal	3,245
Artillery	309
IV Corps – Lieutenant-General Count Milhaud (on 9 June)	
13th Cavalry Division – Lieutenant-General Wathier 14th Cavalry Division – Lieutenant-General Baron Delort	2,556
Artillery	313

Total reserve cavalry	10,942
Total reserve artillery	1,185 and 48 guns

TOTAL INFANTRY . . .	83,753
TOTAL CAVALRY . . .	20,959
TOTAL ARTILLERY . . .	10,028
TOTAL SAPPERS . . .	1,384
GRAND TOTAL . . .	116,124 men and 350 guns

Effective strengths of the Armies at Waterloo

ANGLO-ALLIED ARMY

Infantry	49,608
Cavalry	12,408
Artillery	5,645
TOTAL	67,661 men and 156 guns

PRUSSIAN ARMY

Infantry	at 16.30 hrs	12,043	
	at 18.00 hrs	13,338	
	at 19.00 hrs	15,902	Total 41,283
Cavalry	at 16.30 hrs	2,720	
	at 18.00 hrs	–	
	at 19.00 hrs	6,138	Total 8,858
Artillery	at 16.30 hrs	1,143	
	at 18.00 hrs	–	
	at 19.00 hrs	660	Total 1,803
Guns	at 16.30 hrs	64	
	at 18.00 hrs	16	
	at 19.00 hrs	24	Total 104

TOTAL . . . 51,944 men and 104 guns

FRENCH ARMY

Infantry	47,579
Cavalry	13,792
Artillery	7,529
TOTAL . . .	68,900* men and 246 guns

* This figure excludes Marshal Grouchy's wing; with the troops used to protect the right flank against the Prussian threat, Napoleon had no more than 56,000 men in action at any time against Wellington.

WATERLOO CASUALTIES – KILLED, WOUNDED AND MISSING

British troops
85 officers 1,245 rank and file – killed
365 officers 4,261 rank and file – wounded
10 officers 558 rank and file – missing

King's German Legion
27 officers 306 rank and file – killed
77 officers 931 rank and file – wounded
1 officer 213 rank and file – missing

Hanoverian
20 officers 296 rank and file – killed
77 officers 1,244 rank and file – wounded
6 officers 349 rank and file – missing

Brunswickers
7 officers 147 rank and file – killed
26 officers 430 rank and file – wounded
– officers 50 rank and file – missing

Nassauers
5 officers 249 rank and file – killed
19 officers 370 rank and file – wounded

Appendix III: Chancellorsville and Gettysburg

Federal Army of the Potomac-Chancellorsville Campaign

Army Commander . . . Major-General Joseph Hooker
Chief of Staff . . . Brigadier-General D. Butterfield
Artillery Commander . . . Brigadier-General Henry J. Hunt

I ARMY CORPS – Major-General John F. Reynolds

1st Division Brigadier-General James S. Wadsworth
1st Brigade – Colonel Walter Phelps, Jr.
2nd Brigade – Brigadier-General Lysander Cutler
3rd Brigade – Brigadier-General Gabriel R. Paul
4th Brigade – Brigadier-General Solomon Meredith
Artillery – Captain John A. Reynolds

2nd Division – Brigadier-General John C. Robinson
1st Brigade – Colonel Adrian R. Root
2nd Brigade – Brigadier-General Henry Baxter
3rd Brigade – Colonel Samuel H. Leonard
Artillery – Captain Dunbar R. Ransom

3rd Division – Major-General Abner Doubleday
1st Brigade – Brigadier-General Thomas A. Rowley
2nd Brigade – Colonel Roy Stone
Artillery – Major Ezra W. Matthews

II Army Corps – Major-General Darius N. Couch

1st Division – Major-General Winfield S. Hancock
1st Brigade – Brigadier-General John C. Caldwell
2nd Brigade – Brigadier-General Thomas F. Meagher
3rd Brigade – Brigadier-General Samuel K. Zook
4th Brigade – Colonel John R. Brooke
Artillery – Captain Rufus D. Pettit

2nd Division – Brigadier-General John Gibbon
1st Brigade –
(1) Brigadier-General Alfred Sully
(2) Colonel Byron Laflin
2nd Brigade – Brigadier-General Joshua T. Owen
3rd Brigade – Colonel Norman J. Hall

3rd Division – Major-General William H. French
1st Brigade – Colonel Samuel S. Carroll
2nd Brigade –
(1) Brigadier-General Wm. Hays (captured 3 May)
(2) Colonel Chas. J. Powers
3rd Brigade – Colonel John D. MacGregor

III Army Corps – Major-General Daniel E. Sickles

1st Division – Brigadier-General David B. Birney
1st Brigade –
(1) Brigadier-General Chas. K. Graham (assigned to command of 3rd Division 4 May)
(2) Colonel Thomas W. Egan
2nd Brigade – Brigadier-General J. H. H. Ward
3rd Brigade – Colonel Samuel B. Hayman
Artillery – Captain A. Judson Clark

2nd Division –
(1) Major-General Hiram G. Berry (killed 3 May)
(2) Brigadier-General Joseph B. Carr
1st Brigade –
(1) Brigadier-General Joseph B. Carr
(2) Colonel William Blaisdell
2nd Brigade –
(1) Brigadier-General Joseph W. Revere
(2) Colonel J. Egbert Farnum (assigned to command 3 May)
3rd Brigade –
(1) Brigadier-General Gershom Mott (wounded 3 May)
(2) Colonel Wm. J. Sewell
Artillery – Captain Thomas W. Osborn

3rd Division –
(1) Brigadier-General Amiel W. Whipple (wounded 4 May)
(2) Brigadier-General Charles K. Graham
1st Brigade – Colonel Emlen Franklin
2nd Brigade – Colonel Samuel M. Bowman

3rd Brigade – Colonel Hiram Berdan
Artillery –
(1) Captain Albert A. Von Puttkammer
(2) Captain James F. Huntington

V ARMY CORPS – Major-General George G. Meade

1st Division – Brigadier-General Charles Griffin
1st Brigade – Brigadier-General James Barnes
2nd Brigade –
(1) Colonel James McQuade (disabled 4 May)
(2) Colonel Jacob B. Sweitzer
3rd Brigade – Colonel T. B. W. Stockton
Artillery – Captain Augustus P. Martin

2nd Division – Major-General George Sykes
1st Brigade – Brigadier-General Romeyn B. Ayres
2nd Brigade – Colonel Sidney Burbank
3rd Brigade – Colonel Patrick H. O'Rorke
Artillery – Captain Stephen H. Weed

3rd Division – Brigadier-General Andrew A. Humphreys
1st Brigade – Brigadier-General Erastus B. Tyler
2nd Brigade – Colonel Peter H. Allabach
Artillery – Captain Alanson M. Randol

VI ARMY CORPS – Major-General John Sedgwick

1st Division – Brigadier-General William T. H. Brooks
1st Brigade –
(1) Colonel Henry W. Brown (wounded 3 May)
(2) Colonel William H. Penrose
(3) Colonel Samuel L. Buck (disabled)
(4) Colonel William H. Penrose
2nd Brigade – Brigadier-General Joseph J. Bartlett
3rd Brigade – Brigadier-General David A. Russell
Artillery – Major John A. Tompkins

2nd Division – Brigadier-General Albion P. Howe
1st Brigade – Colonel Lewis A. Grant
2nd Brigade – Brigadier-General Hall
3rd Brigade – Brigadier-General Thomas H. Neill
Artillery – Major J. Watts de Peyster

3rd Division – Major-General John Newton
1st Brigade – Colonel Alexander Shaler
2nd Brigade –
(1) Colonel William H. Browne (wounded 3 May)
(2) Colonel Henry L. Eustis
3rd Brigade – Brigadier-General Frank Wheaton
Artillery – Captain Jeremiah McCarthy

XI Army Corps – Major-General Oliver O. Howard

1st Division –
(1) Brigadier-General Charles Devens, Jr. (wounded 2 May)
(2) Brigadier-General Nathaniel C. McLean
1st Brigade – Colonel Leopold von Gilsa
2nd Brigade – Brigadier-General Nathaniel C. McLean
Artillery – Captain Julius Dieckmann

2nd Division – Brigadier-General Adolph Steinwehr
1st Brigade – Colonel Adolphus Buschbeck
2nd Brigade – Brigadier-General Francis C. Barlow
Artillery – Captain Michael Wiedrich

3rd Division – Major-General Carl Schurz
1st Brigade – Brigadier-General Alexander Schimmelpfennig
2nd Brigade – Colonel W. Krzyzanowski
Reserve Artillery – Lieutenant-Colonel Louis Schirmer

XII Army Corps – Major-General Henry W. Slocum

1st Division – Brigadier-General Alpheus S. Williams
1st Brigade – Brigadier-General Joseph F. Knipe
2nd Brigade – Colonel Samuel Ross (wounded 3 May)
3rd Brigade – Brigadier-General Thomas H. Ruger
Artillery – Captain Robert H. Fitzhugh

2nd Division – Brigadier-General John W. Geary
1st Brigade – Colonel Charles Candy
2nd Brigade – Brigadier-General Thomas L. Kane
3rd Brigade – Brigadier-General George S. Greene
Artillery – Captain Joseph M. Knap

Cavalry Corps – Brigadier-General George Stoneman

1st Division – Brigadier-General Alfred Pleasonton (assumed command of 1st and 2nd Divisions 4 May)
1st Brigade (detached with General Averell to 4 May) – Colonel Benjamin F. Davis
2nd Brigade – Colonel Thomas C. Devin
Artillery –

2nd Division – Brigadier-General William W. Averell (later Colonel Duffié)
1st Brigade – Colonel Horace B. Sargent
2nd Brigade – Colonel John B. McIntosh (later Colonel Irvin Gregg)
Artillery –

3rd Division – Brigadier-General David McM. Gregg
1st Brigade – Colonel Judson Kilpatrick
2nd Brigade – Colonel Percy Wyndham
Reg. Reserve Cavalry Brigade – Brigadier-General John Buford
Artillery – Captain John M. Robertson
Reserve Artillery – Captain Graham
Engineer Brigade – Brigadier-General Benham

	Total Strength	Casualties
General and Staff	60	0
I Corps	16,908	299
II Corps	16,893	1,925
III Corps	18,721	4,119
V Corps	15,824	700
VI Corps	23,667	4,610
XI Corps	12,977	2,412
XII Corps	13,450	2,824
General Artillery Reserve	1,610	0
Engineers and Signal Corps	800	9
Cavalry Corps	11,541	389
Provost Guard	2,217	0
ARMY TOTAL	134,668	17,287

Confederate Army of Northern Virginia – Chancellorsville Campaign

Army Commander . . . General Robert E. Lee
Chief of Staff . . . Brigadier-General Chilton

I ARMY CORPS – Lieutenant-General James Longstreet (absent with two divisions)

Anderson's Division – Major-General Richard H. Anderson
Wilcox's Brigade – Brigadier-General C. M. Wilcox
Mahone's Brigade – Brigadier-General William Mahone
Wright's Brigade – Brigadier-General A. R. Wright
Posey's Brigade – Brigadier-General Carnot Posey
Perry's Brigade – Brigadier-General E. A. Perry
Artillery –
(1) Lieutenant-Colonel J. J. Garnett
(2) Major Chas. Richardson

McLaws' Division – Major-General Lafayette McLaws
Wofford's Brigade – Brigadier-General W. T. Wofford
Kershaw's Brigade – Brigadier-General James D. Kershaw
Semmes's Brigade – Brigadier-General Paul J. Semmes
Barksdale's Brigade – Brigadier-General William Barksdale
Artillery –
(1) Colonel H. C. Cabell
(2) Major S. P. Hamilton
Artillery Reserve, I Corps – not known
Alexander's Battalion – Colonel E. P. Alexander
Washington (La.) Artillery – Colonel J. B. Walton

II ARMY CORPS – Lieutenant-General Thomas J. Jackson

A. P. Hill's Division (Light Division) –
(1) Major-General A. P. Hill
(2) Brigadier-General Henry Heth

(3) Brigadier-General W. D. Pender
(4) Brigadier-General J. J. Archer
Heth's Brigade –
(1) Brigadier-General Henry Heth
(2) Colonel J. M. Brockenbrough
McGowan's Brigade –
(1) Brigadier-General S. McGowan
(2) Colonel O. E. Edwards
(3) Colonel A. Perrin
(4) Colonel D. H. Hamilton
Thomas's Brigade – Brigadier-General E. L. Thomas
Lane's Brigade – Brigadier-General J. H. Lane
Archer's Brigade –
(1) Brigadier-General J. J. Archer
(2) Colonel B. D. Fry
Pender's Brigade – Brigadier-General W. D. Pender
Artillery – Colonel R. L. Walker

D. H. Hill's Division –
(1) Brigadier-General R. E. Rodes
(2) Brigadier-General S. D. Ramseur
Rodes' Brigade –
(1) Brigadier-General R. E. Rodes
(2) Colonel E. A. O'Neal
(3) Colonel J. M. Hall
Doles' Brigade – Brigadier-General George Doles
Colquitt's Brigade – Brigadier-General A. H. Colquitt
Iverson's Brigade – Brigadier-General Alfred Iverson
Ramseur's Brigade –
(1) Brigadier-General S. D. Ramseur
(2) Colonel F. M. Parker
Artillery – Lieutenant-Colonel D. H. Carter

Early's Division – Major-General Jubal A. Early
Gordon's Brigade – Brigadier-General John B. Gordon
Smith's Brigade – Brigadier-General William Smith
Hoke's Brigade – Brigadier-General Robert F. Hoke
Hays' Brigade – Brigadier-General Harry T. Hays
Artillery – Lieutenant-Colonel R. S. Andrews

Trimble's Division – Brigadier-General R. E. Colston
Paxton's Brigade –
(1) Brigadier-General E. F. Paxton
(2) Colonel J. H. S. Funk
Jones's Brigade –
(1) Brigadier-General J. R. Jones
(2) Colonel T. S. Garnett
(3) Colonel A. S. Vanderventer
Colston's Brigade –
(1) Colonel E. T. H. Warren
(2) Colonel T. V. Williams

(3) Lieutenant-Colonel S. T. Walker
(4) Lieutenant-Colonel S. D. Thruston
(5) Lieutenant-Colonel H. A. Brown
Nicholls' Brigade –
(1) Brigadier-General F. T. Nicholls
(2) Colonel J. M. Williams
Artillery – Lieutenant-Colonel H. P. Jones
Artillery Reserve, II Corps –
(1) Colonel S. Crutchfield
(2) Brigadier–General Pendleton
Brown's Battalion – Colonel J. Thompson Brown
McIntosh's Battalion – Major D. G. McIntosh

Reserve Artillery Army of Northern Virginia – Brigadier-General William N. Pendleton

Cavalry Division – Major-General J. E. B. Stuart
Brigade of Brigadier-General W. H. F. Lee
Brigade of Brigadier-General Fitzhugh Lee

	Total Strength	*Casualties*
I Corps		
Anderson's Division	8,370	1,445
McLaws' Division	8,665	1,775
Corps Artillery	720	106
TOTAL I Corps	17,755	3,326
II Corps		
A. P. Hill's Division	11,751	2,940
D. H. Hill's Division	10,063	2,937
Early's Division	8,596	1,346
Trimble's Division	6,989	2,078
Corps Artillery	800	80
TOTAL II Corps	38,199	9,381
General Artillery Reserve	480	3
Cavalry	2,500	111
ARMY TOTAL	58,934	12,821

Federal Army of the Potomac – Battle of Gettysburg

Army Commander . . . Major-General George G. Meade

I Army Corps –
(1) Major-General John F. Reynolds (killed 1 July)
(2) Major-General Abner Doubleday
(3) Major-General John Newton

1st Division – Major-General J. S. Wadsworth
1st Brigade –
(1) Brigadier-General S. Meredith (wounded)
(2) Colonel W. W. Robinson
2nd Brigade – Brigadier-General L. Cutler

2nd Division – Brigadier-General John C. Robinson
1st Brigade –
(1) Brigadier-General Gabriel R. Paul (wounded)
(2) Colonel Richard Coulter
2nd Brigade – Brigadier-General Henry Baxter

3rd Division – Major-General Abner Doubleday
1st Brigade – Brigadier-General Thomas A. Rowley
2nd Brigade –
(1) Colonel Roy Stone (wounded)
(2) Colonel Langhorne Wister (wounded)
(3) Colonel Edmund L. Dana
3rd Brigade – Brigadier-General George J. Stannard
Artillery Brigade – Colonel C. S. Wainwright

II Army Corps –
(1) Major-General Winfield S. Hancock
(2) Brigadier-General John Gibbon

1st Division – Brigadier-General John C. Caldwell
1st Brigade –
(1) Colonel E. E. Cross (killed)
(2) Colonel H. B. McKeen
2nd Brigade – Colonel Patrick Kelly
3rd Brigade –
(1) Brigadier-General S. K. Zook (killed)
(2) Lieutenant-Colonel John Fraser
4th Brigade – Colonel John R. Brooke

2nd Division –
(1) Brigadier-General John Gibbon
(2) Brigadier-General William Harrow
1st Brigade –
(1) Brigadier-General William Harrow
(2) Colonel Francis E. Heath
2nd Brigade – Brigadier-General A. S. Webb
3rd Brigade – Colonel N. J. Hall

3rd Division – Brigadier-General Alexander Hays
1st Brigade – Colonel S. S. Carroll
2nd Brigade –
(1) Colonel Thos. A. Smyth (wounded)
(2) Lieutenant-Colonel F. E. Pierce
3rd Brigade –
(1) Colonel G. L. Willard (killed)
(2) Colonel Eliakim Sherrill (killed)

(3) Lieutenant-Colonel James M. Bull
Artillery Brigade – Captain J. G. Hazard
Cavalry Squadron – Captain Riley Johnson

III Army Corps –
(1) Major-General Daniel E. Sickles (wounded)
(2) Major-General D. B. Birney

1st Division –
(1) Major-General D. B. Birney
(2) Brigadier-General J. H. H. Ward
1st Brigade –
(1) Brigadier-General C. K. Graham (wounded)
(2) Colonel A. H. Tippin
2nd Brigade –
(1) Brigadier-General J. H. H. Ward
(2) Colonel H. Berdan
3rd Brigade – Colonel P. R. Trobriand

2nd Division – Brigadier-General A. A. Humphreys
1st Brigade – Brigadier-General Joseph B. Carr
2nd Brigade – Colonel William R. Brewster
3rd Brigade – Colonel George C. Burling
Artillery Brigade – Captain George E. Randolph

V Army Corps – Major-General George Sykes

1st Division – Brigadier-General James Barnes
1st Brigade – Colonel W. S. Tilton
2nd Brigade – Colonel J. B. Sweitzer
3rd Brigade –
(1) Colonel Strong Vincent (killed)
(2) Colonel James C. Rice

2nd Division – Brigadier-General R. B. Ayres
1st Brigade – Colonel Hannibal Day
2nd Brigade – Colonel Sidney Burbank
3rd Brigade –
(1) Brigadier-General S. H. Weed (killed)
(2) Colonel Kenner Garrard

3rd Division – Brigadier-General S. W. Crawford
1st Brigade – Colonel William McCandless
2nd Brigade – Colonel J. W. Fisher
Artillery Brigade – Captain A. P. Martin
Provost Guard – Captain H. W. Rider

VI Army Corps – Major-General John Sedgwick

1st Division – Brigadier-General H. G. Wright
1st Brigade – Brigadier-General A. T. Torbert
2nd Brigade – Brigadier-General J. J. Bartlett
3rd Brigade – Brigadier-General D. A. Russell

2nd Division – Brigadier-General A. P. Howe
1st Brigade – Colonel L. A. Grant
2nd Brigade – Brigadier-General T. H. Neill

3rd Division – Brigadier-General Frank Wheaton
1st Brigade – Brigadier-General Alexander Shaler
2nd Brigade – Colonel H. L. Eustis
3rd Brigade – Colonel David I. Nevin
Artillery Brigade – Colonel C. H. Tompkins
Cavalry Detachment – Captain William L. Craft

XI Army Corps – Major-General O. O. Howard (during the interval between the death of General Reynolds and the arrival of General Hancock on the afternoon of 1 July, all the troops on the field of battle were commanded by General Howard, General Schurz taking command of XI Corps, and General Schimmelpfennig of the 3rd Division)

1st Division –
(1) Brigadier-General F. C. Barlow (wounded)
(2) Brigadier-General Adelbert Ames
1st Brigade – Colonel Leopold von Gilsa
2nd Brigade –
(1) Brigadier-General Adelbert Ames
(2) Colonel A. L. Harris

2nd Division – Brigadier-General A. von Steinwehr
1st Brigade – Colonel C. R. Coster
2nd Brigade – Colonel Orland Smith

3rd Division – Major-General Carl Schurz
1st Brigade –
(1) Brigadier-General A. Schimmelpfennig (captured)
(2) Colonel George von Amsberg
2nd Brigade – Colonel W. Krzyzanowski
Artillery Brigade – Major T. W. Osborn

XII Army Corps – Major-General H. W. Slocum

1st Division – Brigadier-General Alpheus S. Williams
1st Brigade – Brigadier-General Thomas H. Ruger
2nd Brigade (unassigned during progress of battle; afterward attached to 1st Division as 2nd Brigade) – Brigadier-General H. H. Lockwood
3rd Brigade – Colonel S. Colgrove

2nd Division – Brigadier-General John W. Geary
1st Brigade – Colonel Charles Candy
2nd Brigade –
(1) Colonel George A. Cobham, Jr.
(2) Brigadier-General Thomas L. Kane
3rd Brigade – Brigadier-General George S. Greene
Artillery Brigade – Lieutenant Edward D. Muhlenberg
Headquarters Guard – not known

CAVALRY CORPS – Major-General Alfred Pleasonton

1st Division – Brigadier-General John Buford
1st Brigade – Colonel William Gamble
2nd Brigade – Colonel Thomas C. Devin
3rd Brigade – Brigadier-General Wesley Merritt

2nd Division – Brigadier-General D. McM. Gregg
1st Brigade – Colonel J. B. McIntosh
2nd Brigade (not engaged) – Colonel Pennock Huey
3rd Brigade – Colonel J. I. Gregg

3rd Division – Brigadier-General Judson Kilpatrick
1st Brigade –
(1) Brigadier-General E. J. Farnsworth
(2) Colonel N. P. Richmond
2nd Brigade – Brigadier-General George A. Custer
Horse Artillery – not known
1st Brigade – Captain John M. Robertson
2nd Brigade – Captain John C. Tidball

ARTILLERY RESERVE –
(1) Brigadier-General R. O. Tyler (disabled)
(2) Captain John M. Robertson
1st Regular Brigade – Captain D. R. Ransom
1st Volunteer Brigade – Lieutenant-Colonel F. McGilvery
2nd Volunteer Brigade – Captain E. D. Taft
3rd Volunteer Brigade – Captain James F. Huntington
4th Volunteer Brigade – Captain R. H. Fitzhugh
Train Guard – Major Charles Ewing
Headquarters Guard – Captain J. C. Fuller

DETACHMENTS AT HEADQUARTERS ARMY OF THE POTOMAC
Command of the Provost Marshal-General – Brigadier-General M. R. Patrick
Engineer Brigade – Brigadier-General H. W. Benham
Guards and Orderlies – Captain D. P. Mann

	Total Strength	*Casualties*
Headquarters	2,580	4
Artillery Reserve	2,868	242
I Corps	10,355	6,059
II Corps	13,056	4,369
III Corps	12,630	4,211
V Corps	13,211	2,187
VI Corps	15,710	242
XI Corps	10,576	3,801
XII Corps	8,597	1,082
Cavalry	10,192	852
TOTAL	99,775	23,049

Confederate Army of Northern Virginia – Battle of Gettysburg

Army Commander . . . General Robert E. Lee

I Corps – Lieutenant-General James Longstreet

1st Division – Major-General John B. Hood (wounded)
1st Brigade – Brigadier-General D. R. Anderson
2nd Brigade – Brigadier-General H. L. Bennings
3rd Brigade –
(1) Brigadier-General E. M. Law
(2) Colonel Jas. L. Sheffield
4th Brigade – Brigadier-General J. B. Robertson
Artillery – Major M. W. Henry

2nd Division – Major-General Lafayette McLaws
1st Brigade –
(1) Brigadier-General W. Barksdale (wounded)
(2) Colonel B. G. Humphreys
2nd Brigade – Brigadier-General J. B. Kershaw
3rd Brigade – Brigadier-General W. T. Wofford
4th Brigade –
(1) Brigadier-General P. J. Semmes (wounded)
(2) Colonel Goode Bryan
Artillery – Colonel H. C. Cabell

3rd Division – Major-General George E. Pickett
1st Brigade –
(1) Brigadier-General J. L. Kemper (wounded)
(2) Colonel Joseph Mayo, Jr.
2nd Brigade –
(1) Brigadier-General L. A. Armistead (killed)
(2) Colonel W. R. Aylett
3rd Brigade –
(1) Brigadier-General R. B. Garnett (killed)
(2) Major George C. Cabell
4th Brigade (not engaged at Gettysburg; encamped at Gordonsville 1–8 July) –
Brigadier-General M. D. Corse
Artillery – Major James Dearing
Reserve Artillery, I Corps – Colonel J. B. Walton, Chief of Artillery
Alexander's Battalion – Colonel E. P. Alexander
Corps Artillery – Major B. F. Eschelmann

II Corps – Lieutenant-General Richard S. Ewell

1st Division – Major-General Jubal A. Early
1st Brigade –
(1) Brigadier-General William Smith
(2) Colonel John S. Hoffman
2nd Brigade –
(1) Brigadier-General R. F. Hoke
(2) Colonel Isaac E. Avery (wounded)

(3) Colonel A. C. Godwin
3rd Brigade – Brigadier-General Harry T. Hays
4th Brigade – Brigadier-General J. B. Gordon
Artillery – Lieutenant-Colonel H. P. Jones

2nd Division – Major-General Edward Johnson
1st Brigade –
(1) Brigadier-General John M. Jones (wounded)
(2) Lieutenant-Colonel R. H. Dungan
(3) Colonel B. T. Johnson
2nd Brigade – Brigadier-General James A. Walker
3rd Brigade – Brigadier-General George H. Stewart
4th Brigade –
(1) Colonel J. M. Williams
(2) Brigadier-General A. Iverson (assigned 19 July)
Artillery – Lieutenant-Colonel R. S. Andrews

3rd Division – Major-General R. E. Rodes
1st Brigade –
(1) Brigadier-General E. A. Neal
(2) Colonel C. A. Battle
2nd Brigade (temporarily consolidated 10 July) – Brigadier-General S. D. Ramseur
3rd Brigade – Brigadier-General George Doles
4th Brigade (temporarily consolidated 10 July) –
(1) Brigadier-General Alfred Iverson
(2) Brigadier-General S. D. Ramseur
5th Brigade – Brigadier-General Junius Daniel
Artillery – Lieutenant-Colonel Thomas H. Carter
Reserve Artillery, II Corps – Colonel J. Thompson Brown, Chief of Artillery
Brown's Battalion (1st Virginia Artillery) – Captain W. J. Dance
Nelson's Battalion – Lieutenant-Colonel William Nelson

III Corps – Lieutenant-General Ambrose P. Hill

1st Division – Major-General R. H. Anderson
1st Brigade – Brigadier-General William Mahone
2nd Brigade –
(1) Brigadier-General A. R. Wright
(2) Colonel William Gibson
(3) Colonel E. J. Walker
(4) Captain B. C. McCurry
(5) Captain C. H. Anderson
3rd Brigade –
(1) Colonel David Lang
(2) Brigadier-General E. A. Perry
4th Brigade – Brigadier-General Carnot Posey
5th Brigade – Brigadier-General C. M. Wilcox
Artillery (Sumter Battalion) – Major John Lane

2nd Division –
(1) Major-General William D. Pender (wounded)

(2) Brigadier-General James H. Lane
1st Brigade –
(1) Brigadier-General S. McGowan
(2) Colonel A. Perrin
2nd Brigade – Brigadier-General James H. Lane (under Trimble's command 3 July)
3rd Brigade – Brigadier-General E. L. Thomas
4th Brigade (under Trimble's command 3 July) –
(1) Brigadier-General A. M. Scales (wounded)
(2) Colonel W. Lee J. Lawrence
Artillery – Major William T. Poague

3rd Division –
(1) Major-General Henry Heth
(2) Brigadier-General J. J. Pettigrew
1st Brigade (temporarily consolidated 10 July, under Pettigrew's command) –
(1) Brigadier-General J. J. Pettigrew (wounded)
(2) Major J. Jones
(3) Lieutenant-Colonel W. J. Martin
(4) Colonel J. K. Marshall
(5) Colonel T. C. Singeltary
2nd Brigade –
(1) Brigadier-General Charles W. Field
(2) Colonel J. M. Brockenbrough
(3) Brigadier-General H. H. Walker (assigned 19 July. Appears in return for 31 July as commanding both 2nd and 3rd Brigades)
3rd Brigade (temporarily consolidated 10 July, under Pettigrew's command) –
(1) Brigadier-General James J. Archer
(2) Colonel B. D. Fry
(3) Colonel S. G. Shepard
(4) Brigadier-General H. H. Walker
4th Brigade – Brigadier-General Joseph R. Davis
Artillery –
(1) Lieutenant-Colonel John J. Garnett
(2) Major Charles Richardson
Reserve Artillery, III Corps – Colonel R. L. Walker, Chief of Artillery
McIntosh's Battalion – Major D. G. McIntosh
Pegram's Battalion –
(1) Major W. J. Pegram
(2) Captain E. B. Brunson

Cavalry Division – Major-General J. E. B. Stuart
1st Brigade – Brigadier-General B. H. Robertson
2nd Brigade –
(1) Brigadier-General Wade Hampton
(2) Colonel L. S. Baker
3rd Brigade – Brigadier-General Fitzhugh Lee
4th Brigade – Brigadier-General W. H. F. Lee
5th Brigade – Brigadier-General William E. Jones
6th Brigade – Brigadier-General A. G. Jenkins

	Total Strength	Casualties
General Staff	47	—
Artillery*	4,703	—
I Corps	29,158	7,539
II Corps	30,279	5,937
III Corps	21,000	6,735
Cavalry	10,292	240
TOTAL	95,479	20,451

Appendix IV: Palestine
First and Second Battles of Gaza
Eastern Force, April 1917

General Officer Commanding – Lieutenant-General Sir C. M. Dobell
B.G.S. – Brigadier-General G. P. Dawnay
B.G.R.A. – Brigadier-General A. H. Short
Chief Engineer – Brigadier-General R. L. Waller

DESERT COLUMN – Lieutenant-General Sir Philip Chetwode

Australian and New Zealand Mounted Division – Major-General Sir H. G. Chauvel
1st Australian Light Horse Brigade † – Brigadier-General C. F. Fox
2nd Australian Light Horse Brigade – Brigadier-General G. de L. Ryrie
New Zealand Mounted Rifle Brigade – Brigadier-General E. W. C. Chaytor
22nd Mounted Brigade – Brigadier-General F. A. B. Fryer
3rd (T.F.) Brigade R.H.A. – two batteries
4th (T.F.) Brigade R.H.A. – two batteries

Imperial Mounted Division – Major-General H. W. Hodgson
3rd Australian Light Horse Brigade – Brigadier-General J. R. Royston
4th Australian Light Horse Brigade* – Brigadier-General J. B. Meredith
5th Mounted Brigade – Brigadier-General E. A. Wiggin
6th Mounted Brigade – Brigadier-General T. M. S. Pitt
Two batteries R.H.A. and two batteries H.A.C.

53rd (Welsh) Division – Major-General A. G. Dallas
158th Infantry Brigade – Brigadier-General S. F. Mott
159th Infantry Brigade – Brigadier-General J. H. du B. Travers
160th Infantry Brigade – Brigadier-General W. J. C. Butler
265th, 266th and 267th Brigades R.F.A.

* Casualties reflected in corps casualties.
† Not present at First Battle of Gaza.

FORCE TROOPS

52nd (Lowland) Division – Major-General W. E. B. Smith
155th Infantry Brigade – Brigadier-General J. B. Pollok-M'Call
156th Infantry Brigade – Brigadier-General A. H. Leggett
157th Infantry Brigade – Brigadier-General C. D. H. Moore
261st, 262nd and 263rd Brigades R.F.A.

54th (East Anglian) Division – Major-General S. W. Hare
161st Infantry Brigade – Brigadier-General W. Marriott-Dodington
162nd Infantry Brigade – Brigadier-General A. Mudge
163rd Infantry Brigade – Brigadier-General T. Ward
270th, 271st and 272nd Brigades R.F.A.

74th (Yeomanry) Division – Major-General E. S. Girdwood
229th Infantry Brigade – Brigadier-General R. Hoare
230th Infantry Brigade* – Brigadier-General A. J. McNeill
231st Infantry Brigade* – Lieutenant-Colonel W. J. Bowker
Artillery – Nil

Imperial Camel Brigade – Brigadier-General C. L. Smith, V.C.
Three battalions
Hong Kong and Singapore Camel Battery

No. 7 Light Car Patrol
Nos. 11 and 12 Armoured Motor Batteries
5th Wing R.F.C.

Third Battle of Gaza
Egyptian Expeditionary Force

Commander-in-Chief – General Sir Edmund H. V. Allenby G.C.B., G.C.M.G.
Chief of the General Staff – Major-General L. J. Bols, C.B., D.S.O.
B.G.S. – Brigadier-General G. P. Dawnay, D.S.O., M.V.O.

DESERT MOUNTED CORPS – Lieutenant-General Sir H. G. Chauvel, K.C.M.G., C.B.
B.G.S. – Brigadier-General R. G. H. Howard-Vyse, D.S.O.

Australian and New Zealand Mounted Division – Major-General E. W. C. Chaytor, C.B., C.M.G.
1st Australian Light Horse Brigade – Brigadier-General C. F. Cox, C.B.
2nd Australian Light Horse Brigade – Brigadier-General G. de L. Ryrie, C.M.G.
New Zealand Mounted Rifles Brigade – Brigadier-General W. Meldrum, C.B., D.S.O.
Artillery – commander not known
18th Brigade R.H.A. Engineers – commander not known
Australian and New Zealand Field Squadron

* These two brigades were not complete at the time of the First Battle of Gaza.

Australian Mounted Division – Major-General H. W. Hodgson, C.B., C.V.O.
3rd Australian Light Horse Brigade – Brigadier-General L. C. Wilson, C.M.G.
4th Australian Light Horse Brigade – Brigadier-General W. Grant, D.S.O.
5th Mounted Brigade – Brigadier-General P. D. Fitzgerald, D.S.O.
Artillery – 14th Brigade R.H.A.
Engineers – Australian Mounted Division Field Squadron

Yeomanry Mounted Division – Major-General G. de S. Barrow, C.B.
6th Mounted Brigade – Brigadier-General C. A. C. Godwin
8th Mounted Brigade – Brigadier-General C. S. Rome
22nd Mounted Brigade – Brigadier-General F. A. B. Fryer
Artillery – 20th Brigade R.H.A.
Engineers – No. 6 Field Squadron R.E.

Corps Troops
Machine-Gun Corps Attached 7th Mounted Brigade – Brigadier-General J.T. Wigan, D.S.O.

XX Corps
G.O.C. – Lieutenant-General Sir P. W. Chetwode, K.C.M.G., D.S.O.
B.G.S. – Brigadier-General W. H. Bartholemew, C.M.G.

53rd (Welsh) Division – Major-General S. F. Mott
158th Brigade – Brigadier-General H. A. Vernon, D.S.O.
159th Brigade – Brigadier-General N. E. Money, D.S.O.
160th Brigade – Brigadier-General V. L. N. Pearson
Artillery –
265th Brigade R.F.A.
266th Brigade R.F.A.
267th Brigade R.F.A.
Engineers – 437th and 439th Field Companies R.E.

60th (London) Division – Major-General J. S. M. Shea, C.B., D.S.O.
179th Brigade – Brigadier-General FitzJ. M. Edwards, C.M.G., D.S.O.
180th Brigade – Brigadier-General C. F. Watson, C.M.G., D.S.O.
181st Brigade – Brigadier-General E. C. Da Costa
Artillery –
301st Brigade R.F.A.
302nd Brigade R.F.A.
303rd Brigade R.F.A.
Engineers – 519th, 521st and 522nd Field Companies R.E.

74th (Yeomanry) Division – Major-General E. S. Girdwood
229th Brigade – Brigadier-General R. Hoare
230th Brigade – Brigadier-General A. J. M'Neill
231st Brigade – Brigadier-General C. E. Heathcote, D.S.O.
Artillery –
44th Brigade R.F.A.
117th Brigade R.F.A.
268th Brigade R.F.A.
Engineers – 5th (R. Monmouth) and 5th (R. Anglesey) Field Companies R.E.

Corps Troops
Mounted Troops – 1/2nd County of London Yeomanry
Artillery – XCVI Heavy Artillery Group
Attached 10th (Irish) Division – Major-General J. R. Longley, C.B.

XXI Corps
G.O.C. – Lieutenant-General E. S. Bulfin, C.B., C.V.O.
B.G.S. – Brigadier-General E. T. Humphreys, D.S.O.

52nd (Lowland) Division – Major-General J. Hill, D.S.O.
155th Brigade – Brigadier-General J. B. Pollok-M'Call, C.M.G.
156th Brigade – Brigadier-General A. H. Leggett, D.S.O.
157th Brigade – Brigadier-General C. D. H. Moore, D.S.O.
Artillery –
261st Brigade R.F.A.
262nd Brigade R.F.A.
264th Brigade R.F.A.
Engineers – 410th, 412th and 413th Field Companies R.E.

54th (East Anglian) Division – Major-General S. W. Hare, C.B.
161st Brigade – Brigadier-General W. Marriott-Dodington
162nd Brigade – Brigadier-General A. Mudge
163rd Brigade – Brigadier-General T. Ward, C.M.G.
Artillery –
270th Brigade R.F.A.
271st Brigade R.F.A.
272nd Brigade R.F.A.
Engineers – 484th and 486th Field Companies R.E.

75th Division – Major-General P. C. Palin, C.B.
232nd Brigade – Brigadier-General H. J. Huddleston, D.S.O.
233rd Brigade – Brigadier-General the Hon E. M. Colston, D.S.O.
234th Brigade – Brigadier-General F. G. Anley, C.B., C.M.G.
Artillery –
37th Brigade R.F.A.
172nd Brigade R.F.A.
1st S. African F.A. Brigade
Engineers – 495th and 496th Field Companies R.E.

Corps Troops
Mounted Troops – Composite Regiment
Artillery –
XCVII Heavy Artillery Group
C Heavy Artillery Group
102nd Heavy Artillery Group
Machine-Gun Corps –
E Company, Tank Corps
211th Machine-Gun Company

General Headquarters Troops – Brigadier-General W. S. Brancker, D.S.O.
Palestine Brigade R.F.C.
Artillery –
8th Mountain Brigade R.G.A.
9th Mountain Brigade R.G.A.
Attached to Desert Mountain Corps – 7th Mounted Brigade and Imperial Camel Corps
Attached to XX Corps – 10th Division

Order of Battle for Yilderim October 1917
Commander-in-Chief – Marshal E. von Falkenhayn
Chief of Staff – Oberst von Dommes

SEVENTH ARMY
G.O.C. – General Fevzi Pasha
III Corps – Colonel Ismet Bey

EIGHTH ARMY
G.O.C. – General Freiherr Kress von Kressenstein
XX Corps – Colonel Àli Fuad Bey
XXIII Corps – Colonel Refet Bey

G.H.Q. TROOPS
19th Division
German Flying Corps (301st, 302nd, 303rd and 304th Flight Detachments)

Appendix V: Anzio

VI U.S. CORPS –
(1) Major-General John P. Lucas
(2) Major-General Lucian K. Truscott

British Troops

1st Infantry Division – Major-General W. R. C. Penney
24th Guards Brigade – relieved by 18th Infantry Brigade, 7 March
2nd Infantry Brigade
3rd Infantry Brigade
2nd, 19th, 24th, 67th Field Regiments, R.A.
80th Medium Regiment, R.A. (The Scottish Horse)
81st Anti-Tank Regiment, R.A.
90th Light Anti-Aircraft Regiment, R.A.
46th Royal Tank Regiment
1st Reconnaissance Regiment
2nd/7th Battalion The Middlesex Regiment (M.G.)
2nd Special Service Brigade

56th Infantry Division – Major-General G. W. R. Templer
(Arrived at the beachhead between 3 and 16 February: departed 11 March)
167th Infantry Brigade
168th Infantry Brigade
169th Infantry Brigade

5th Infantry Division – Major-General P. G. S. Gregson-Ellis
(Arrived at the beachhead 7 March)
13th Infantry Brigade
15th Infantry Brigade
17th Infantry Brigade

United States Troops

1st Armored Division – Major-General Ernest N. Harmon
Combat Command A (Combat Command B only came to the beachhead in May)
6th Armored Infantry Regiment

3rd Infantry Division –
(1) Major-General Lucian K. Truscott
(2) Brigadier-General John W. O'Daniel
7th, 15th, 30th Infantry Regiments
504th Parachute Infantry Regiment
509th Parachute Infantry Battalion
751st Tank Battalion
1st, 3rd and 4th Battalions of 6615th Ranger Force
36th Combat Engineer Regiment

45th Infantry Division – Major-General William W. Eagles
(Arrived end of January/beginning of February)
157th, 179th, 180th Infantry Regiments
158th, 160th, 171st, 189th Field Artillery Battalions
645th Tank Destroyer Battalion

1st Special Service Force – Brigadier-General Robert I. Frederick
(American/Canadian Force: arrived beachhead 3 February)

34th Infantry Division – Major-General Charles W. Ryder
(Arrived beachhead end of March)
133rd, 135th and 168th Infantry Regiments

36th Infantry Division – Major-General F. L. Walker
(Arrived beachhead after break-out)
141st, 142nd and 143rd Infantry Regiments

German Troops

The Germans fed bits and pieces of units into the beachhead area with great speed in the early days, and thereafter there were many changes in divisions and lower formations. The table below gives a full list of the units involved at one time or another in the Anzio operation, either as a complete unit or more often in part. The commanders of the formations principally engaged in the beachhead battles are as follows:

Fourteenth Army . . . Generaloberst Eberhard von Mackensen
LXXVI Panzer Corps . . . Generalleutenant Traugott Herr
I Parachute Corps . . . General der Flieger Alfred Schlemm
114th Jaeger Division . . . Generalleutnant Eglseer
362nd Infantry Division . . . Generalleutnant Heinz Greiner
715th Infantry Division . . . Generalleutnant Hildebrandt
Hermann Göring Panzer Division . . . Generalleutnant Conrath
4th Parachute Division . . . Colonel Heinrich Tiettner (Gazetted Major-General, I, July 1944)
3rd Panzer Grenadier Division . . . Generalleutnant Fritz-Hubert Graser
65th Infantry Division . . . Generalleutnant Pfeiffer
71st Infantry Division . . . Generalleutnant Wilhelm Raapke
26th Panzer Division . . . Generalleutnant Luttwitz
29th Panzer Grenadier Division . . . Generalleutnant Fries

Divisions	Type	Infantry Regiments*	Panzer (Tank) Units	Anti-tank Battalions	Artillery Regiments
1st	Parachute	1st, 3rd, 4th		1st	1st
3rd	Panzer Grenadiers	8th, 29th	103rd Battalion	3rd	3rd
4th	Parachute	10th, 11th, 12th		4th	4th
5th	Mountain	85th, 100th		95th	95th
8th	Mountain	296th, 297th		—	1057th
15th	Panzer Grenadiers	104th, 115th	115 Reconnaissance	33rd	33rd
16th	S.S-Panzer Grenadiers	35th, 36th	16th Battalion	16th	16th
26th	Panzer	9th, 67th	26th Regiment	51st	93rd
29th	Panzer Grenadiers	15th, 71st	129th Battalion	29th	29th
44th	Infantry	131st, 132nd, 134th		46th	96th
65th	Infantry	145th, 146th, 147th	1165th Battalion	165th	165th
71st	Infantry	191st, 194th, 211th		171st	171st
90th	Panzer Grenadiers	200th, 361st	190th Battalion	90th	190th
92nd	Infantry	1059th, 1060th		192nd	192nd

*Included in these are the parachute regiments of the parachute division, and the panzer grenadier regiments of the panzer and panzer grenadier division.

Divisions	Type	Infantry Regiments*	Panzer (Tank) Units	Anti-tank Battalions	Artillery Regts
94th	Infantry	267th, 274th, 276th		194th	194th
114th	Jaeger	721st, 741st		114th	661st
162nd*	Infantry	303rd, 314th, 329th		236th	236th
278th	Infantry	992nd, 993rd, 994th		278th	278th
305th	Infantry	576th, 577th, 578th		305th	305th
334th	Infantry	754th, 755th, 756th		334th	334th
356th	Infantry	869th, 870th, 871st		356th	356th
362nd	Infantry	954th, 955th, 956th		362nd	362nd
715th	Infantry	725th, 735th, 1028th		715th	671st
Herman Göring	Panzer	1st, 2nd, Hermann Göring	H.G. Regiment	1st H.G.	1st H.G.

* Composed mainly of 'volunteers' from Soviet Central Asia. Full designation is: 162nd Infantry Division (Turk).

List of Books Consulted

Chapter 1: HASTINGS

Belloc, Hilaire, *William the Conqueror*, Peter Davies, 1933.
Bryant, Arthur, *The Story of England*, Collins, 1953.
Burne, A. H., *The Battlefields of England*, Methuen, 1950.
Compton, Piers, *Harold the King*, Robert Hale, 1961.
Douglas, David C., *William the Conqueror*, Eyre and Spottiswoode, 1964.
Freeman, E. A., *The Norman Conquest*, Clarendon Press, 1869.
Fuller, J. F. C., *The Decisive Battles of the Western World*, Eyre and Spottiswoode, 1954.
Lemmon, C. H., *The Field of Hastings*, Budd and Gillatt, 1957.
Oman, Charles, *A History of the Art of War in the Middle Ages, Vol. I*, Methuen, 1924.
Seymour, William, *Battles in Britain, Vol. I*, Sidgwick and Jackson, 1975.
Young, Peter, and Adair, John, *Hastings to Culloden*, G. Bell, 1964.

Chapter 2: CRÉCY

Belloc, Hilaire, *British Battles (Crécy)*, Stephen Swift, 1912.
Burne, A. H., *The Crécy War*, Eyre and Spottiswoode, 1955.
Churchill, Winston S., *A History of the English-Speaking Peoples, Vol. I*, Cassell, 1956.
George, H. B., *Battles of English History*, Methuen, 1895.
Green, Howard, *Famous Engagements*, 1969 – printed by J. Tiehen & Co.
Lodge, E. C., *Gascony Under English Rule*, Methuen, 1926.
Oman, Charles, *A History of the Art of War in the Middle Ages, Vol. II*, Methuen, 1924.
Perroy, Edouard, *The Hundred Years War*, Eyre and Spottiswoode, 1962.
Ramsay, Sir James, *Genesis of Lancaster, Vol. I*, Clarendon Press, 1913.
Seward, Desmond, *The Hundred Years War*, Constable, 1978.

Chapter 3: AGINCOURT

Burne, A. H., *The Agincourt War*, Eyre and Spottiswoode, 1956.
George, R. H., *Battles of English History*, Methuen, 1895.
Green, Howard, *Famous Engagements*, 1969.
Hibbert, Christopher, *Agincourt*, Batsford, 1964.
Hutchison, Harold F., *Henry V*, Eyre and Spottiswoode, 1967.
Jacob, E. F., *Henry V and the Invasion of France*, Hodder and Stoughton, 1947.
Oman, Charles, *A History of the Art of War in the Middle Ages, Vol. II*, Methuen, 1924.
Perroy, Edouard, *The Hundred Years War*, Eyre and Spottiswoode, 1962.
Ramsay, J. H., *Lancaster and York, Vol. I*, Clarendon Press, 1892.
Seward, Desmond, *The Hundred Years War*, Constable, 1978.
Wylie, James Hamilton, *The Reign of Henry the Fifth, Vol. II*, Cambridge University Press, 1919.

Chapter 4: THE THIRD CIVIL WAR

Abbott, W. C., *The Writings and Speeches of Oliver Cromwell, Vol. II*, Harvard University Press, 1939.
Bund, J. H., Willis, *The Civil War in Worcestershire and the Scottish Invasion*, London and Birmingham, 1905.
Burne, A. H., *Battlefields of England*, Methuen, 1950.
Clarendon, Edward, Earl of, *The History of the Rebellion, Book XIII*, Clarendon Press, 1704.
Firth, C. H., *Oliver Cromwell*, London, 1924.
Firth, C. H., 'The Battle of Dunbar', *Transactions of the Royal Historical Society*, Vol. XIV, Longmans, 1900.
Fraser, Antonia, *Cromwell – Our Chief of Men*, Weidenfeld and Nicolson, 1973.
Rogers, H. C. B., *Battles and Generals of the Civil Wars*, Seeley Service, 1968.
Seymour, William, *Battles in Britain, Vol. II*, Sidgwick and Jackson, 1975.
Young, Peter, and Adair, John, *Hastings to Culloden*, G. Bell, 1964.
Young, Peter, and Holmes, Richard, *The English Civil War*, Eyre, Methuen, 1974.

Chapters 5–7: THE SARATOGA CAMPAIGN

Alden, J. R., *The American Revolution 1775–1783*, Harper and Row, 1954.
Bird, Harrison, *March to Saratoga 1777*, Oxford University Press, N.Y., 1963.
Blackmore, Howard L., *British Military Firearms 1650–1850*, Herbert Jenkins, 1961.
Burgoyne, John, *A State of the Expedition from Canada as Laid Before the House of Commons*, London, 1780
Carrington, Henry B., *Battles of the American Revolution*, New York Times and Arno Press, 1877.
Clark, Jane, 'Responsibility for the Failure of the Burgoyne Campaign', *American Historical Review*, Vol. XXXV, April 1930.
Cumming, P., and Rankin, Hugh, *The Fate of a Nation: the American Revolution Through Contemporary Eyes*, Phaidon Press, 1975.
Digby, William, *The British Invasion from the North*, Joel Munsell's Sons, 1887.
Elting, John R., *The Battles of Saratoga*, Philip Freneau Press, 1977.
Fortescue, the Hon. J. W., *A History of the British Army, Vol. III, 1763–1793*, Macmillan, 1902
Fuller, J. F. C., *The Decisive Battles of the United States*, Hutchinson, 1942.
Fuller, J. F. C., *The Decisive Battles of the Western World, Vol. II*, Eyre and Spottiswoode, 1955.
Furneaux, Rupert, *Saratoga: the Decisive Battle*, George Allen and Unwin, 1971.
Higginbotham, Don, *The War of American Independence*, Macmillan, N.Y., 1971.
Hudleston, F. J., *Gentleman Johnny Burgoyne*, Bobbs-Merrill Company, 1927.
Lewis, Paul, *The Man Who Lost America*, The Dial Press, 1973.
Lunt, James, *John Burgoyne of Saratoga*, Harcourt Brace Jovanovich, 1975.
Luzader, John, *Decision on the Hudson*, National Parks Service, 1975.
Mackesy, Piers, *The War for America 1775–1783*, Harvard University Press, 1964.
Nickerson, Hoffman, *The Turning Point of the Revolution, Vols. I and II*, Kennikat Press, 1967.
Paine, Lauran, *Gentleman Johnny: the Life of General John Burgoyne*, Robert Hale, 1973.
Peterson, Harold L., *The Book of the Continental Soldier*, Stackpole Books, Harrisburg, 1963.
Smith, Paul H., *Loyalists and Redcoats: a Study in British Revolutionary Policy*, W. W. Norton, 1964.
Ward, Christopher, *The War of the Revolution, Vols. I and II*, Macmillan, N.Y., 1952.
Willcox, William B., *Portrait of a General, Sir Henry Clinton in the War of Independence*, Knopf, N.Y., 1964.
Willcox, William B., 'Too Many Cooks: British Planning Before Saratoga', *The Journal of British Studies*, May 1962.

Chapters 8–9: WATERLOO

Becke, A. F., *Napoleon and Waterloo*, London, 1936.
Bryant, Arthur, *Jackets of Green*, London, 1972.
Chalfont, Lord, ed., *Waterloo: Battle of Three Armies*, Sidgwick and Jackson, 1979.
Clausewitz, General C. von, *La Campagne de 1815 (en France)*.
Jomini, Baron de, *The Campaign of Waterloo*, New York, 1853.
Kennedy, General Sir James Shaw, *Notes on the Battle of Waterloo*, London, 1865.
Longford, Elizabeth, *Wellington: The Years of the Sword*, London, 1969.
Müffling, Baron von, *Passages from my Life*, second edition, London, 1853.
Oman, C. W. C., *Wellington's Army*, London, 1913.
Siborne, W., *History of the War in France and Belgium in 1815*, London, 1848.
Sloane, William, *Life of Napoleon (4 vols.)*, New York, 1901.
Weller, Jac, *Wellington at Waterloo*, London, 1967.

Chapter 10: BACKGROUND TO CHANCELLORSVILLE

Barker, Alan, *The Civil War in America*, A. and C. Black, 1961.
Catton, Bruce, *This Hallowed Ground*, Victor Gollancz, London, 1957.
Haythornthwaite, Philip J., *Uniforms of the Civil War 1861–65*, Macmillan, N.Y., 1976.
Henderson, G. F. R., *Stonewall Jackson and the American Civil War*, Longmans, Green, 1936.
Johnson, Rossiter, *A Short History of the War of Secession*, Boston, 1888.
Lee, G. C., *The True History of the Civil War*, London, 1903.
McClellan, George B., *McClellan's Own Story*, New York, 1887.
Miers, Earl Schenck, *Robert E. Lee*, New York, 1956.
Mitchell, Joseph B., *Decisive Battles of the Civil War*, Putnam's, New York, 1955.
Palfrey, Francis Winthrop, *The Antietam and Fredericksburg*, New York, 1882.
Paxson, Frederic L., *The American Civil War*, London, 1911.
Rogers, H. C. B., *The Confederates and Federals at War*, Ian Allan, 1973.
Thomas, Benjamin P., *Abraham Lincoln*, New York, 1952.
Tyler, Mason Whiting, *Recollections of the Civil War*, New York, 1912.

Chapter 11: CHANCELLORSVILLE

Battles and Leaders of the Civil War, Vols. I–IV, New York, 1956.
Bigelow, John, *The Campaign of Chancellorsville*, Yale, 1910.
Catton, Bruce, *This Hallowed Ground*, Victor Gollancz, London, 1957.
Cullen, Joseph P., *Where a Hundred Thousand Fell*, Washington, 1966.
Espasito, Vincent J., ed., *The West Point Atlas of American Wars*, New York.
Freeman, Douglas Southall, *Lee's Lieutenants, Vols. 2 and 3*, Scribner's, N.Y., 1943.
Gough, Colonel J.E., *Fredericksburg and Chancellorsville*, London, 1913.
Hatchless, Jed and Allan, William, *The Battlefields of Virginia. Chancellorsville*, New York, 1967.
Henderson, G. F. R., *Stonewall Jackson and the American Civil War*, Longmans, Green, 1936.
Johnson, Rossiter, *A Short History of the War of Secession*, Boston, 1888.
Lee, G. C., *The True History of the Civil War*, Philadelphia and London, 1903.
Miers, Earl Schenck, *Robert E. Lee*, New York, 1956.
Mitchell, Joseph B., *Decisive Battles of the Civil War*, Putnam's, New York, 1955.
Paris, the Comte de, *History of the Civil War in America, Vol. 3*, London, 1883.
Paxson, Frederic L., *The American Civil War*, London, 1911.
Rogers, H. C. B., *The Confederates and Federals at War*, Ian Allan, 1973.
Selby, John, *Stonewall Jackson*, Batsford, 1968.
Stackpole, Edward J., *Chancellorsville*, The Stackpole Company, 1958.

Chapter 12: GETTYSBURG

Burnage, H. S., *Gettysburg and Lincoln*, Putnam's, 1906.
Catton, Bruce, *Gettysburg: The Final Fury*, London, 1975.
Catton, Bruce, *Glory Road*, Doubleday, 1952.
Catton, Bruce, *This Hallowed Ground*, Victor Gollancz, London, 1957.
Coddington, Edwin, *The Gettysburg: A Study in Command*, New York, 1968.
Freeman, Douglas Southall, *Lee's Lieutenants, Vols. 2 and 3*, Scribner's, N.Y., 1943.
Fox, Charles K., ed., *Gettysburg*, A. S. Barnes, 1969.
Gettysburg, National Park Service Historical Handbook Series 9, Washington D.C., 1954.
Haskell, Frank A., *The Battle of Gettysburg*, Eyre and Spottiswoode, London, 1957.
Johnson, Robert U., and Buel, C. C., *Battles and Leaders of the Civil War, 4 vols.*
Johnson, Rossiter, *A Short History of the Civil War of Secession*, Boston, 1888.
Lee, G. C., *The True History of the Civil War*, Philadelphia and London, 1903.
Miers, Earl Schenck, *Robert E. Lee*, New York, 1956.
Mitchell, Joseph B., *Decisive Battles of the Civil War*, Putnam's, N.Y., 1955.
Paris, the Comte de, *History of the Civil War in America, Vol. 3*, London, 1883.
Paxson, Frederic L., *The American Civil War*, London, 1911.
Rogers, H. C. B., *The Confederates and Federals at War*, Ian Allan, 1973.
Stackpole, Edward J., *They Met at Gettysburg*, The Stackpole Company, 1951.
War of the Rebellion, U.S. War Department, Official Records.

Chapter 13: GAZA

Australian Official History, Vols. I and VII. Powles, Lieutenant-Colonel C. G.
Garsia, Lieutenant-Colonel Clive, *A Key to Victory*, Eyre and Spottiswoode, 1940.
History of the Great War, Egypt and Palestine 1914–17, H.M.S.O., 1928.
History of the Great War, Egypt and Palestine 1917–18, Parts I and II, H.M.S.O., 1930.
Massey, W. T., *How Jerusalem Was Won*, Constable, 1919.
New Zealand Official History, Vol. III.
Preston, Lieutenant-Colonel R. M. P., *The Desert Mounted Corps*.
Robertson, Field-Marshal Sir William, *Soldiers and Statesmen 1914–1918*, Cassell, 1926.
Wavell, Colonel A. P., *The Palestine Campaigns*, Constable, 1928.

Chapter 14–16: ANZIO

Blumenson, Martin, *Anzio: the Gamble that Failed*, Weidenfeld and Nicolson, 1963.
Clark, General Mark, *Calculated Risk*, George G. Harrap, 1951.
Fisher, Ernest F. Jr., *United States Army in World War II: The Mediterranean Theater of Operations, Cassino to the Alps*, Washington D.C., 1977.
The German Operation at Anzio, Military Intelligence Division, U.S. War Department, Maryland.
Greenfield, K. R. (general editor), *Command Decisions*, Harcourt, Brace, 1960.
Hibbert, Christopher, *Anzio: The Bid for Rome*, Hazell Watson and Viney, 1970.
Jackson, Major-General W. G. F., *Alexander of Tunis*, Batsford, 1971.
Jackson, Major-General W. G. F., *The Battle for Rome*, London, 1969.
Kesselring, Albert, *The Memoirs of Field-Marshal Kesselring*, William Kimber, 1953.
Malony, Brigadier C. J. C., *History of the Second World War: the Mediterranean and Middle East, Vol. V*, H.M.S.O., 1973.
Nicolson, Nigel, *Alex*, Weidenfeld and Nicolson, 1973.
Truscott, Lucian K., *Command Mission*, New York, 1954.
Vaughan-Thomas, Wynford, *Anzio*, Longmans, 1961.
Verney, Peter, *Anzio 1944*, Batsford, 1978.
Westphal, General Siegfried, *The German Army in the West*, Cassell, 1951.

Index

Bold figures in brackets refer to numbers of maps or plans.